Happ

Read + Learn — Then find
the land and give me
a ring — partners!

love Pete x

The Wines and Vineyards of New Zealand

To
ROBIN MORRISON

Whose unique photography brought the wine industry to
life to a degree no writer could ever hope to rival.

— Michael Cooper

The Wines and Vineyards of New Zealand

MICHAEL COOPER

PHOTOGRAPHY BY
Robin Morrison

Hodder & Stoughton
AUCKLAND LONDON SYDNEY TORONTO

Photographic Acknowledgements
All photographs in this book are by Robin Morrison except those
attributed otherwise below:
Marti Friedlander: pp 8, 11, 12, 13, 16
Wine Institute: p 17
New Zealand Wine Guild: p 25

Cartography
Maps reproduced by permission of the Department of Survey
and Land Information.

Book design by Barbara Nielsen
Cartography by Sue Gerrard
Production services by Graeme Leather
Typeset by Catherine Haigh, Typocrafters Ltd, Auckland
Printed and bound by Kyodo Printing Co. Ltd, Singapore,
for Hodder & Stoughton Ltd, 44–46 View Road, Glenfield,
Auckland, New Zealand

CONTENTS

PREFACE

This book is the fruit of an unforgettable journey. I flew into Central Otago at the start of a marathon, six week-long national vineyard tour. From Alexandra in the south to Matakana in the north, I visited all the key winemakers, toured their vineyards and cellars, learned of their achievements and dreams — and tasted a dazzling array of wines.

Robin Morrison, prior to his premature death in 1993, also ventured on the vineyard trail. *The Wines and Vineyards of New Zealand*, from the first 1984 edition to this fourth edition, has been a labour of love for the Cooper/Morrison team. For the final time in these pages, Robin Morrison's camera captures the rich diversity of the New Zealand wine industry — the stunning landscapes in which the vineyards flourish and the absorbing gallery of personalities behind the labels.

Few days pass in the New Zealand wine industry without the arrival of a new label on the shelves. New wineries are proliferating too — sixty companies were featured in the third edition of this book; in this edition there are over a hundred. The historical and technical chapters of the book have therefore been trimmed to enable the burgeoning number of wineries and labels to be comprehensively chronicled.

The book begins with the recent history of New Zealand wine, and profiles several crucial figures in the flowering of the industry since mid-century. My thanks are due to Alex Corban and David Lucas for their willing co-operation, and to Philip Gregan, executive officer of the Wine Institute, for reading the latter part of the history chapter in manuscript.

Chapter Two takes an in-depth look at the country's principal grape varieties, pinpointing the top producers of each variety. Dr David Jordan, New Zealand's viticultural scientist, gave invaluable assistance with this chapter.

The heart of the book follows: the third chapter is a comprehensive region-by-region tour of every significant winery in the land. I have visited most of the producers several times, to meet the winemakers, view their vineyards and wineries — and taste their wines. For the first time, this chapter features brief profiles of some of the most influential winemaking personalities and top vineyard sites. If you dip into this chapter — glass in hand — I hope you find almost everything you want to know about New Zealand wine.

Sue Gerrard prepared the regional maps with her usual precision. For this edition, the maps pinpoint not only the winery locations, but also where the major concentrations of vines are found. Here David Hoskins, Ross Spence, Peter Babich, Kim and Jeanette Goldwater, Dr Rainer Eschenbruch, Steve Smith, Phyllis Pattie, Tim and Judy Finn, Petter Evans and Rob Hay all made valuable contributions.

For permission to use material first published in my monthly column in *North and South* and my feature articles in *Cuisine*, I am indebted to those magazines' publishers. The designer's job is critically important to a richly illustrated book; my thanks to Barbara Nielsen. Tom Beran of Hodder and Stoughton steered the project throughout. To Linda, my wife, my deepest thanks.

MICHAEL COOPER

CHAPTER ONE

The Modern Era of New Zealand Wine

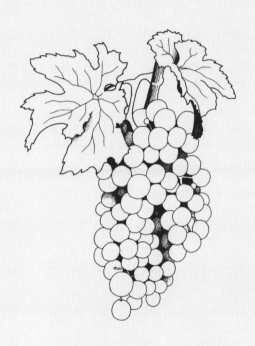

The grapevine is a hardy plant, able to survive climatic extremes of heat, drought and frost. Those who make their living from the vine are also a resilient breed. Throughout the first century of winemaking in New Zealand, winemakers needed all the passion and tenacity they could muster.

Samuel Marsden, the Anglican missionary, made the first recorded planting of grapevines in New Zealand at Kerikeri on 25 September 1819. The honour of being the country's first recorded winemaker, however, belongs to the first British Resident in New Zealand, James Busby. When the French explorer Dumont d'Urville visited Busby at Waitangi in 1840, he saw 'a trellis on which several flourishing vines were growing . . . with great pleasure I agreed to taste the product of the vineyard that I had just seen. I was given a light white wine, very sparkling, and delicious to taste, which I enjoyed very much.'[1]

French priests and peasants, Hawke's Bay pastoralists — who made and marketed table wine from classic *vinifera* grape varieties — Dalmatian gum-diggers and others kept the flame ignited by Busby alight throughout the nineteenth century. Romeo Bragato's far-sighted 'Report on the Prospects of Viticulture in New Zealand', published in 1895, stimulated a flurry of vineyard plantings. But the assaults of oidium (powdery mildew), the vine-destroying phylloxera aphid and prohibitionist zealots (who secured the passage of increasingly restrictive liquor legislation between 1881 and 1918) together ensured that Bragato's dream of 'an industry that will by far eclipse any other that has hitherto been prosecuted here . . . [making] wine of the finest quality, both red and white and champagne . . .'[2] was soon forgotten.

Antonio ('Tony') Zame ran Capri Vineyards in Gisborne during the 1950s and 1960s. An Italian from the island of Stromboli, near Sicily, Zame worked as a restaurateur and part-time winemaker before he plunged into full-time winemaking.

The 1920s and 1930s witnessed gradual but unspectacular growth in the wine industry. By 1932 a hundred licensed winemakers were producing a stream of fortified 'sherries' and 'ports' from low-grade hybrid, native American and virused *vinifera* vines. The industry boomed during World War Two, when duties were raised on overseas wines and import licences for wine halved, but when import restrictions were eased in the late 1940s, the growers' fortunes again faded.

At mid-century the industry was failing to capitalise on New Zealand's potential for the production of world-class table wine; the vast majority of the winemakers were not even aware that the potential existed. The modern era of New Zealand wine, however, was about to begin.

A new impulse was felt when George Mazuran was elected president of the Viticultural Association — composed mainly of small-scale growers of Dalmatian origin — in 1950. Convinced early that the future prosperity of the wine industry hinged on relaxation of the country's restrictive licensing laws, Mazuran subsequently carved out a long career for himself as one of the most successful political lobbyists that New Zealand has known. The efforts of Mazuran and the Viticultural Association yielded an impressive string of legislative concessions from successive governments that laid the foundation for the industry's phenomenal growth rates of the 1960s and 1970s.

The early 1950s under a new National Government brought a series of measures designed to boost the ailing wine industry. The wartime forty percent sales tax was halved. Then in 1953 separate licences were created for the manufacture of grape and fruit wines and no person or company was allowed to hold both. The idea was to prevent the sale of fruit wines masquerading as grape wine.

The winegrowers' annual dinner and field-day for parliamentarians and government officials was launched in 1952 and subsequently brought the wine industry to the favourable attention of a host of politicians. The field-day transformed the traditional European harvest celebrations into a superbly effective public relations exercise. MPs gained the background knowledge, and industry leaders the social contacts that together assured the wine industry of an accessible and responsive legislature.

Mazuran's assiduous lobbying soon paid off. A crucial breakthrough came in 1955 when Parliament reduced the minimum quantities of wine that could be sold by winemakers and wine resellers, from two gallons (9.1L) to a quart (1.14L) for table wines and — temporarily — to a half-gallon (2.3L) for fortified wines.

An even more important contribution to the resurgence of interest in the wine industry was made by the Winemaking Industry Committee, set up in 1956 to investigate all aspects of the manufacture and sale of New Zealand wine. For several years the breweries had succeeded in preventing the spread of wine resellers' licences (introduced in 1948 to make New Zealand wine more freely available to the public), on the grounds that such outlets were unnecessary where New Zealand wine could already

George Mazuran — Lobbyist

The modern era of New Zealand wine dawned with the legislative break-throughs of the 1950s. Those early political successes are largely attributable to the efforts of one man — George Mazuran.

The power of Mazuran's political influence was admired by Hugh Watt, MP: 'There are not eighty MPs but really eighty-one. The extra one [Mazuran] who comes down to sit in the House . . . has almost as much influence as a Member.'[3] Mazuran himself claimed to have been able to sit in the gallery during debates and 'signal to the MPs at the division bells to tell them which way to vote'.[4]

Mazuran made a poor fit for the popular image of a political lobbyist as a smooth, professional operator. Dr Martyn Finlay observed that MPs often regarded Mazuran 'with affection, tinged with amusement at his quaint ways'.[5] What Mazuran lacked in sophistication, however, he more than compensated for with his strong conviction of the justice of his case. Other political assets included an amiable personality, a well-armoured persistence that refused to accept 'no' for an answer, and an exhaustive knowledge of legislation relevant to the wine industry.

Mazuran realised in the early 1950s that the Viticultural Association — of which he was the president — would have to scratch and claw to make progress. 'We had people who would make the wine, but couldn't sell it [in less than two-gallon lots]. I saw it right away — how can you exist if you can't sell your product? Parliament was the only place that could change the laws; if they tied the knot they could undo it.'[6]

While other Dalmatian winemakers helped to tend his Henderson vine-yard, Mazuran immersed himself in the politics of wine. He went 'deeper and deeper. I took an interest in overseas licensing laws; went to Australia and could quote their legislation.'[7]

To foster the sales of the smaller, principally Dalmatian growers belonging to the Viticultural Association was the raison d'être of Mazuran's lobbying career. 'All my life,' he reflected in the 1970s, 'I have been protecting the family winemaker from the [corporate] monster getting in behind.'[8] In contrast to 'the big growers who only represent the breweries',[9] Mazuran stressed the value to the community of 'the wine-growing families and children I was speaking for'.[10]

Frequent contact with those most in a position to benefit the wine industry was the key source of Mazuran's political influence. He care-fully cultivated the friendships of MPs and government officials, spent days at a time on the floor of the House, and canvassed MPs widely before a bill came to the vote. The payoff from such diligent lobbying was, as Peter Wilkinson, the MP for Rodney, put it in 1975: 'It has been well recognised by the Legislature for many years that if New Zealand is to have a healthy wine industry it is vital that the interests of the smaller growers should be constantly kept in mind.'[11]

George Mazuran was awarded the OBE in 1971, in his twenty-first year as president of the Viticultural Association. He died in 1982, aged seventy-three. For decades, in the eyes of many, he had been 'Mr Wine'.

Tom McDonald — Lord of the Bay

Tom McDonald, *Cooks Wine Bulletin* declared in April 1971, was 'like a benevolent grizzly bear . . . big, bald. [He] lords it over the Bay. He rules by force of personality, his views influencing those outside his domain, his insistence on tradition admired there and mostly copied.'[12] Today, Tom McDonald is honoured as the driving force behind New Zealand's first prestige red.

McDonald's first involvement with the wine industry was in a humble role — washing bottles at the Mission during the 1918–19 influenza epidemic. There wasn't room for all the sons on his father's farm. In 1921, fourteen-year-old Thomas Bayne McDonald started work at Bartholomew Steinmetz's Taradale Vineyards.

'We made only fortified wines, such as sherries, ports and muscats then . . . I was very fortunate that the old fellow was a talented wine-maker brought up in the European school, and it was from him I got my basic principles.'[13] In 1926, when Steinmetz, born in Luxembourg, decided to return to Europe for a few years, he leased his business to McDonald and George Hildred. When Steinmetz later decided not to return to New Zealand, and Hildred pulled out of the partnership, McDonald took over the business, carrying on trading as Taradale Vineyards.

The 1931 earthquake ravaged McDonald's winery but in 1932, and again in 1938, he extended his land holdings. During the Depression, McDonald 'really had to sell . . . Often I had to spend ten shillings getting an order. I had to "shout" for the place and make two or three visits to collect the money . . . Table wines were just a sideline then. The main grapes were Pinot Meunier, Chasselas and Cabernet Sauvignon. White wines were made from Pinot Meunier fermented off the skins.'[14]

In 1944, McDonald sold out to the Christchurch-based brewers and merchants, Ballins. 'I had two reasons for selling. The first was that I had two daughters and no sons to hand the business on to, and the other was that it brought in the necessary finance to allow for expansion.'[15] McDonald stayed on, however, as manager; new vineyards were planted and a modern winery built. The company's name was finally changed to McDonald's Wines.

There now flowed a rivulet of Cabernet Sauvignon under the McDonald label that pioneered the modern red-winemaking era in Hawke's Bay. McDonald himself was inspired by the past. Born in Taradale, he was acquainted with Henry Tiffen's daughter and heir, Mrs Amelia Randall, and keenly aware of the *vinifera* table wines produced from the mid-1890s onwards at Tiffen's Greenmeadows Vineyard.

When André Simon, the famous writer on gastronomy and wine, visited Hawke's Bay in 1964, McDonald served him cheek by jowl the 1949 vintages of his own 'Cabernet' and Chateau Margaux. To Simon, this was 'a rare and convincing proof that New Zealand can bring forth table wines of a very high standard of quality . . . The Margaux had a sweeter finish and a more welcoming bouquet, greater breed, but it did not shame the New Zealand cousin of the same vintage.'[16] When he left the country, André Simon took a bottle of McDonald Cabernet 1949 'to show my Australian friends what New Zealand can do'.[17]

At a 1989 dinner to celebrate Montana's purchase and restoration of the old McDonald winery (see The McDonald Winery, Chapter Three), two bottles of McDonald Cabernet 1951 were served. Discovered when the buildings were stripped for renewal, the wine was packaged in slim, 'hock'-shaped bottles. 'After 38 years', wrote J.C. Graham in the *New Zealand Herald*, 'the wine had lost some fruit but retained clean, lingering port-like bouquet. It had lightened in colour to deep rosé, with a nutty taste and distinct Cabernet character.'[18]

In 1962 McDonald's Wines merged with the dominant force in the Bay, McWilliam's Wines (NZ). McDonald climbed to the influential post of McWilliam's production director. In 1965 came the first vintage of the great string of white-labelled McWilliam's Cabernet Sauvignons that were so excitingly superior to anything else in the country.

Tom McDonald's 1965 Cabernet Sauvignon under the McWilliam's label was praised in 1971 by author Frank Thorpy as 'the finest commercial wine made in my lifetime in New Zealand and the forerunner of what I think will be the grand naissance of New Zealand wines'.[19]

John Buck was equally excited by the 1965. 'Aptly described as "Tom McDonald's baby", this wine is of darkish, Cabernet colour and has a lovely varietal nose which, although rather strong in new oak, should improve with age. It also has good body (perhaps still fractionally light but a great improvement by New Zealand standards) and a palate well balanced in tannin and acid . . . it looks a certain winner.'[20]

With McWilliam's financial resources, McDonald and his winemaker Denis Kasza — who made a vital contribution — were not only able to purchase French oak barriques in which to mature their Cabernet Sauvignon for a year; it was then bottle-aged until it was four years old. Each vintage was snapped up by wine lovers who put their names on allocation lists; few bottles ever reached retail shelves.

Awarded the OBE in 1971 — the year of his fiftieth vintage — McDonald retired from McWilliam's in 1976, but stayed on as a consultant and director until 1982, and even chaired the Wine Institute from 1980 to 1982.

A burly man with a powerful intellect, gentle manner and shrewd, sparkling eyes, Tom McDonald died in 1987, aged seventy-nine. 'We remember him as Dad,' his daughter, Mrs Janet Gordon, two years later told a gathering at The McDonald Winery, 'but to most people he was one of the Grand Old Men of New Zealand wine.'[21]

be bought from hotels. In a decision of cardinal significance, the Committee recommended that the existence of other forms of licence should not affect the spread of wineshops and that wine-resellers' licences should be more freely granted. The outcome was the doubling in number of New Zealand wineshops by 1965.

Encouraging the spread of wineshops across the nation was the 1957 committee's essential achievement. There were other useful proposals. The committee suggested the formation of a Viticultural Advisory Committee consisting of winegrowers and departmental officials; that restaurants should be licensed to serve New Zealand (and no other) wine; and that the provision allowing the sale of dessert wine in half-gallon quantities should be made permanent. Together, the recommendations received widespread legislative support.

The incoming second Labour Government rendered further assistance to the winegrowers. Imports of wines and spirits dropped in 1958 and 1959 to half their former volume. Another shot in the arm came with the high taxes slapped on beer and spirits in the 'Black Budget' of 1958. *The Weekly News* in September 1958 declared that the beer drinker 'was rocked on his heels by the sharp bump upwards in beer tax that came with the Nordmeyer Budget. Today, New Zealanders who wend their way homeward after 6 p.m. with brown parcels under arm will often have a bottle of wine as well as the traditional nut-brown brew.'[22]

The effects of Labour's moves are well described in the 1959 Annual Report of the Department of Agriculture. The tax and licensing adjustments had 'created an immediate and unprecedented demand for New Zealand wines. The market position for New Zealand wines changed from one of difficult and competitive trading to a buoyant market capable of absorbing all the wine that producers could supply.'[23] Shortly after, the law was eased to remove restrictions on wine sales in no-licence districts and to allow single bottle sales of fortified wine.

Our southern outpost of the earth's millions of hectares of vines presently covers an area revealed by the New Zealand Vineyard Survey in 1992 to be 6099 hectares — more than fifteen times larger than that surveyed in 1960, when there were 387 hectares planted. This rapid growth rate has involved the New Zealand wine industry in a number of drastic changes.

One outstanding feature, especially of the 1960s, was the heavy investment by overseas companies in New Zealand wine. Foreign interests often staked their claims through investment in previously family-controlled wineries. McWilliam's led the way, establishing vineyards and a winery in Hawke's Bay between 1947 and 1950. In 1961 McWilliam's joined forces with McDonald's at Taradale to form what then became the largest winemaking group in the country.

Takeovers soon became common. Like McWilliam's, Penfolds of Australia decided that the establishment of vineyards in New Zealand would best serve their interests in the local market. In a new company, Penfolds Wines

Brothers Frank (left) and Cecil Vidal, whose Claret and Burgundy were praised by John Buck in the 1960s as 'the two finest, freely available dry reds on the New Zealand market'.

(NZ) Ltd, founded in 1963, the parent company in Australia owned sixty-two percent of the capital and local brewers and merchants held the rest. Later Gilbey's moved into Nobilo at Huapai and Seppelt's of Australia became involved with Vidal at Hastings.

All three companies reverted to New Zealand ownership in the 1970s as foreign investors pared their overseas operations. Control of Nobilo passed to a triumvirate of the Public Service Investment Society, Reid Nathan Ltd and the Development Finance Corporation; George Fistonich of Villa Maria acquired Vidal; and Penfolds was bought by Frank Yukich, formerly head of Montana. Meanwhile, Rothmans entered the wine industry through the purchase of a controlling share in Corbans.

The greatest impact by foreign capital on New Zealand wine was made through Montana. A crash expansion programme launched in the 1960s culminated in 1973 when Seagram of New York acquired a forty percent share in the company. American finance and expertise subsequently enabled Montana to emerge as the dominant force in the New Zealand wine industry.

Having failed to block the emergence of an indigenous wine industry, local wine merchants and brewers, who formerly had favoured overseas labels, began to display a more positive attitude towards New Zealand wine. Hotel bottlestores throughout the country now stocked and promoted the products of the vineyards in which the breweries had a financial stake. Their involvement in wine could be incestuous; in 1980, for example, two-thirds of McWilliam's shares were held by New Zealand Breweries, Ballins and Dominion Breweries between them.

The wine industry also benefited from a proliferation of new forms of liquor licences. From the 1960s the trend towards liberalisation of the licensing laws, evident since 1948, grew much more decisive. Restaurants were licensed in 1960 and taverns in 1961. Theatres, airports and cabarets became licensed between 1969 and 1971, offering new avenues for wine sales. The creation of a permit system in 1976 gave belated legislative recognition to the BYO (bring

Frank Yukich — Visionary

He took the helm of an obscure, fortified-producing winery and with a cyclonic explosion of energy constructed the largest wine company in the land. In 1973 he led Montana into the almost virgin wine territory of Marlborough, triggering the emergence of New Zealand's most internationally acclaimed wine region. Ambitious, ruthless and far-sighted, in the 1960s and 1970s Frank Yukich revolutionised the New Zealand wine industry.

'He has . . . a tack-sharp mind and a squared-away muscularity which would make you think three times before tangling with him,'[24] declared *The Dominion* in 1972. To call Yukich hard-driving would be an understatement — he drank two bottles of wine and slept no more than three or four hours daily.

Montana was founded by Ivan Yukich, a Dalmatian immigrant who planted his vineyard in the Waitakere Ranges west of Auckland and marketed his first wine in 1944. Under the direction of Ivan's sons, Frank and Mate, in the 1960s Montana admitted new shareholders to build up its financial clout, and then plunged into a whirlwind of expansion.

'This virile young company has now risen to the top bracket of the industry,' observed *Wine Review* in 1966. 'Through the medium of major distributors and the company's own wine-shops throughout both islands, Montana wines, sherries, ports, cocktails and liqueurs have become some of the best known.'[25] Montana led the way with such novelties as Cold Duck and Montana Pearl, and was the first to put classical labels in quantity on the market. The winery's output soared from 13,500 litres in 1961 to 5.85 million litres in 1972. A year later, Frank Yukich bought New Zealand's first mechanical grape-harvester.

Above all else, Frank Yukich is remembered as the driving force behind the foundation of the Marlborough wine industry. Alerted to the region's viticultural potential by a favourable report from Wayne Thomas, then a young DSIR scientist, Yukich, according to *Marlborough Wines and Vines*, 'set out, under the cover of the company Cloudy Bay Developments, to buy 1,600 hectares of flat land on the southern side of the Wairau Plain.

'He paid the deposit on the land out of his own pocket, then told the [Montana] board what he had done. His plan was rejected. The board, still convinced that the North Island was the only place to be, argued that Thomas's research was inconclusive. Professor Berg [of California] subsequently studied Thomas's report, agreed with the conclusions, the board then reversed its decision, and gave Yukich the go-ahead to start planting in Marlborough.'[26] The first vine in the modern era of Marlborough viticulture was planted on 24 August 1973.

Bold in action, Yukich was also an indefatigable planner. His five-year plan was the springboard for Montana's initial flurry of growth; 1966 brought an 800-page, ten-year plan. *The Dominion* was agape at 'such planning and analysis as one would regard as being far beyond the capacity of a young man who reached form two at a country school'.[27]

In 1973 Seagram, the multinational liquor giant, took a forty percent shareholding in Montana. The investment was trumpeted as 'basically an export deal . . . [to] export three million gallons [13.64 millionL] a year within five years'[28] to the United States. Yukich's relationship with Seagram, however, proved stormy. In 1974, with the winery reporting a $178,000 loss, Yukich stepped down from the top post and cut his ties with Montana, although his brother, Mate, stayed on as production director.

In 1977 Frank Yukich rebounded with the purchase of a controlling sixty-four percent share in Penfolds (NZ). By buying most of the Australian parent company's stake in Penfolds, the man responsible for heavy overseas investment in Montana turned Penfolds into a New Zealand company. The large Henderson winery, Yukich declared, had been 'just standing still . . . nobody has been pushing to expand the company.'[29] Penfolds' traditional emphasis on sherry and port was swiftly shifted to table wines, and by January 1978 Yukich was back in Marlborough looking for new contract growers.

'The stormy, often controversial rise of Penfolds,' wrote the *Auckland Star* in 1984, 'has been accompanied by bloodletting . . . the rush to become big . . . brought so many problems . . .'[30] Short of grapes, in 1981 Penfolds was apprehended importing concentrated grapejuice, in an effort to satisfy its burgeoning market for restaurant bulk wines. When its hock and moselle catering kegs were seized by the Health Department, the ensuing controversy severely damaged the company's reputation. Penfolds' profitability nose-dived. A proposed $2.5 million share-issue had to be withdrawn, and in 1982 Yukich bowed out by selling his shares in Penfolds to Lion Breweries. As a condition of the sale, Yukich agreed to have no involvement with the wine industry for at least five years. His tumultuous career in New Zealand wine was at an end.

In 1993, in recognition of his mammoth contribution to the industry, Frank Yukich was elected a Fellow of the Wine Institute. 'I take off my hat to Frank Yukich,' says David Lucas, the founder of Cooks. 'He always wanted to be the biggest — and for a long time he succeeded.'[31]

Frank Yukich (right) watches Sir David Beattie, then chairman of Montana, gingerly plant the company's first vine in Marlborough on 24 August 1973.

your own) wine phenomenon by allowing the consumption of wine in unlicensed restaurants. Another amendment that year introduced vineyard bar licences, to enable the sale of wine by the glass or bottle at vineyards for consumption on the premises.

The emergence, relatively recently, of contract grapegrowing has reshaped the structure of the viticultural industry in New Zealand. Traditionally, wineries had grown all their own grape requirements. Viticulture and winemaking formed integral parts of each winery's activities. This pattern altered in the late 1960s when several companies, seeking to avoid the heavy capital expenditure required to establish new vineyards, persuaded farmers to plant their surplus acres in grapevines. The wine companies provided vines, viticultural advice and assistance with financial arrangements in return for guaranteed access to the fruit of the new vineyards.

Vineyard acreages tripled between 1965 and 1970 as contract grapegrowing swept the Gisborne plains. An average winery bought in four percent of its grape requirements in 1960. In 1992, 460 independent growers produced seventy-seven percent of all New Zealand grapes, including eighty-eight percent of bulk-wine grapes and sixty-three percent of premium-wine varieties.

Since 1970 Auckland has lost its former pre-eminence as New Zealand's major grapegrowing region. Auckland's share of the national vineyard area dropped between 1970 and 1992 from nearly fifty percent to four percent. Shifts occurred within the province in this period as many Henderson wineries developed new vineyards further north, in the more rural Huapai-Kumeu area.

Although vineyard expansion was slow in Auckland and the Waikato, further south in Marlborough, Hawke's Bay and Poverty Bay the pace has been hot. Corbans' plantings, for example, spread from Henderson to Kumeu and Taupaki, and then to the East Coast and finally Marlborough. Cooks in the late 1960s established vineyards and a winery at Te Kauwhata, later contracted growers in Poverty Bay and then acquired vineyards at Riverhead (Auckland) and in Hawke's Bay. Montana planted at Mangatangi, south of Auckland, before contracting large acreages in Poverty Bay and pioneering the spread of viticulture to Marlborough.

According to the then Viticultural Association chairman, George Mazuran, quoted in *The Weekly News* in April 1971, the wine boom of the 1960s 'was achieved at the expense of quality'. During the 'Cold Duck' era twenty-five years ago an undiscriminating and unsuspecting public snapped up large quantities of cheap adulterated sherries and table wines. Charged Mazuran: 'Some growers have been getting away with blue murder.'[32]

The 1970s brought an overall improvement in wine quality and heavy emphasis on the production of table wines. Wine production rose between 1960 and 1983 from 4.1 million litres to 57.7 million litres. This, said Alex Corban, means that New Zealand had 'probably the fastest growing wine industry in the world'.[33] The growth area was table wines, which captured twelve percent of the

Josip (Joe) Babich, the founder of Babich Wines, welcomes visitors to his vineyard for the Viticultural Association's field-day in 1966. Born in Dalmatia, he died in 1983, after sixty-seven vintages in the new land.

market in 1962. Today that figure stands at over ninety-two percent and fresh, fruity white table wines dominate the market.

The predominance of these wines reflects the sweeping changes in the composition of New Zealand vineyards. Thirty years ago fewer than one-third of the vines planted in New Zealand were of classical European varieties — the most common varieties were the heavy cropping but poor wine-producing Baco 22A and Albany Surprise. As a Cooks publication has noted: 'The first is prohibited in most European winemaking districts. The second would be if anyone proposed to plant it.'[34]

By 1992 the classical Chardonnay variety was more heavily planted in New Zealand than any other variety. Cabernet Sauvignon is the major variety for red wine. Classic varieties now constitute over ninety-eight percent of all vines in New Zealand.

Thirty years ago New Zealanders each drank an average of two bottles of wine annually — today the average is over twenty bottles. This increase in wine consumption reflects the much greater awareness of wine in the community.

Government viticultural inspector B.W. Lindeman had observed back in 1939 that New Zealanders showed 'not only a lamentable ignorance of wine, but also a very conservative attitude toward it'.[35] Soon after, thousands of New Zealanders stationed in European wine districts during the Second World War had their first fumbling encounters with wine. An anonymous 'Kiwi Husband' writing in the magazine *Here and Now* in January 1952 recalled that most New Zealand soldiers made their first acquaintance with wine only when the supply of beer ran dry. 'It was a rough and ready meeting, and many of us dealt with wine in the

J.C. ('Jock') Graham, who embarked on his love affair with wine while billeted in Tuscany — Chianti country — at the end of the Second World War, is today regarded as the doyen of New Zealand wine writers.

manner to which we had become accustomed. We drank it from the bottle, and by the bottleful, often with sad results to ourselves and a total absence of respect for the vintage . . . It was consumed in quantities that horrified the inhabitants and tortured our stomachs. We drained the countryside of mature stocks and caught up with the harvest. We collected our wine in water carts that held some hundreds of gallons and imparted a taint of chlorine and foul lime sediments; we dispensed it in jerrycans designed for petrol and drank it from the mugs we used for hot tea. And we abandoned it for beer whenever we had the chance . . .'[36]

Some, like 'Kiwi Husband', later developed a more appreciative understanding of wine. The migration of thousands of continental Europeans to New Zealand introduced large groups of Italian, Yugoslav and Greek wine drinkers into our midst. Countless New Zealanders passing through Europe during the post-war boom in overseas travel were exposed to the traditional European enthusiasm for wine.

Yet in the mid 1950s, except in hotel dining-rooms, it was still extraordinarily difficult to savour wine with a restaurant meal. 'The custom was to smuggle in a bottle and hide it under the table . . . some establishments took away the bottle and decanted it into soft-drink bottles, to look more innocent in the event of a police raid.'[37]

'Today may go down as a gastronomic landmark,' the *New Zealand Herald* observed on 8 February 1954, 'for at 7.30 tonight a little group of food enthusiasts . . . will inaugurate a Wine and Food Society in New Zealand, dedicated to the cause of better eating — and drinking.' Dr T.D.C. Childs was interviewed about 'neglected' practices in New Zealand, including 'the habit of drinking wine moderately and intelligently with meals'.[38] Alex Corban, and Dudley Russell of Western Vineyards, were among the society's inaugural officers.

Rising affluence at home encouraged more and more New Zealanders to seek new experiences in food and drink. The mushrooming restaurant trade finally promoted wine as an essential aspect of 'the good life'. No longer, as in wine-

maker Paul Groshek's day, was wine viewed as 'plonk, to be consumed in shame behind hedges and bullrushes'. Wine has become fashionable. The industry's own marketing efforts, the improved availability of quality wine and the emergence of wine competitions, clubs and courses have combined to raise the level of public wine awareness in New Zealand to new heights.

Several wine writers played crucial roles in arousing enthusiasm for wine. Dick Scott's masterly historical treatise, *Winemakers of New Zealand*, published in 1964, was the first book devoted to the subject and remains one of the best. Scott also published the industry's first journal, the quarterly *Wine Review*, between 1964 and 1978.

J.C. (Jock) Graham launched his influential wine column, 'Cellar Book', in the *New Zealand Herald* in 1970. The first regular wine column to be run by a New Zealand daily newspaper, after twenty-three years it is still appearing, a model of good sense and objectivity. Michael Brett started writing about wine for the *Auckland Star* in the early 1970s; today his lively, frank column still appears in the *Sunday Star*.

Frank Thorpy's *Wine in New Zealand*, published in 1971, also made a strong impact. In his book Thorpy delivered broadsides at such abuses as excessive sugaring and wine-watering, and condemned the winegrowers' heavy reliance on hybrid grape varieties. Thorpy was the modern wine industry's first outspoken critic.

Yet the turmoil in the wine industry in the 1970s produced a number of casualties. To switch from the manufacture of fortified wine to the production of classical table wines required heavy expenditure. While some companies acquired the necessary technical and marketing abilities, others fell behind in the race to expand and improve.

The failure of Western Vineyards, Spence's and Eastern Vineyards to lift their winemaking standards beyond those prevailing in the 1960s contributed to their demise in the late 1970s.

Western Vineyards — founded by nineteen-year-old Dudley Russell in the Waitakere Ranges in 1932 — had helped pioneer the modern era of classical table wines and enjoyed formidable competition success in the sixties. At its height of production, the winery put out 35–40,000 gallons (157,500–180,000L) of wine a year, with its Cabernet Sauvignon, 'Gamay de Beaujolais', Pinot Chardonnay and private bin Flor Sherry all enthusiastically received.

The stunning forty-acre (15ha) terraced vineyard in 1964 excited André Simon to jot in the visitors' book: 'I have never seen a more picturesque vineyard anywhere but in Tuscany'. For many years thereafter, Western Vineyards' advertisements brandished this memorable Simon quote — minus the last three words.

But in his later years, Dudley Russell began to lose interest in winemaking; beef farming increasingly absorbed his energies and investments. After his eldest son chose not to enter the family wine business, Russell announced in 1979 that the 1978 vintage had been his last; the vineyard

Alex Corban — Innovator

As Corbans' production manager from 1952 to 1976, he was the first winemaker in the country to adopt such modern technical wizardry as pressure fermentations, stainless steel, refrigeration and yeast starter cultures. The first chairman of the Wine Institute, from 1975 to 1979, Alex Annis Corban cut a powerful figure in the wine industry's affairs for more than thirty years.

Alex (68) — a grandson of the winery's founder, Assid Abraham ('A.A.') Corban — recalls that his 'first job, when I could walk, was to follow the hand-harvesters [through the vineyard] and pick up all the berries that had fallen to the ground'.[39] Only the second New Zealander (after Frank Berrysmith) to gain a Diploma in Oenology from Roseworthy College, South Australia, Corban started work at the family winery in 1949.

Although exposed during his Australian studies to winemaking equipment and techniques unknown to an older generation, Alex Corban initially had to work under the direction of Wadier Corban, his winemaker uncle. 'When W.A. [Wadier] and his brothers were presented with different ways of doing things they were hard men to shift,'[40] recalls Alex.

At the fiftieth anniversary of the company in 1952, sweet fortified wines still dominated Corbans' output, and Jubilee Port and Jubilee Sherry were marketed as the Golden Anniversary specials. Yet there soon flowed an impressive series of winemaking 'firsts' from the Corbans cellars.

In 1949 the entire Corbans range was fermented with cultured yeasts — the first time natural, 'wild' yeasts were not used in New Zealand. At

harvest, instead of using the grapes' sugar levels as the sole measure of ripeness, Corban concentrated on a balance between their total acidity and pH. At the first Easter Show Wine Competition in 1953, the champion wine was Corbans Sauternes. Stainless steel tanks made their New Zealand wine industry debut in 1958 at Corbans. Then the company's Dry White 1962, based on hybrid Baco 22A grapes, was pressure-fermented at controlled temperatures — another first.

Corbans' 1964 vintage was New Zealand's first commercial Pinotage. The company's Riverlea Riesling 1965, according to Alex Corban, was the country's first white wine of that era made '100 percent' from classic *vinifera* grapes. Refrigeration, centrifuges, freeze concentration, pasteurisation, cold sterile bottling — the array of equipment and techniques Corban introduced to New Zealand is vast.

Another breakthrough by Alex Corban was the production of the country's first 'Charmat' method (tank-fermented) sparkling wine. 'The decision had been made that Corbans Premiere Cuvée, rouge, medium white and dry white, would reach the bottling chamber in 1962 — in time for the main event of the company's diamond jubilee celebrations. All the directors lined up to watch the first batch emerge. There were no bubbles.'[41] Contracted by the chilled wine, the joints along the pipeline had opened, allowing the gas to escape. Spanner in hand, Corban set to work, and soon New Zealand had its first 'naturally' fermented sparkling wine.

The year 1962 also witnessed the launch of the country's first commercial 'flor' sherry. Made from classic Palomino and Pedro Ximenes grapes, Corbans Palomino Flor Sherry swept all other dry sherries before it at the 1963 Industries and Commerce judging, and in the same year was even served to Queen Elizabeth II.

In 1965 Corbans gained a notable commercial accolade, by arranging with John Harvey and Sons of Bristol, England, to market a selection of Corbans sherries in New Zealand under a Harvey's Cream and three other labels. This was a breakthrough — the first time a European wine interest had formed a production and marketing link with an established New Zealand producer.

A politician as much as a winemaker, Alex Corban was president of the New Zealand Wine Council (which represented primarily large-scale non-Dalmatian companies) from 1952 to 1976. Looking back, he draws deep satisfaction from having 'stuck up for the industry on organisational bodies. When I was chairman of the Wine Institute we worked out the 1978 Industry Study and Development Plan, and I was forever immersed in the industry's tariff battles.'[42] In 1978 Corban was awarded an OBE, and in 1983 he was elected a Fellow of the Wine Institute for 'long, distinguished and signal service to the wine industry of New Zealand'.

Alex Corban is now retired, although he still acts as a consultant to small wineries. His son Alwyn is a partner in the Ngatarawa winery in Hawke's Bay. Alex's unpublished autobiography, *Behind the Wine Curtain*, has earned praise from Hugh Johnson, the leading British wine author.

Corban wants this verse for his epitaph:

I hope they say, when in the ground I'm sunk,
He was sober but his wines were drunk.[43]

Harold Innes — Wine Promoter

'Go anywhere in the wine industry in the Auckland province these days and his name crops up — this idiosyncrasy, this brewer turned wine promoter, a craggy-voiced, white-haired, hustling man of 62, who more than anyone has boosted sales of New Zealand table wines.'[44] The anonymous writer of the December 1970 edition of *Cooks Wine Bulletin* was clearly much impressed with Harold Hirst Innes.

In an era when most New Zealanders viewed wine with slight suspicion, Innes was a self-confessed wine lover. His family had been involved in the beer trade since his grandfather sold beer during the nineteenth century New Zealand Wars. Yet Harold Innes was fond of saying: 'There is more wine drunk in my house than beer. A bottle of whisky lasts three months, a case of wine only a week.'[45]

Innes joined his family's brewing business in 1950. Appointed managing director of a company of wine and spirit merchants in Hamilton, he sought advice on wine from Denis Kasza, Kevin Schollum and Paul Hitchcock at the Te Kauwhata viticultural research station. 'They took an interest in me, we collected New Zealand wines and started our own panel for judging.'[46] He also began a life-long friendship with Mate Brajkovich of San Marino (Kumeu River).

In 1962, when Innes Industries amalgamated with L.D. Nathan, Harold Innes shifted to Auckland. Disappointed with the lack of interest in New Zealand wines, in 1965 he launched his masterstroke — the Connoisseurs Club. A wine and cheese club based at Newmarket Wines and Spirits, with Joan Ricca as the popular secretary, from the start the Connoisseurs Club was a roaring success.

The aim of the club, Innes declared in 1970, was to bring together people, wine and food. 'The title was presumptuous then, and perhaps still is now. But we have brought the average man into contact with wine and food. And we have developed the idea of bringing New Zealand table wines into the home.'[47]

The Connoisseurs Club staged its first function in May 1965. Over three evenings, 700 members tasted award-winning entries from that year's Easter Show Wine Competition. A year later, 2000 Aucklanders consumed almost 5000 bottles of wine while dancing on the green at Redwood Park, Swanson. 'Never before has the New Zealand vintage been so splendidly heralded,'[48] enthused *Wine Review*. In 1974 — the first year the event was held at Kumeu — the 4000 festival-goers drained 8000 bottles of New Zealand wine.

As L.D. Nathan's wine interests multiplied, Innes's influence spread. He became chairman of John Reid and Co., a major wine and spirit company, and was also deeply involved with Reid's chain of twenty-five wineshops. His enthusiasm for the initially unfashionable cause of New Zealand wine undiminished, Innes's impact was strongly felt on Auckland retail shelves.

Harold Innes died in 1985, at the age of seventy-five. 'I practise what I preach,' he was fond of saying. 'No guest sits down at my table without wine.'[49]

Harold Innes in his favourite role — spreading the gospel of New Zealand wine at an early Connoisseurs Club festival.

The historic first meeting of the provisional executive committee of the Wine Institute, held on 1 October 1975. From left: Terry Dunleavy (acting executive officer), Montana; Peter Fredatovich, Lincoln; Russell Gibbons, Montana; George Mazuran (deputy chairman); Alex Corban (chairman); Mate Selak; Tom McDonald, McWilliam's; Stan Chan, Totara SYC; Mate Brajkovich, San Marino; Peter Babich.

would be uprooted and all wine stocks auctioned off. He died less than a year later, aged sixty-seven.

Other old-established wineries in the Henderson Valley have suffered drastic declines in production. Vineyards seeking a permanent place had to acquire the sophistication necessary in an increasingly competitive market or face an uncertain future.

Many years' negotiations finally achieved in 1975 the formation of a single, united wine organisation to represent all New Zealand winemakers. Previously, the internal politics of the wine industry had been dominated by an extreme divisiveness rooted in the contrasting economic fortunes of large and small-scale growers. When the Viticultural Association on the one hand — composed mainly of small-scale growers of Dalmatian origin — and the Wine Council and Hawke's Bay Grape Winegrowers' Association on the other — representing primarily large-scale non-Dalmatian companies — were not pursuing individual paths on matters affecting the wine industry, they spent rather less time in co-operation than at each others' throats.

Management of the new Wine Institute was vested in a ten-member executive committee — three of whom are non-voting — consisting of the elected representatives of three categories of winemakers grouped according to annual levels of production. The formula agreed upon — three representatives each of the small and medium-sized growers and four representatives of the big companies — ensured that, at least initially, a majority of executive committee members would belong to the Viticultural Association. The large companies were relying upon their belief that the wine industry fell 'into three different categories rather than two, and that the kind of representatives who would emerge from the middle group would tend towards the view of the larger companies rather than the smaller growers'.[50]

The Wine Institute's existence is recognised by statute in the Wine Makers Levy Act 1976, which requires all licensed grape winemakers to fund the Institute's activities via a compulsory annual levy based on sales, and also the Wine Makers Act 1981, which governs the grant and renewal of licences. In 1993 the Wine Institute had 180 members.

Fundamentally a pressure group, the Wine Institute has summarised its functions as 'representational, regulatory and promotional, in approximately that order of importance'.[51] Apart from lobbying, a multitude of tasks are accomplished, from behind-the-scenes subcommittee work on such areas of special concern as viticulture, winemaking regulations and tariffs, to more visible promotional activities, including the co-ordination of export campaigns and administration of the Air New Zealand Wine Awards.

Responsible for the day-to-day conduct of the Institute from its inception was its indefatigable executive officer, Terry Dunleavy. Extremely extroverted, and with a powerful political bent, he had a pronounced personal impact on the wine industry. Following Dunleavy's retirement in late 1990, Philip Gregan (33), his former deputy, stepped into the top post. Gregan is a quieter, more studious figure than Dunleavy. 'With the modern accent on fact and dispassionate analysis,' wrote Dunleavy, 'Philip Gregan is better qualified than I ever was to find acceptable solutions . . . to the new challenges facing the industry.'[52] Gregan has a clear view of his role: 'to get the big picture right for winemakers, so they can get on and do what they do best — which is to produce wine'.[53]

The wine industry encountered sustained criticism in 1979 when it became widely known that the illegal practice of wine-watering was common in New Zealand. Many wineries had taken advantage of the continuing shortage of wine in the market place to 'stretch' their products.

'The fact is,' wrote David Lucas of Cooks in 1980, 'the New Zealand wine industry has been following thoroughly fraudulent winemaking practices for many years, and if this news gets into the overseas press, our chances of getting into the world's wine markets will be irreparably damaged.'[54] Several scientific studies yielded evidence suggesting that consumers annually had been paying for up to fifteen million litres of tap water masquerading as wine.

Amendments to the Food and Drug Regulations in 1980 dropped the previous prohibition of water addition and set a scale of minimum grapejuice levels: ninety-five percent for premium or varietal wines, eighty percent for non-premium table wines and sixty percent for dessert wines. A pledge by the Health Department to tighten its surveillance of the regulations led to a marked lift in the quality of 1980 vintage wines.

Yet the whole issue flared again in 1981 when several firms released large volumes of 'flavoured wine', especially in casks. 'Flavoured wine' by law was able to contain as low as forty percent grapejuice. Confronted by a heavy barrage of adverse publicity — and an impending grape glut — the winegrowers finally agreed in 1982 to support moves to prevent watering. Amendment No. 7 to the Food and Drug Regulations dropped altogether the 'flavoured wine' category (although the launch of wine coolers in the mid–late 1980s effectively re-introduced flavoured wine by another name). From 1983 table wines were permitted to contain only fifty millilitres of water per litre of wine, where the water had been used as a processing aid for legal additives. Now, table wines of any description must be produced almost wholly from grapejuice, although the sixty percent minimum juice level for dessert wines remains.

In 1979 the wine industry had been referred for study to the Industries Development Commission as part of the Government's policy on economic 'restructuring'. In its 'Wine Industry Development Plan to 1986', published in December 1980, the Commission sought to 'assess the potential of the wine-producing industry to contribute to the future growth of the economy, taking into account the interests of the wine-producing industry, consumers, and the distributive trades, and recommend a strategy for future development'.[55]

The Commission's essential conclusion was that the wine industry deserved special encouragement. The industry employed 3000 people and turned out a product with eighty-five percent domestic content. Nevertheless, the IDC strongly criticised the price of most New Zealand wine as being too high for the future welfare of the industry. Soaring costs were threatening to push the price of a bottle of wine beyond most New Zealanders' reach.

The IDC produced a series of valuable recommendations designed to contain escalating costs through the encouragement of stiffer competition in the market. In the Commission's view, over-protection of the local industry from imported wine had placed a burden on the consumer unjustified by the wine industry's 'poor' export performance. Distortion of competition in the wine market also derived from the commercial dominance of a select cluster of wineries, merchants and resellers described by the IDC as 'a highly cartelised group characterised by their oligopolistic influence in the market'. The lack of real competition had encouraged an unhealthy 'cost-plus attitude to escalating costs . . . to a point where consumer resistance to price shows incipient signs of developing into a major constraint upon consumption . . .'[56]

Several, although not all, of the IDC's recommendations won government acceptance. Foreign wines were freed from import licensing in 1981, although not from tariff restrictions. In an effort to stimulate greater competition in the wholesale distribution of wine, a new class of wine-distributor licence was created. And sales tax on wine was altered from a value basis to a volume rate, producing a drop in the retail price for the better class wines.

Yet as early as 1982 serious doubts arose about the successful achievement of the aims of the Wine Industry Development Plan. The plan had projected an annual per capita consumption of New Zealand wine of fifteen litres by 1986, but consumption eased from 12.2 litres in 1981 to 12.1 litres in 1982, then to 12.0 litres in 1983. The drop in sales, largely linked to heavier imports and consumer price resistance, proved a serious setback to an industry geared to rapid growth.

Heavy overplanting of new vineyards in the early 1980s helped raise the spectre of a wine glut. For over two decades, from the early hints of the wine boom in the late fifties until the early eighties, wine grapes were in short supply. Winemakers, understandably eager to cater for the burgeoning demand for New Zealand wine, reacted by committing themselves to a multitude of long-term contracts with grapegrowers. The industry needed 55,000 tonnes from the bumper 1985 vintage to satisfy the current thirst for wine — but crushed 78,000 tonnes. Production climbed by forty-three percent on 1984, sales rose only two percent, and a surplus resulted of record proportions.

David Lucas — Entrepreneur

The word 'entrepreneur' crops up repeatedly in David Lucas's conversation. That's not surprising, for in the late 1960s Lucas was the driving force behind the flowering of a high-profile, major new winery in New Zealand — Cooks.

Eloquent and extroverted, Lucas is an extremely fit-looking 54 year-old. Born in Stratford, he abandoned farm labouring at an early age ('the hard slog was not for me') to move behind retail counters. Vinka, his former wife, poured her design flair into an Auckland city boutique, Maree de Maru, with Lucas as managing director. The success of Maree de Maru's bridal outfits gave Lucas his financial springboard into wine.

In the 1960s Lucas noticed that Dalmatian friends in the Henderson area 'had large cars, small vineyards and went for trips back home. At about the same time, licensed restaurants were being opened. A waiter at La Boheme in Auckland suggested a bottle of wine with our meal. Like most people in those days, I thought he was having me on. But I ordered it, drank it and put two and two together. Two years later, I was writing feasibility studies for a winemaking business using classical grape varieties.'[57]

The first report, on 17 May 1967, circulated by Lucas and Simon Ujdur of Birdwood Vineyards, called for the purchase of a fifty-acre (20ha) vineyard site at Te Kauwhata. 'The industry has only a limited number of vines and these in the main are large bearers, planted by pioneers who knew and cared little about quality wine . . . Many of the leading wine connoisseurs in New Zealand recognise the need for a model vineyard, devoted only to quality, that will raise the standard of New Zealand wines.'[58]

Lucas's frequent barbs about the wine standards prevailing in the late 1960s won him few friends within the industry. 'Everyone predicted we would go broke — that classical grapes would not grow in New Zealand.'[59] Lucas, however, saw himself as 'an idealist — one who sets a goal and goes for it . . . My major concern was to beat the sceptics.'[60]

One early masterstroke was to name the infant company Cooks, thereby instantly evoking thoughts of age and respectability. By knocking on wine merchants' and private investors' doors all over the country, Lucas raised $35,000 — enough to set the ball rolling. A talented team of directors was assembled: Lucas, wine and spirit company director Jim Wallace, lawyer John Fernyhough, accountant Graham Goodare, and sharebroker David Smythe.

The first vine — a Müller-Thurgau — was planted at Te Kauwhata in May 1969. By the end of the year the planned fifty-acre (20ha) vineyard of Pinot Gris, Chasselas, Müller-Thurgau, Cabernet Sauvignon, Pinotage and Meunier had been established. In 1970 Cooks went public, raising its issued capital to $432,000.

The year 1972 brought Cooks' first vintage, with winemaker Kerry Hitchcock (now chief winemaker at Corbans) at the helm. Four bronze medal-winning table wines emerged: two 1972 'Rieslings' (Müller-Thurgaus) from Te Kauwhata and Gisborne fruit, a Chasselas, and a sweetish blend of Chasselas and Müller-Thurgau labelled 'Chasseur'. Two years later, when Cooks' 1972–73 Cabernet Sauvignon scooped the trophy for champion wine at the Easter Show Wine Competition, the winery emerged as one of New Zealand's leading table wine producers.

Cooks' early fortunes were boosted by Lucas's enormous flair for publicity. *Wine Review*, the industry's quarterly magazine, noted slightly disapprovingly in 1973 that 'Mr Lucas, who once operated an advertising agency, has himself shown no mean ability at winning the spotlight. As the vineyard progressed toward maturity, his pronouncements on wine industry matters occupied more column inches than the views of established industry spokesmen.'[61]

Yet Lucas's early vision of Cooks — a quality-orientated winery producing premium table wines from classic *vinifera* grape varieties — was to be betrayed. With 'Bragatto', a sweet fortified wine flavoured with quinine and herbs (and labelled as Italian Blend — Riserva Superiore) the company departed from its lofty intentions, used the water hose, and made a lot of money.

'Once we were in profitability and had gained recognition,' recalls Lucas, 'in the mid–late '70s things went wrong. The accountants [answering to a changing parade of big investors] took over and made the marketing decisions.'[62] To generate cashflow, Cooks' marketing became more aggressive. 'After 1976 we were forced to "stretch" some of our wines just to maintain our share of the market.'[63] Watery-thin 'Party Packs' involved Cooks in the very winemaking practices it had once so vehemently condemned.

In 1982 Lucas resigned from Cooks' board, having 'lost interest completely'.[64] Four years earlier, after circulating a report strongly critical of the breweries' stranglehold on liquor distribution in New Zealand, he had been forced to step down from Cooks' deputy chairmanship. Defiance, doggedness and cheek — the qualities Lucas had so strikingly displayed in Cooks' formative years — were no longer in demand.

Has Lucas ever contemplated re-entering the wine industry? 'No. I'm a pioneer. I've done my bit, and can't see the same marketing opportunities now.' Looking back, what gives him the greatest satisfaction? 'Chasseur — an enormous amount still goes out. Over the years, you're talking about vast, vast quantities.'[65]

David Lucas is remembered in the wine industry as an extraordinarily talented entrepreneur. In the late 1960s Lucas set out to create a brand-new wine company, producing large volumes of New Zealand table wines based on *vinifera*, rather than hybrid, grape varieties — and pulled it off.

TOTAL VINE AREA IN NEW ZEALAND
(Hectares)

1983	5927
1984	5998
1985	6000
1986	4500
1987	Not Available
1988	Not Available
1989	5437
1990	6000+
1991	6000+
1992	6099

SOURCE: Wine Institute of N.Z. Annual Report 1992

NEW ZEALAND WINE PRODUCTION AND SALES
(incl. exports)
(Million litres)

	Production	Sales
1983	57.7	38.8
1984	41.7	42.6
1985	59.6	43.4
1986	42.4	51.8
1987	37.7	47.6
1988	39.2	47.8
1989	45.6	41.7
1990	54.4	43.3
1991	49.9	46.7
1992	41.6	51.1

SOURCE: Wine Institute of N.Z. Annual Report 1992

The winemaking industry in New Zealand has been on a century-old rollercoaster ride, soaring and plunging through successive periods of growth and optimism, decline and disillusionment. Between mid-1985 and early 1986, following the sustained prosperity of the 1960s and 1970s, the rollercoaster once again turned groundwards, carrying three privately owned wineries — the Villa Maria/Vidal stable and Delegat's — into receivership and, by their own admission, Nobilo and Glenvale to the brink of it.

Two committees of government officials which studied independently and in depth the issue of the wine industry grape surplus reached identical conclusions on its principal cause. In the view of the 1983 'Hartevelt' Committee, 'both growers and companies appear to have ignored the mounting evidence that trends since 1980 have been unsustainable . . . the surplus is seen as an inevitable result of a production system which lost touch with the market for the final product.'[66]

Two years later, the interdepartmental overview committee which conducted a 'mid-term' review of the Wine Industry Development Plan was concerned that 'despite all the recent debate on the subject, significant overproduction of wine in any given year is still possible . . . The problem arises from the industry's ability to divorce itself from market realities . . .'[67]

Discontent within the wine industry focused most angrily on the Labour Government's November 1984 Budget, which lifted the sales tax on fortified sherries and ports by fifty-four percent, from $1.05 to $1.62 per bottle, and on table wines from fifty-four cents to ninety-nine cents per bottle — a searing hike of eighty-three percent. Finance Minister Roger Douglas and Trade and Industry Minister David Caygill claimed they were remedying a position whereby the industry had 'received very favoured treatment with regard to alcohol tax'. In concert with the over-enthusiastic expansion of vine plantings and the slow-down in consumption growth, this brutal tax hike brought the wine industry to its knees.

The decision by the giant Cooks/McWilliam's company — formed by a merger in September 1984 — to unload its surplus stocks by cutting prices then set alight the ferocious price war of 1985–86. Villa Maria/Vidal, Penfolds, Montana, Delegat's, Corbans and Glenvale all chopped their ex-winery trade prices to as low as one-half the production cost. The cut-throat discounting brought an explosion in the sale of cheap sparkling wines and casks, sliding demand for the wineries which stayed aloof from the price slashing, and a battle for life for the family vineyards lacking corporate financial backing.

In February 1986 the Government intervened. According to Douglas Myers, chief executive of Lion Corporation, which then owned Penfolds, 'We came together with Mr Douglas, the major companies in the wine industry — Brierley, Rothmans, Montana and ourselves — and the Government asked us what the problems were and what do you want us to do? Within a fortnight a package was announced.'[68]

The Government's decision to provide adjustment assistance partly stemmed from its belief that, without some form of direct intervention, the prospects for an efficient and lasting solution to the oversupply problem were dim. It also accepted that, in hindsight, the Wine Industry Development Plan had encouraged the industry to expand to an unrealistic extent, and that the sales tax increases of 1984 had had a strong negative impact on wine consumption. The looming prospect of another bumper harvest in 1986 added urgency to the issue.

The Government's offer, up to ten million dollars to fund a vine-uprooting programme, had initially been opposed as too short-term by the Wine Institute, which had campaigned for reform of the sales tax. Following the Government's decision to pay grapegrowers $6175.00 per hectare to eradicate up to twenty-five percent of the national vineyard, 1517 hectares of vines were uprooted.

The heaviest vine pulls were on the East Coast: Hawke's Bay lost 534 hectares of vineyards and Poverty Bay/East Cape 586 hectares. In the South Island, 210 hectares were removed; in Auckland, 161 hectares. Müller-Thurgau, the grape variety upon which most bulk white wines are based, suffered the severest losses: 507 hectares, notably in Hawke's Bay and Gisborne. According to George Fistonich of Villa Maria/Vidal, more than sixty grapegrowers quit the industry.

The industry's travails also resulted in a severe loss of investor confidence and sweeping ownership changes at Montana, Penfolds, Villa Maria/Vidal, Cooks/McWilliam's and Delegat's. Peter Masfen gained control of Montana through Corporate Investments, in which he has a majority stake. Seagram's decision to divest itself of its forty-three percent shareholding in Montana is not difficult to understand — in the half-year to 31 December 1985 the company chalked up a net loss of $4,621,000. In August 1986 Montana succeeded in a takeover of Penfolds Wines (NZ), thereby creating the country's biggest wine company, commanding about forty percent of the market for New Zealand wines.

Corbans bought the vineyard and winery assets of Cooks/McWilliam's in early 1987 for $20 million. This move surprised few observers: Corbans is a subsidiary of DB Group Limited (formerly Magnum Corporation), itself a subsidiary of Brierley Investments — as was Cooks/McWilliam's. The enlarged Corbans has now clearly emerged as the second largest wine company in the country, holding over thirty-five percent of the market.

A highlight of the past decade has been the emergence of Canterbury, Wairarapa and latterly Central Otago as successful new wine regions. Of the principal established regions, Marlborough has clearly emerged as the most heavily planted in the country, with 2071 hectares under vines, according to the 1992 vineyard survey. Hawke's Bay, with 1614 hectares, is a shade ahead of Poverty Bay's 1503 hectares. The only other regions with more than 100 hectares in vines are Auckland (241ha), Canterbury (198ha), Wairarapa (189ha), and Waikato/Bay of Plenty

(161ha). Given the high fertility of Gisborne and Hawke's Bay compared to Marlborough, the two North Island regions produce over sixty percent of the nation's grape crop. New Zealand wine, therefore, is still primarily of Hawke's Bay and Gisborne origin.

Hard on the heels of the chronic oversupply problem of 1985–86, 1987 and 1988 brought what was inconceivable a year or two earlier — a wine shortage. The 1986 — following the vine-pull scheme — and low-cropping 1987 and 1988 vintages each yielded up to twenty percent (or ten million litres) less wine than the current level of sales. By mid-1988, so acute had become the shortage of New Zealand wine that the two industry giants, Montana and Corbans, turned to the production of cask wines based on apples or kiwifruit. 'In blind tastings, people have great difficulty differentiating fruit wines from the real thing,' observed Bryan Mogridge, then managing director of Montana.

In mid-1988 the Labour Government again angered winemakers by holding out its hand for $23 million. By abolishing the 'standard values' system of stock valuation — which since 1968 had allowed wineries to pay income tax on their wine stocks when they were sold, rather than when they were still maturing — the Government severely strained the winemakers' financial resources, and removed a key incentive to improve wine quality by maturing stocks.

The supply/demand position again reversed in 1989–1990. Wine consumption had nose-dived from 1986 to 1989, when the domestic demand for New Zealand wines reached its lowest level for a decade. The fall-off in demand alarmed winemakers, whose cellars were awash with stocks from the bumper 1989 and 1990 vintages.

The Wine Institute, citing such factors as the country's economic recession and emigration, saw the causes of the decline as 'many and complex'. Resistance to rising prices was part of the story. Following the low-yielding 1986–1988 vintages, a number of wineries took advantage of the temporary wine shortage to inflate their prices, in some cases by up to fifty percent overnight. Uppermost among the Wine Institute's fears, however, must have been the rising tide of imports. Overseas wines, which in 1987 only cornered 5.2 percent of the total New Zealand wine market,

by mid-1992 had grabbed 17 percent. New Zealand brands continued to dominate bulk cask wine sales — the largest market segment — but their grip on the premium bottled wine market had been broken.

The Government, although funding the vine-uprooting scheme, had also moved to speed up the removal of barriers against overseas wines. A radically liberalised tariff structure was unveiled in late 1985. Part of the Labour Government's Wine Industry Assistance Package — which also featured the Grapevine Extraction Scheme — this new regime was centred on a series of annual tariff reductions and quota increases.

By the middle of 1990 there were no quantitative restraints on the volume of overseas wine which could be brought into New Zealand. The heavy duties with which cheap imported wines valued at less than $2 per litre FOB (free on board) were previously saddled, were slashed. Imports from countries other than Australia carried a twenty-five percent tariff. A diverse array of wines from Italy, Chile and Eastern Europe flooded in. By 1995 the tariff will have shrunk to only thirteen percent.

By mid-1990 the Government had also entirely abolished trans-Tasman tariffs on wine, allowing Australian wineries to contest the New Zealand market on an equal footing with local producers. The Australian invasion was spearheaded by its buxom red wines, which have long been popular in New Zealand. Less conspicuously, Australia also dominated the imported still white, sparkling and fortified wine markets.

Never before had New Zealand consumers been treated to such an enthralling pageant of overseas labels. The local industry's determination to claw back some of the lower-priced market it had forfeited to cheap imports was soon signalled by the launches of such budget-priced labels as Penfolds' Clive River (from Montana), Villa Maria's St Aubyns, and Corbans' 'white label' range of sub-$10 varietals. Mission, Robard and Butler, and deRedcliffe slashed their prices across the board.

The steady decline in the domestic sales of New Zealand wine in the mid–late 1980s plateaued in 1989–90, and between mid-1990 and mid-1992 sales rose by more than

WINE CONSUMPTION
(Litres of New Zealand wine, per capita)

Year	Litres
1983	12.0
1984	13.0
1985	13.1
1986	15.4
1987	13.7
1988	12.9
1989	11.7
1990	11.7
1991	12.1
1992	12.9

SOURCE: Wine Institute of N.Z. Annual Report 1992

WINE IMPORTS
(Millions of litres)

Year	Millions
1983	2.50
1984	3.18
1985	3.85
1986	2.79
1987	3.73
1988	4.38
1989	6.80
1990	7.99
1991	11.40
1992	8.40

SOURCE: Wine Institute of N.Z. Annual Report 1992

A symbol of the much-discussed grapevine extraction scheme — a mound of uprooted vines in an Auckland vineyard.

Wine arrived on most supermarkets' shelves only in late 1990, but already the grocery trade commands over a quarter of the New Zealand retail wine market.

ten percent. The trend reflected a key liberalising influence in the Sale of Liquor Act 1989, which licensed supermarkets and grocery stores to sell wine.

With liquor in New Zealand a complex and emotion-charged issue, it is hardly surprising that the road to supermarket and grocery store wine sales was long and tortuous. Conflicting advice given to successive governments by various committees had clouded the issue. The position of the winemakers themselves also shifted.

The Wine Institute's Study and Development Plan submitted to the Government in early 1979 conspicuously omitted any request for wine sales in supermarkets and grocery stores. The reasons are not hard to isolate: smaller winemakers often had ownership links with the wine reseller outlets, and larger wineries with the hotels and 'wholesale' outlets, which stood to lose market share if supermarkets and grocery stores stocked wine.

Confronted with surplus wine stocks in the mid-1980s, however, and the declining competitiveness of wine reseller outlets, the Wine Institute began vigorously crusading for supermarket wine sales. 'The law says to women: if you want wine, make a special trip to a shop which specialises in alcoholic beverages only; you are not allowed, or not to be trusted, to buy it where you buy the household groceries . . . In view of the close association, especially of table wine, with food, we believe that there is no justification for legal denial of the opportunity for its sale alongside the other food products wine is intended to accompany . . .'[69]

Parliament's epochal 1989 decision to allow food stores (except in licensing trusts and no-licence districts) to sell table wine for off-premise consumption nearly went the other way. The provision was initially lost in a June vote by a margin of 33:37; only an improved attendance in the House of yes-voting MPs when the issue was recommitted in July — rather than MPs changing their minds — carried the day by 41:34. Fourteen months later, the Devonport New World in Auckland became the first new food store to sell wine in more than thirty years (a few Hamilton grocers had long had licences).

Much debate continues around the issue of the rate of sales tax (now called excise) on New Zealand wines. A decade ago, a New Zealand Institute of Economic Research report on 'The Impact of Taxation on Alcohol Consumption' found that wine was more price 'elastic' (sensitive) than beer or spirits, and predicted that any increase in the real price of wine would lead to a corresponding fall-off in demand. When Labour virtually doubled the excise on table wines in 1984 from 72 cents to $1.32 per litre, and also raised the excise on fortified wine from $1.40 to $2.16, the ensuing slump in sales was inevitable and immediate.

For five years thereafter, despite soaring inflation, the excise rate was not increased by Labour. Finally in the 1989 Budget the excise on table wine rose to $1.50 per litre, and was thereafter adjusted on a six-monthly basis, according to movements in the Consumer Price Index. In late 1992, the Wine Institute welcomed an undertaking by the National Government not to increase wine excise in real terms in the foreseeable future.

Successive governments have sought to justify alcohol taxes on the grounds that they provide funds with which to tackle the social costs of excessive drinking. High excise taxes also force the vast majority of responsible drinkers to subsidise the irresponsible ones. In mid-1993 the excise on table wine was $1.73 per litre — equivalent to $1.30 per bottle. About $3 of the cost of a $10 bottle of wine is now attributable to excise, tax-related margins and GST.

Yet despite the widely condemned excise and the industry's long-running inability to match its supply of grapes to market demand, new wineries are mushrooming from Northland to Central Otago. In 1984 the Wine Institute had 93 members; in 1990, 131; in 1993, 180. This increase in wineries has not been matched by a major increase in domestic wine sales, but the wine industry is forever abuzz with the excitement of new companies, new faces, new labels.

Attracted by New Zealand's recently proven ability to produce world-class wines, a wave of French and Australian producers has recently swept into the industry. One expression of this burgeoning international interest has been the purchase of New Zealand grapejuice and wine in bulk: the Australian companies Wolf Blass, Andrew Garrett, Thomas Hardy and Domaine Chandon have all produced trans-Tasman blends. Yalumba, another Australian company, buys Hawke's Bay and Marlborough-grown Sauvignon Blanc and Chardonnay fruit, produces the wines in New Zealand, and markets them under its Nautilus label on both sides of the Tasman. Cape Mentelle, a high-flying Western Australian winery, founded Cloudy Bay in Marlborough and set the wine world instantly alight with the magical quality of its 1985 Sauvignon Blanc. (Both Cape Mentelle and Cloudy Bay are now principally owned by the French Champagne house, Veuve Clicquot.) Mildara Wines of Australia from 1988 to 1993 owned Morton Estate, the Bay of Plenty-based winery. Wolf Blass, now merged with Mildara, from 1989 to 1993 owned vineyards in Marlborough and a joint venture winery — Marlborough Cellars — with Corbans.

Montana's premium bottle-fermented sparkling, Deutz Marlborough Cuvée, is produced under the guidance of the French Champagne house of Deutz. The deRedcliffe winery in the Waikato and the Highfield winery in Marlborough are wholly or principally Japanese-owned. New Zealand is a new star on the international wine scene — and increasingly foreign investors want part of the action.

To ensure international confidence in New Zealand's wine labels, a Certified Origin (C.O. for short) system will shortly be introduced. In the past, North Island wines are known to have masqueraded as Marlborough's; undoubtedly other abuses of regional labelling have occurred. The new system is designed to achieve 'honesty and integrity' in labelling.

The raison d'être of the Certified Origin system is 'to guarantee the geographic origin of the grapes from which a wine is produced, when such an origin is claimed on the label'. For a wine to be part of the C.O. system, at least eighty-five percent of its grapes must come from the geographic area, vintage and variety stated on the label. To enforce the system, wineries will be required to keep records that enable auditors to trace back from a wine to its geographic area. Wines not labelled with a geographic descriptor are not eligible for inclusion in the scheme.

The C.O. system is based on a hierarchy of geographical denominations. The broadest designations — designed to accommodate regional blends — are New Zealand, followed by North Island or South Island. Next in the hierarchy are sixteen regions, such as Canterbury or Auckland. Then come localities, such as Fernhill, defined as 'identifiable localised areas'. Finally, and most precisely, come individual vineyards.

The Certified Origin system makes no judgments on the quality of wines produced in denominated areas; it seeks only to guarantee their authenticity. The development of the system is long overdue — but no less welcome for that. The winemakers benefit by the protection of their well-known regional, locality and vineyard names, and wine lovers have a greater chance of getting the truth about the contents of the bottles they buy.

At mid-1993 the outlook for the New Zealand wine industry was mixed. Overseas demand for New Zealand wine was soaring, but a near-halt in vine planting since 1990 (in response to false predictions of yet another wine surplus) was threatening its ability to expand exports. The low-cropping nature of the 1992 and 1993 vintages exacerbated the problem; in both years Montana and Corbans were forced in import Australian bulk wine to maintain their cask-wine supply.

In a forthright November 1992 report entitled 'New Zealand Wine 1993–2000: A Working Paper', the Wine Institute predicted that 'export sales, which have already risen from $4.5 million in 1987 to $34.7 million in 1992, can achieve a further three-fold increase to at least $100 million by the year 2000 . . . To meet the production requirements of a new $100 million export market, the industry requires at least another two thousand hectares of grapes — an increase of thirty-three percent from the current vineyard area . . . the total new investment [needed] is an estimated $280 million.'[70]

Samuel Marsden observed 174 years ago that 'New Zealand promises to be very favourable to the vine'.[71] That distant prediction has lately been brilliantly fulfilled — leading one overseas enthusiast to even suggest that New Zealand should be planted in a sea of Sauvignon Blanc vines stretching from North Cape to Stewart Island.

Following the successful invasion of the New Zealand market by foreign labels, the wine industry's most urgent task is to ensure that export sales strongly surge ahead on the rising tide of international applause for New Zealand wines. In the 1980s New Zealanders discovered the fresh, fragrant, flavour-packed quality of their wines — what will the 1990s bring?

Wine Exports

'An international impact with white wines of startling quality'

New Zealand's white wines are frequently praised as amongst the finest in the world, yet the country commands a meagre 0.2 percent of the international wine trade. The recent growth trend in New Zealand wine exports has been impressively steep; the future promises an even sharper ascent.

For over a century enthusiasts have predicted a buoyant overseas trade in New Zealand wine. Walter Brodie, for instance, ignoring the almost total absence of wines in the colony, declared in 1845 that 'New Zealand, in a few years, will export much wine'.[72] Brodie was wrong. Andrew Tod, a Scotsman who made wine in Wanganui as early as 1873, later shipped a batch home, and in December 1874 the *Dundee Advertiser* reported: 'We have seen some samples of the New Zealand wine . . . it appears to be of excellent quality, and has improved rather than deteriorated in bouquet by the long voyage.'[73]

This, the first recorded export of New Zealand wine, came to nought. In 1934 the British newspaper the *Daily Mail* enquired whether there were any 'New Zealand or West Indian wines that could be offered in this country?' Observed the paper generously: 'We Englishmen are prepared to try anything once.'[74]

A small, steady overseas trade in New Zealand wine has been maintained since 1963, when Corbans received a trial order from the British Columbia Liquor Control Board for twenty-five cases of medium-sweet sherry. One early flirtation with export was a 1971 480-case shipment to Britain. Organised by Graeme Reid, a leading Auckland wine merchant, the selection included four Corbans labels — Montel Sauternes, Riverlea Riesling, Riverlea White and Medium Sherry — and other wines from Aspen Ridge, Penfolds, San Marino, McWilliam's, McDonald's and Western Vineyards.

The result was a 'disaster'.[75] No interest was displayed by New Zealand firms operating in the United Kingdom. Four months after the wines arrived in the U.K., 104 cases had been sold, 16 cases had been broken in transit, 133 cases were unsaleable because of inadequate labelling, and the other 227 cases were unsold.

The export sales of New Zealand wine climbed from a minuscule $43,000 in the year to June 1972, to over 500,000 litres worth $1.25 million in the year to June 1982. By June 1992, wine exports had surged to 7.1 million litres worth $34.7 million.

There is no doubt that New Zealand's top wines possess the quality to compete on world markets. The country's cool-climate growing conditions — an advantage enjoyed by few other viticultural regions — yield fragrant, naturally crisp, deep-flavoured wines that rival the premium wines of France and Germany.

'Over the past decade,' Hugh Johnson wrote in 1991, 'New Zealand has made an international impact with white table wines of startling quality.'[76] *Decanter*, the prestigious British wine magazine, enthused in 1990 that 'New Zealand can justly claim to produce some of the best Sauvignon Blanc wines in the world.'[77] Such critical praise has done much to foster the growing international respect for New Zealand wines.

The difficulty until recently lay in converting this stream of applause into concrete sales. An interdepartmental committee in 1985 described the winemakers' export performance as 'lacklustre'[78] and the industry narrowly failed to achieve its own goal, set in 1978, of over $5 million per year in export earnings by the mid-1980s. Since then, however, wine exports have grown swiftly: between 1985 and 1992 the volumes shipped rose by 800 percent, and earnings by over 1000 percent.

With the New Zealand market now opened up to increased competition from abroad, especially from Australian wines under CER, a powerful commitment to export is evident in many wineries. But only since the mid-1980s have the vintners possessed sufficient volumes of premium table wine to think seriously of exporting. Corbans until the early 1980s dominated activity with their sales of Select Medium Sherry to Canada. In 1992 such fortified wines accounted for only 3.9 percent of New Zealand wine exports; sparkling wines accounted for 4.3 percent; red wines 9.9 percent; white wines almost 82 percent.

Sweden, Finland and Japan have emerged as important markets, although principally for low-priced white wine supplied in bulk. The key markets for premium, bottled New Zealand wines are the United Kingdom and Australia. In the year to June 1992, the United Kingdom absorbed the equivalent of 400,000 cases, and Australia 73,778 cases.

The United Kingdom is one of the toughest wine markets in the world to crack. Nevertheless, acceptance by the British wine trade, and its internationally influential wine press, has been pivotal to the worldwide acceptance of New Zealand wines. The annual New Zealand tasting in London has generated enormous favourable publicity and now stands as a popular event on the crowded London wine calendar. Three hundred wine buyers, distributors and writers attended the January 1992 tasting, where almost 200 New Zealand wines were on show.

'New Zealand has a good quality image,' according to the wine buyer at Selfridge's, the London department store. 'Many of our customers now realise that for half or even a third of the cost of white Burgundy, they can buy a decent New Zealand Chardonnay, or your even better known dry white, Sauvignon Blanc.'[79]

To heighten consumer acceptance, in 1991 the key wine exporters to the United Kingdom joined forces to form the New Zealand Wine Guild. The move — recommended by the British wine trade — involved the exporters' voluntary agreement to the funding of 'a London-based office and administration, generic promotional materials, and activities to take the taste of New Zealand wine to target consumer groups'.[80]

The New Zealand Wine Guild's initial marketing plan calls for $2.8 million to be spent by 1995, with forty-five percent of the funds to flow from Tradenz (formerly the New Zealand Trade Development Board). Vicky Bishop — an Englishwoman who previously promoted Australian wines — was appointed the Guild's United Kingdom manager in mid-1991.

A string of triumphs in international wine competitions has also boosted the image of New Zealand wine in the United Kingdom. At the International Wine and Spirit Competition in London, for instance, New Zealand wines have won the Marquis de Goulaine Trophy for the champion Sauvignon Blanc three years in a row: in 1990 with Montana Marlborough Sauvignon Blanc 1989; in 1991 with Delegat's Oyster Bay Sauvignon Blanc 1990; and in 1992 with Hunter's Sauvignon Blanc 1991.

The 'International Wine Challenge' 1991, staged by *Wine* magazine in London, awarded the trophy for the best Chardonnay to Kumeu River 1989; Te Mata Elston Chardonnay 1989 scooped the trophy for the best white wine of the show. Greater kudos often attaches to these awards in New Zealand than in the United Kingdom — where reputations are built over decades and competition results are much less influential — but these and other victories have clearly helped to build New Zealand's quality image.

A striking export coup was announced in October 1991. The House of Nobilo won a $3 million contract to supply 1.4 million 200ml bottles of Poverty Bay Chardonnay for service on British Airways' European routes. The United Kingdom absorbed 50.7 percent by volume, and 62 percent by value, of New Zealand's wine exports in the year to mid-1992.

New Zealand's other most valuable market for wine lies across the Tasman. Critics in both countries agree that New Zealand's explosively flavoured Sauvignon Blanc is markedly superior to Australian versions, its Gewürztraminers more intensely spicy, its Chardonnays of equally high interest. The market is close, transport costs are reasonable, tariff barriers are non-existent and the locals are relatively sophisticated about wine. According to Corbans' former general manager, Paul Treacher, however, 'the difficulty is persuading Australians that New Zealand wine is fit to drink'.

Montana established a bridgehead in Australia in 1984 with a carefully researched launch on the Queensland market. Its Marlborough Sauvignon Blanc and bottle-fermented sparkling, Lindauer, soon proved successful. The company pushed into Sydney and Melbourne, and later enjoyed instant recognition of the quality of its Deutz Marlborough Cuvée. In the face of a depressed Australian wine market, however, in 1992 Montana announced it was scaling down, although not abandoning, its

NEW ZEALAND WINE EXPORTS

	Million litres	($ million) Value
1983	.55	1.24
1984	.72	2.17
1985	.79	3.00
1986	1.09	3.92
1987	1.03	4.46
1988	2.90	11.63
1989	2.70	11.63
1990	4.00	18.38
1991	5.60	25.28
1992	7.10	34.74

SOURCE: Wine Institute of N.Z. Annual Report 1992

promotions across the Tasman, in favour of intensifying its export efforts to Europe.

Other vineyards — Corbans and Nobilo (with Montana the three big players of New Zealand wine exports), Coopers Creek, Delegat's, Cloudy Bay, Te Mata, Matua Valley, Hunter's, Villa Maria, Selaks and Babich — have also established a significant presence on Australian wineshop shelves. As Terry Dunleavy, former executive officer of the Wine Institute, put it: 'Our winemakers have now broken through their own confidence barrier.'

The Wine Institute has recently predicted that 'the New Zealand wine industry is likely to become a net exporter (exports greater than imports) within the next three to four years, and ultimately more New Zealand wine will be consumed off-shore than in New Zealand'.[81] A bullish forecast — but who would argue?

Lord Porritt, patron of the New Zealand Wine Guild, speaking at the guild's London launch in May 1991. Left to right: Hon. George Gair, New Zealand High Commissioner and the guild's vice-patron; Bill Spence of Matua Valley Wines, guild chairman; Don Walker, New Zealand Trade Commissioner.

NOTES FOR CHAPTER ONE

1. Wright, Olive, 'Voyage of the Astrolabe 1840', Wellington, 1955. See Thorpy, F., 'Wine in New Zealand', Collins, Auckland, 1st edn. 1971, p. 20.
2. Bragato, R., 'Report of the Prospects of Viticulture in New Zealand', Department of Agriculture, 1895, pp. 8–10.
3. Cooper, M., 'The Wine Lobby', University of Auckland M.A. thesis, 1978, p. 41.
4. *Ibid.*
5. *Ibid.*
6. Cooper, M., *op. cit.*, p. 43.
7. *Ibid.*
8. Cooper, M., *op. cit.*, p. 44.
9. *Ibid.*
10. *Ibid.*
11. Cooper, M., *op. cit.*, p. 45.
12. *Cooks Wine Bulletin*, April 1971, p. 6.
13. *Wine Review*, Summer 1971, p. 39.
14. *Wine Review*, Autumn 1976, p. 22.
15. *Wine Review*, Summer 1971, p. 39.
16. Thorpy, F., *op. cit.*, p. 87.
17. *Ibid*, p. 105.
18. *New Zealand Herald*, 12 September 1989.
19. Thorpy, F., *op. cit.*, p. 117.
20. Buck, J., *Take a Little Wine*, Whitcombe & Tombs, Christchurch, 1969, p. 109.
21. 'Wine Pioneer's Name to Live On', Montana Wines press release, 31 August 1989.
22. Bolster, T.N., 'Lucky Break for New Zealand Winemakers', *Weekly News*, 24 September 1958, p. 3.
23. Annual Report, Department of Agriculture and Fisheries, 1959.
24. 'Man Who Produces the Food of the Sun', *Dominion*, 3 June 1972.
25. *Wine Review*, Spring 1966, p. 11.
26. Brooks, C., *Marlborough Wines and Vines*, C. & B. H. Brooks, Blenheim, 1992.
27. *Dominion*, 3 June 1972.
28. *Wine Review*, Summer 1972, p. 5.
29. *Auckland Star*, 8 August 1978.
30. *Ibid*, 31 August 1984.
31. Lucas, D., personal interview, March 1992.
32. *Weekly News*, 5 April 1971, p. 4.
33. Corban, A., address to an International Wine Symposium, Auckland 1978.
34. Company Brochure, 1979.
35. Lindeman, B.W., Report on the Possibilities of Expansion in the Wine Industry of New Zealand, 3 March 1939. Unpublished. National Archives.
36. 'Observations at the Shrine of Bacchus', *Here and Now*, 2, No. 4, January 1952, p. 2.
37. Graham, J.C., *Silver Jubilee History of the Wine and Food Society of Auckland*, 1979.
38. *Ibid.*
39. Corban, A., Behind the Wine Curtain, unpublished manuscript, November 1987.
40. Scott, D., *A Stake in the Country*, Southern Cross Books, Auckland,
1977, p. 158.
41. *Ibid*, p. 160.
42. Corban, A., personal interview, March 1992.
43. Corban, A., Behind the Wine Curtain, *op. cit.*, p. 221.
44. *Cooks Wine Bulletin*, December 1970, p. 2.
45. *Ibid.*
46. *Ibid.*
47. *Ibid.*
48. *Wine Review*, April 1966, p. 11.
49. *Cooks Wine Bulletin*, December 1970, p. 2.
50. Minutes of a Meeting of the New Zealand Wine Council, 16 July 1974. See Cooper, M. 'The Wine Lobby: Pressure Group Politics and the New Zealand Wine Industry'. University of Auckland M.A. thesis, 1977, p. 150.
51. Wine Institute submission to the Working Party on Liquor, 1986, p. 2.
52. Wine Institute of New Zealand, letter to wine writers, 3 September 1990.
53. 'The Wine Thief', July 1992, p. 6.
54. Lucas, D., '1980 Vintage: Influence of Wine Quality Requirements', unpublished paper, 1980.
55. Report of the Industries Development Commission: The Wine Industry Development Plan to 1986, 1980.
56. *Ibid.*
57. *New Zealand Company and Director*, October 1973, p. 23.
58. Lucas, D. and Ujdur S., 'New Zealand Wine', unpublished report, 17 May 1967.
59. Lucas, D., personal interview, March 1992.
60. Lucas, D., *ibid.*
61. *Wine Review*, Spring 1973, p. 29.
62. Lucas, D., *ibid.*
63. Lucas, D., personal interview, March 1992.
64. Lucas, D., *ibid.*
65. Lucas, D., *ibid.*
66. Report To Prime Minister On Investigation of Grape Surplus by Officials Committee, 1983.
67. The Inter-Departmental Overview Committee, *op. cit.*, p. 34.
68. Quoted in *Southern Horticulture*, February 1986, p. 10.
69. *Grocers' Review*, March 1991, p. 33.
70. Wine Institute of New Zealand, 'New Zealand Wine 1993–2000: A Working Paper', November 1992.
71. Rawson-Elder, J., 'Letters and Journals of Samuel Marsden', Otago University Council, 1932.
72. Quoted in Scott, D., *Winemakers of New Zealand*, Southern Cross Books, Auckland 1964, p. 14.
73. *Listener*, 8 February 1971, p. 49.
74. *New Zealand Herald*, 12 July 1935.
75. *Cooks Wine Bulletin*, September 1971, p. 1.
76. Johnson, H., *Hugh Johnson's Pocket Wine Book 1992*, p 196.
77. *Decanter*, August 1990, p. 42.
78. 'The Inter-Departmental Overview Committee Mid Term Review of the Wine Industry Development Plan to 1986', 1985, p. 94.
79. *Cuisine*, No. 11, Oct./Nov. 1988, p. 115.
80. Wine Institute of New Zealand, Annual Report, 1991, p. 16.
81. *Ibid*, p. 8.

CHAPTER TWO

The Principal
Grape Varieties

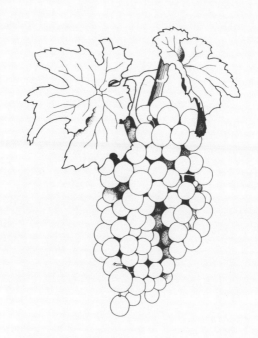

Where is wine made — in the vineyard or the winery? Take a bow if you answered 'in both places'. A common saying in the wine industry is that 'you can't make good wine from bad grapes'. Although in New Zealand the media spotlight is invariably focused on winemakers rather than grape-growers, it is in the vineyards that the raw materials of wine are cultivated, and there that each wine's basic potential is set.

'The cult of the winemaker is essentially a New World phenomenon,' observes Michael Brajkovich of Kumeu River. 'The cellar master where I worked in France was not a noted celebrity; accolades were given to the vineyard and its management, or the proprietor.' Brajkovich finds he can more easily manipulate wine style and quality in the vine-yard than in the winery.

'There is no short cut to quality,' according to Brian Croser of Petaluma, one of Australia's top small wineries. 'It is only achieved by the long term, teeth-gritting commit-ment in the vineyard to moderate crop yields and the risk-fraught decision to hang out for high grape maturity. Real fruit concentration, implying mouth-filling flavour and substantial texture, can only be achieved in the vineyard . . .'

Several basic factors influence the emergence of all wine styles — climate, the soil, grape varieties, vine management and the winemaker.

Grapes are more responsive to climate than most other fruits, and during the growing season the amount of rain-fall, hours of sunshine and degree of heat all have an even-tual effect on the quality of the crop. Variations in soil types also influence the character and quality of wine: although vines grow in a wide variety of soils, heavy clays and poorly drained soils are less suitable than gravelly or sandy soils.

The careful selection of soils and climatic zones, how-ever, must be matched by the planting of suitable grape varieties. In New World wine countries, the selection of grape varieties involves a very considered judgment about grape quality, hardiness and yield.

Every five years between 1960 and 1980, and every three years since 1983, a comprehensive survey of New Zealand's vineyards has been undertaken. The most up-to-date survey, conducted by Dr David Jordan and Sherwyn Veldhuizen in 1992, like its forerunners reveals drastic changes in the varietal composition of our vineyards.

In 1960 the total vineyard area of 388 hectares was most heavily planted in Albany Surprise (sixty hectares), Baco 22A (forty-five hectares) and Seibel 5455 (twenty-nine hectares); these were all hybrid (crossings of classic *vinifera* and native American) or *labrusca* (a species of native American) varieties, offering no scope to the winemaker in pursuit of quality.

The rush of hybrid and *vinifera* plantings in the late 1960s showed up in the survey conducted in 1970. Palo-mino, the Spanish sherry grape, topped the list with 243 hectares, heading Baco 22A, at 217 hectares still in its ascendancy, and Müller-Thurgau rising from obscurity to

third place with 194 hectares. In 1975 Müller-Thurgau emerged well on top, at 649 hectares far ahead of Palomino (338 hectares) and the resilient Baco 22A (208 hectares).

Vine plantings more than doubled between 1975 and 1980, from 2351 to 4853 hectares. The 1980 vineyard survey also revealed that Müller-Thurgau had maintained its ascendancy with an almost 200 percent increase in plantings to 1819 hectares — thirty-eight percent of the total vineyard area.

Then Cabernet Sauvignon rose, between 1980 and 1983, from the relative obscurity of sixth place, to become New Zealand's second most heavily planted vine. By 1986 Chardonnay had moved past Cabernet Sauvignon and was clearly New Zealand's second most extensively planted variety. In 1989 Sauvignon Blanc emerged in third position, trailing only Müller-Thurgau and Chardonnay.

The latest, 1992 survey reveals that Müller-Thurgau, after a reign of fifteen years, has finally been displaced as New Zealand's most extensively planted variety by Chardonnay. Since the 1989 survey, plantings of Char-donnay, Sauvignon Blanc, Pinot Noir and Merlot have forged ahead, but the area devoted to Müller-Thurgau has dropped by fifteen percent (see chart below).

NATIONAL TOTAL AREA OF PLANTED VINES

	1992		1989	
	Hectares	% of Total Plantings	Hectares	% of Total Plantings
Chardonnay	1283	21.0	769	15.0
Müller-Thurgau	1083	17.7	1277	24.9
Sauvignon Blanc	857	14.1	552	10.8
Cabernet Sauvignon	522	8.6	432	8.4
Pinot Noir	402	6.6	199	3.9
Riesling	281	4.6	284	5.5
Muscat Dr Hogg	229	3.8	229	4.5
Merlot	220	3.6	104	2.0
Chenin Blanc	176	2.9	214	4.2
Sémillon	169	2.8	138	2.7
Gewürztraminer	143	2.3	169	3.3

Other varieties covering between fifty and 100 hectares are Palomino, Reichensteiner, Chasselas, Cabernet Franc and Pinotage.

Relatively rare grapes that are occasionally produced as varietal wines or named on labels are Pinot Blanc (a white mutation of Pinot Noir which can yield rewardingly weighty and savoury wines); Grey Riesling (a source of plain, slightly earthy white wines); Flora (a cross of Sémillon and Gewürztraminer); Morio-Muskat (a very blowsy cross of Sylvaner and Pinot Blanc); Ehrenfelser (a cross of Riesling and Sylvaner with very Riesling-like qualities); Scheurebe (another Riesling and Sylvaner cross, highly rated in Germany); Osteiner (yet another crossing of Riesling and Sylvaner, thus far not impressive in New Zealand); and Malbec (a dark-hued red wine variety grown in the Gironde and Cahors in France).

Overall, in 1992 surveyed vines covered 6099 hectares — a steep rise from the 5125 hectares recorded in 1989.

White-grape varieties predominate in New Zealand's vineyards. Chardonnay, Müller-Thurgau and Sauvignon Blanc — the big three — account for over fifty percent of the country's total vine plantings.

Of the 'classic' white-grape varieties listed in Jancis Robinson's *Vines, Grapes and Wines*, all five — Riesling, Chardonnay, Sémillon, Sauvignon Blanc and Chenin Blanc — are well established in New Zealand. Since 1975 the proportion of the national vineyard devoted to white-grape varieties has risen from 65.9 percent to 77.5 percent. New Zealand, despite its burgeoning plantings of red-wine grapes, is still largely white-wine country.

WHITE WINE VARIETIES

BREIDECKER

1992 plantings: 26 hectares

Breidecker is a crossing of Müller-Thurgau with the white hybrid Seibel 7053. It is named in honour of Heinrich Breidecker, the nineteenth-century German winemaker at Kohukohu in the Hokianga.

Although Breidecker displays the typical hybrid resistance to rot, its wine is bland and unmemorable: tasted blind, it is usually mistaken for a rather ordinary quality Müller-Thurgau. A couple of varietal wines are marketed, by Soljans and Larcomb, but the grape is principally of value as a blending variety. Plantings are limited.

CHARDONNAY

1992 plantings: 1283 hectares

Weighty, rich-flavoured, absorbingly complex — the wood-matured Chardonnays of Burgundy and the New World — New Zealand, Australia and California — rank among the greatest dry white wines of all. In New Zealand, plantings of this currently most fashionable of all grapes have outstripped those of every other variety.

Chardonnay's success here is hardly surprising, for, unlike its Burgundian stablemate, Pinot Noir, it is a versatile variety that thrives in many parts of the world. In its homeland, France, where in the famous Côte d'Or of Burgundy its plantings cover only one-sixth of the area devoted to Pinot Noir, it reaches its full glories in the vineyards of Puligny-Montrachet and Meursault. Elsewhere, the Californians are infatuated with Chardonnay, as are the Australians, with both countries now boasting wines capable of challenging the Burgundians at the highest level.

Chardonnay used to be commonly referred to as Pinot Chardonnay, but it is not a true Pinot. The clones imported into New Zealand in the late 1920s never grew well, being infected with leaf-roll virus, and the vines languished, raising doubts — now proved unfounded — as to whether the variety was the same vine as the classic Chardonnay of Burgundy.

Of the eleven Chardonnay clones established in New Zealand, the Mendoza (or McCrae) clone of Chardonnay, imported from Australia in 1971, is now the most widely planted. Relatively low-yielding because of its characteristically poor grape 'set', which produces the 'hen and chicken' (large and small berries) effect, Mendoza is nevertheless favoured by most Chardonnay producers because its smaller berries give a higher skin to juice ratio and richer flavour.

Chardonnay vines are spread throughout all the major wine regions, particularly in Marlborough, Hawke's Bay (where it is now the number one grape) and Poverty Bay. The variety is as popular in the vineyard as it is in the market, adapting well to a wide range of climates and soil types. Chardonnay's early bud burst renders it vulnerable to damage from spring frosts in colder regions, but the grapes ripen mid-season in small bunches of thick-skinned, yellow-green berries harbouring high sugar levels (hence the sturdy alcohol typical of its wine). Yields, although moderate, are consistent at about ten tonnes per hectare.

New Zealand's Chardonnays are full-bodied, with fruit flavours ranging from crisp, flinty apples and lemons, through to the lush stone-fruit — peach and apricot — flavours of very ripe grapes. Styles produced range from fresh, unwooded wines like Nobilo Poverty Bay Chardonnay through to mouth-filling, multi-faceted wines like Morton Estate Black Label.

The hallmark of New Zealand Chardonnays is their delicious varietal intensity. The leading wines display such concentrated aromas and flavours, supported by crisp, authoritative acidity, that they have rapidly emerged on the world stage. Vidal Reserve Gimblett Road 1989 won the trophy for the champion Chardonnay at the 1991 International Wine and Spirit Competition in London; Te Mata Elston 1989 was selected as the top white wine at the 1991 International Wine Challenge in London.

Winemakers delight in the multitude of options available to them in Chardonnay production. Most are experimenting with different clones, varying periods of skin contact, different yeasts, fermentation temperatures, barrel fermentation and lees aging, types of oak (usually French and often Nevers), barrel sizes (usually barriques), length of wood maturation and malolactic fermentations. As a result, the standard of our Chardonnay is improving virtually from one vintage to the next.

LEADING LABELS:
Ata Rangi, Babich Irongate, Brookfields Reserve, Church Road, Cloudy Bay, Collards Hawke's Bay and Rothesay Vineyard, Cooks Winemakers Reserve, Corbans Private Bin, Delegat's Proprietors Reserve, Dry River, Esk Valley Reserve, Giesen Reserve, Hunter's, Kumeu River, Martinborough Vineyard, Matua Valley Ararimu, Mills Reef, Montana Ormond Estate and Renwick Estate, Morton Estate Black Label, Nautilus, Neudorf, Ngatarawa Alwyn, Nobilo Marlborough Reserve, Palliser, Robard & Butler Gisborne, Rongopai Reserve, Te Mata Elston, Vavasour, Vidal Reserve, Villa Maria Reserve Barrique Fermented and Reserve Marlborough, Waipara Springs

CHASSELAS

1992 plantings: 92 hectares

Chasselas, often cultivated overseas as a table grape, as a wine variety achieves some prominence in cool-climate regions, where its low-acid, early-ripening qualities are of value. In Switzerland Chasselas is a major grape variety and its wine is also known in Alsace, the Loire, Germany and Austria.

Chasselas was a significant white-wine grape in New Zealand a decade ago, but since then plantings have shrunk and the vine is now almost entirely confined to Poverty Bay and Hawke's Bay. Maturing early, at about the same time as Müller-Thurgau, Chasselas produces bountiful crops of large, low-acid grapes.

As a white table wine, Chasselas has a light bouquet and pleasant, unmemorable, low-acid flavour. The sales response to Chasselas wines marketed as varietals in New Zealand was slow. Easily mistaken for Müller-Thurgau in a blind tasting, it is now almost invariably used as a blending variety.

CHENIN BLANC

1992 plantings: 176 hectares

At their best, New Zealand's Chenin Blancs are full in body, with a fresh, buoyant, pineappley flavour and mouth-watering acidity. At their worst, they are searingly tart and totally devoid of charm.

Chenin Blanc is only a workhorse variety in Australia and California — where it wins favour for its abundant yields of fresh, medium-dry wines which usefully retain an invigorating acidity — but in the Loire it achieves greatness. The finest dry Vouvrays are substantial, fruity wines with tongue-curling acidity and an ability to unfold in the bottle for decades.

In New Zealand the wine industry is less enthusiastic than it used to be about this vigorous variety, which is largely concentrated in Hawke's Bay and Poverty Bay. Although Chenin Blanc ripens early in warm climates, in New Zealand the grapes tend to ripen late. Yields are high for a premium variety, at twelve to fifteen tonnes per hectare. Nevertheless, the grapes are vulnerable to wet weather and to botrytis. Many growers have recently discarded their vines and replanted with other varieties; over the past decade plantings have more than halved.

Chenin Blanc reveals its full personality only in cooler growing conditions, and should therefore feel very much at home in New Zealand. It demands careful vineyard site selection and long hours of sunshine, however, to build up its potentially high sugar levels. In New Zealand, as in France, its acid levels can often be searingly high.

Chenin Blanc's most common role here is as a blending variety, adding body and 'spine' to its blends with Müller-Thurgau in casks; it also has a pervasive presence in many low-priced dry whites labelled as White Burgundy, Chablis or Dry White. To its mid-priced blends with Chardonnay, Chenin Blanc contributes a splash of vigorous fruitiness.

Only Collards and The Millton Vineyard — and to a lesser extent Esk Valley — have consistently mastered Chenin Blanc as a strong, distinctive varietal with seductive tropical-fruit flavours.

LEADING LABELS:
Collards, Esk Valley, The Millton Vineyard

GEWÜRZTRAMINER

1992 plantings: 143 hectares

Gewürztraminer, which is at its ravishingly perfumed and pungently seasoned best in Alsace, has also been a success story in New Zealand's cool climate. Our top wines burst with Gewürztraminer's heady scents and flavour-packed spiciness.

Pronounced Ge-vertz-truh-meen-uh, with the stress on the 'meen', the name of the wine is sometimes shortened to Traminer. 'Gewürz' means spicy. In Germany it was customary to call the wine Gewürztraminer if it was spicy, Traminer if it was not. In Alsace the current practice is to label all the wines Gewürztraminer.

In New Zealand, Gewürztraminer had an inauspicious debut: in 1953 the Department of Agriculture imported a strain known as Roter Traminer which, after McWilliam's established a plot at Tukituki, bore poorly and produced disappointing wine. As a result many winemakers became convinced that Gewürztraminer could not successfully be cultivated here.

Healthier vines are now established in most regions — including Central Otago — with the heaviest plantings in Poverty Bay, the source of most champion wines. Hawke's Bay and Marlborough also have substantial plantings. Gewürztraminer has never risen to the popularity of Chardonnay or Sauvignon Blanc, however, and in the past decade plantings have halved.

On the vines the grapes are easily identified by their distinctive rosy colour; pink and white berries are often found nestling in the same bunch. This vine is notoriously temperamental, ripening its grapes easily in New Zealand with plenty of sugar and fragrance, but highly susceptible to adverse weather during flowering — which can dramatically reduce the crop — and also vulnerable to powdery mildew and botrytis. To plant a vineyard exclusively in Gewürztraminer is a risk, one that few local viticulturists would contemplate.

Gewürztraminer is a wine to broach occasionally, when you're in the mood to delight in its overwhelming aroma — which can often have nuances of gingerbread, cinnamon, lychees and mangoes — and lingering spiciness.

LEADING LABELS:
Collards, Corbans Private Bin, Dry River, Martinborough Vineyard, Matawhero, Matua Valley, Pacific Phoenix, Revington Vineyard, Robard & Butler, Seifried Estate Gewürztraminer Ice Wine, Te Whare Ra, Vidal Reserve, Villa Maria Reserve

MÜLLER-THURGAU

1992 plantings: 1083 hectares

It's amazing how rarely wine buffs talk about this country's second most widely planted grape variety. Although New Zealanders drink oceans of Müller-Thurgau every year, this once-ubiquitous variety is rarely exported and is more often packaged in casks than in bottles. Yet a cool glass of refreshingly fruity and flowery Müller-Thurgau makes delicious summer sipping.

Müller-Thurgau (pronounced Mooler-Ter-gow) is the world's most famous vine crossing. It was bred at Geisenheim in Germany in the 1880s by Professor Hermann Müller, a native of the Swiss canton of Thurgau. Although Müller wrote of his ambition to combine 'the superb characteristics of the Riesling grape with the reliable early maturing qualities of the Sylvaner', it has never been proven that the variety named after him is indeed a crossing of Riesling and the more humble Sylvaner; he may have crossed two different Riesling clones. It makes good sense, therefore, that in New Zealand the variety's older name of Riesling-Sylvaner is gradually being phased out.

Müller-Thurgau was originally regarded as a bulk producer of low merit. Later, German growers unable to ripen Riesling grapes on less favoured sites discovered that the new vine could produce large quantities of attractive wine, with less susceptibility to weather conditions. The early-ripening Müller-Thurgau offered growers the prospect of a reasonable crop every year, and in poor years better quality wine than Riesling.

The drawback is that Müller-Thurgau as a wine cannot match the flavour and aroma intensity or longevity of Riesling. German Müller-Thurgaus are described by one authority, S.F. Hallgarten, as 'mild, aromatic and pleasant with a slight Muscatel flavour'. In Dr Helmut Becker's view the wines are 'elegant, palatable, harmonious and mild, although Riesling drinkers often find them too mild'. Clearly, the Müller-Thurgau lacks the greatness of Riesling.

In the 1930s, Government Viticulturist Charles Woodfin imported the vine into New Zealand. The commercial value of Müller-Thurgau became apparent much later, when the demand for white table wines escalated in the 1960s. Then the vine spread rapidly, prized for its early ripening ability and high yields. A rush of plantings in the early 1970s rapidly established Müller-Thurgau as New Zealand's leading grape variety.

Countless Müller-Thurgau vines were uprooted in the 1986 vinepull scheme, however, and between the 1983 and 1992 vineyard surveys plantings have declined by 42 percent. Müller-Thurgau's plantings have recently been outstripped by Chardonnay, although not yet by Sauvignon Blanc. Its strongholds are Poverty Bay (where over forty percent of the vines are concentrated), Marlborough and Hawke's Bay.

Müller-Thurgau grows vigorously in New Zealand and on most soils yields good crops of about twenty tonnes per hectare. The berries, yellow-green and flecked with small brown spots, ripen early, and Müller-Thurgau is generally the first variety to be picked. The grapes are susceptible to wet weather at vintage and to fungous diseases.

Müller-Thurgau does not usually achieve very high sugar levels on the vines (hence its low-alcohol wine). During ripening, the fruit loses acidity rapidly; this is often corrected by the addition of tartaric acid at the crusher. Most Müller-Thurgaus are backblended with a small amount of unfermented grapejuice to produce an elegant, slightly sweet style that is very similar to, and often better than, most German commercial white wines.

What does a good Müller-Thurgau taste like? It is light bodied, with a garden-fresh bouquet and a mild, distinctly fruity flavour. It reveals flashes of the classic Riesling variety's citric-fruit aroma and flavour, although to a less memorable degree, and lacks Riesling's acid backbone. Müller-Thurgau's natural role is as a beverage wine: an easy-drinking, enjoyable, low-priced white wine to consume anywhere, any time.

LEADING LABELS:
Babich Müller-Thurgau and Dry Riesling-Sylvaner, Giesen Botrytised, Montana Wohnsiedler, Seifried Estate, Villa Maria Private Bin

---

MUSCAT DR HOGG

1992 plantings: 229 hectares

Muscat varieties form a large, instantly recognisable family of white and red grapes notable for their almost overpowering musky scent and sweet grapey flavour. The vines grow all over the Mediterranean and in the New World wine regions, yielding a diversity of styles ranging from delicate dry whites in Alsace through to the full-bloomed sweet Asti sparklings of Italy and the delicious fortified wines of Victoria.

Muscat Dr Hogg, an old English table grape, is by far the most common Muscat variety cultivated in New Zealand, and the country's fifth most widely planted white-wine variety. Only in New Zealand is it used to produce wines. Nearly three-quarters of all plantings are in Poverty Bay, with the rest virtually confined to Hawke's Bay and Marlborough.

The vines crop well in New Zealand, ripening late with good wet-weather resistance, and produce large fleshy berries with an intense Muscat aroma and flavour. Marketed in the 1980s as a fruity, slightly sweet varietal wine labelled Muscat Blanc, it never found favour with consumers.

Often, following the German practice, Muscat is blended with Müller-Thurgau to enhance the wine's bouquet. It also gives a lift — again, often as a minor partner of Müller-Thurgau — to charmingly light and perfumed Asti-type sparklings.

LEADING LABELS:
Corbans Italiano Spumante, Matua Valley Late Harvest, Montana Bernadino Spumante and Fricanté

PALOMINO

1992 plantings: 95 hectares

Palomino is the leading New Zealand 'sherry' variety. The grape is traditionally used to produce the famous sherries of the Jerez region of Spain and at first glance would appear ill-suited to New Zealand's cooler climate.

The vine was largely unknown in New Zealand until its ability to produce very large crops was demonstrated at Te Kauwhata in the early 1950s. Thereafter the vine spread rapidly through all the wine districts. Palomino emerged by 1970 as the main grape variety in the country, with its heaviest concentrations in Auckland and Hawke's Bay. Today, following a seventy-seven percent drop in plantings in the past decade — reflecting the plummeting demand for 'sherry' — it ranks as the fourteenth most important grape variety, with the surviving vines concentrated in Hawke's Bay, Auckland and Poverty Bay.

The Palomino vine grows with much vigour, yielding twenty to thirty tonnes per hectare of large, thick-skinned, fleshy yellow-green grapes that make good eating. The grapes ripen mid to late season with a relatively low acidity and without the high sugars achieved in warmer climates. Palomino withstands wet weather reasonably well, but if there are persistent rain and high humidities as vintage approaches, the grapes are susceptible to botrytis.

Palomino grapes are the foundation of the better New Zealand sherries, dry and sweet, and the best of these are reminiscent of their Spanish counterparts.

LEADING LABELS:
Babich Henderson's Mill Flor Fino and Rare Old Sherry, Corbans Extra Special Sherry, Montana Private Bin Pale Dry Sherry, Pleasant Valley Amontillado, Amoroso and Oloroso Sherry, Soljan's Pergola and Reserve Sherry

PINOT GRIS

1992 plantings: 19 hectares

An outstanding Chardonnay substitute, weighty and deep-flavoured, Pinot Gris is still rare in New Zealand vineyards. The variety belongs to the Pinot family of vines and is cultivated in Italy, Germany and various regions of France. In Alsace — where it is known as Tokay d'Alsace — Pinot Gris produces good wine, dry, full-flavoured and flinty.

Although Bragato praised the variety in 1906 ('in the far north [it] bears heavily and produces an excellent white wine'), Pinot Gris later fell out of favour with most growers because of its tendency to crop erratically. Today plantings are concentrated in Canterbury and Hawke's Bay.

The vines grow with moderate vigour, bearing an average crop of seven to ten tonnes per hectare of small, thin-skinned, reddish-pink berries. The grapes mature early with fairly low acidity and high sugar levels.

Savoury, with an earthy stone-fruit flavour, Pinot Gris offers an underrated alternative to the higher profile dry whites. Dry River has lately proved just how concentrated and savoury a wine Pinot Gris can be persuaded to yield in this country.

LEADING LABELS:
Brookfields, Dry River, Larcomb, Mission, St Helena

REICHENSTEINER

1992 plantings: 98 hectares

Created by Dr Helmut Becker, Reichensteiner is a crossing of Müller-Thurgau with the French table grape Madelaine Angevine and the Italian Caladreser Fröhlich; Becker called it the first 'EEC crossing'. Issued by the Geisenheim Institute in 1978, it swiftly achieved solid support in the vineyards of Germany — largely in the Rheinhessen — and England. The variety takes its name from a castle, Schloss Reichenstein, in the Mittelrhein.

Ripening early, about the same time as Müller-Thurgau,

it enjoys slightly higher acidity and must weights and also, due to its looser branches, less rot.

Reichensteiner wine shares the fruitiness and mildness of Müller-Thurgau and displays good sugar-acid balance. Late-harvest wines are a possibility. In New Zealand plantings are restricted to Poverty Bay. Reichensteiner is the country's tenth most widely planted white wine variety, but, almost invariably, its role is that of an anonymous blending variety.

RIESLING
(Rhine Riesling)

1992 plantings: 281 hectares

Riesling, New Zealand's fourth most widely planted white wine variety, is the greatest and most famous grape variety of Germany. In the Rheingau and the Mosel its wine is strongly scented, the flavour a harmony of honey-like fruit and steely acid. Riesling also performs well in Alsace, Central Europe, California and Australia, and recently it has made its presence strongly felt in New Zealand.

The proper name of the variety is Riesling. In New Zealand it has often been called Rhine Riesling to avoid confusion with Riesling-Sylvaner (Müller-Thurgau). The current trend — led by the smaller wineries — is towards increasing use of the classic name Riesling.

Obtaining a reasonable yield from Riesling has long been recognised as a difficulty in New Zealand. Bragato declined to recommend the vines of this variety, 'being only fair bearers'. The 1975 vineyard survey revealed the scarcity of Riesling vines in the country; eight hectares in Hawke's Bay and half a hectare in Auckland. Since then planting has gathered momentum; the acreage devoted to Riesling almost doubled between 1983 and 1989. Over half of the vines are concentrated in Marlborough, with significant plantings in Hawke's Bay, Canterbury and Poverty Bay.

Riesling is a shy bearer, yielding only about eight tonnes per hectare. The grapes ripen late in the season but hang on well, resisting frosts and cold. In Marlborough, where it has enjoyed the most eye-catching success, it is one of the last white varieties to be picked, and in Germany it is said every vintage is a cliff-hanger.

This variety needs a long slow period of ripening to fully develop its most intricate flavours. Thus the finest Rieslings tend to be grown in cooler regions enjoying long dry autumns — Canterbury and Central Otago are outstanding prospects. In poor years, when the fruit harbours a low level of sugar and high acidity, the wine tends to be unappealingly sharp and thin, but given a good summer and settled autumn, a fragile, luscious wine emerges of unparalleled elegance and perfume.

Noble rot, a beneficial form of *Botrytis cinerea*, can transform Riesling's quality. Riesling produced from nobly rotten grapes usually has a deeper, more golden colour. Depending on the extent of botrytis infection, the bouquet of the grape variety itself is replaced by an aroma strongly reminiscent of honey. The wine tastes richer, more luscious, smoother and honeyed. Even in a dry Riesling, a touch of noble rot can add immensely to the wine's character and style.

Baby-fresh, with tight acidity and unevolved flavours, young Rieslings cry out for time to unleash their beguiling flowery scents and flavour richness. Most Riesling needs two years to achieve a satisfactory level of development: a further spell in the cellar builds up the often breathtaking attributes of its full maturity.

LEADING LABELS:
Allan Scott, Collards Rhine Riesling and Marlborough Riesling, Coopers Creek, Corbans Stoneleigh Vineyard and Private Bin Noble, Cross Roads, Dry River, Giesen Reserve and Botrytised, Grove Mill, Martinborough Vineyard, Montana Marlborough, Neudorf, Ngatarawa Penny Noble Harvest, Palliser, Robard & Butler Amberley, Rongopai Reserve Botrytised, Seifried Late Harvest, Te Whare Ra Botrytis Bunch/Berry Selection, The Millton Vineyard Opou Vineyard and Late Harvest Individual Bunch Selection, Villa Maria Reserve Noble

With one variety New Zealand can claim to have its nose in front of all other wine-producing countries — Sauvignon Blanc.

Sauvignon Blanc is one of the world's noblest white wine grapes. In Bordeaux, where it is widely planted, traditionally it has been blended with the more neutral Sémillon, to produce dry white Graves and sweet Sauternes. But in the regions of Sancerre and Pouilly in the upper Loire Valley, the Sauvignons are unblended and here the wines are assertive, cutting and flinty, in a style readily recognisable as cool-climate Sauvignon Blanc. In California, the grape is second only to Chardonnay as the most fashionable white-wine variety.

Sauvignon Blanc, although a vigorously growing vine, yields only a moderate crop of about ten tonnes of small, yellow-green berries per hectare. The grapes ripen mid to late season, harbouring adequate sugars and a high level of acidity. Viticulturists have to combat two problems, however: the vines are tough-stemmed, making the bunches difficult to harvest mechanically, and in wet weather the grapes are prone to split, causing rot.

Sauvignon Blanc has been grown in New Zealand for just over twenty years. The clone known as UCD1, imported in the early 1970s from the University of California, Davis, is thought to be the source of all current plantings. Matua Valley produced the first trial wine in 1974. 'It showed pronounced varietal character and promised to add a new dimension to the flavour profiles available to our winemakers,' recalls Peter Hubscher of Montana. 'The initial promise was such that Montana planted twenty-three hectares of the variety in Marlborough. This represented a huge financial risk; grapes were new to the region and the variety unproven. I am able to report the decision was a correct one.'

Sauvignon Blanc has rapidly emerged as the third most widely planted variety in the country. Between 1986 and 1992 plantings soared by almost 240 percent. Over sixty percent of vines are in Marlborough (where Sauvignon Blanc is the number one variety), but Sauvignon Blanc also has a strong presence in Hawke's Bay and, to a lesser extent, Poverty Bay.

A smidgen of an organic compound is believed to be the secret of the unrivalled intensity of New Zealand's Sauvignon Blancs. Methoxypyrazine has a green, grassy scent, and humans can detect a difference of just two parts per trillion of methoxypyrazine in wine. New Zealand's Sauvignon Blancs have about three times as much methoxypyrazine as Australia's.

Two distinct methods of handling Sauvignon Blanc are practised in New Zealand wineries. By far the most common involves bottling the wine directly out of stainless steel tanks. These wines, placing their emphasis squarely on the grape's tangy, piquant varietal character, are most often labelled as Sauvignon Blanc.

By contrast, those labelled as Fumé Blanc (or sometimes Reserve Sauvignon Blanc) usually tone down Sauvignon's natural ebullience by maturing, and sometimes fermenting, the wine in oak casks. The result is a broader, potentially more complex wine, more costly to produce.

The pungent, 'nettly' bouquet of Sauvignon Blanc, traditionally described as gunflint — the smell of sparks after a flint strikes metal — leaps from the glass with a forcefulness some criticise as unsubtle. Others adore its distinctiveness. According to Jancis Robinson in *Vines, Grapes and Wines*, '"Cat's pee on a gooseberry bush" is an oft-quoted description [of the wine's aroma] from an impeccable source.' The flavour ranges from a sharp, green capsicum character — stemming from a touch of unripeness in the fruit — through to a riper, fruity gooseberry style and, finally, to the tropical-fruit (melons and passionfruit) overtones and lower acidity of very ripe fruit.

'In New Zealand it is not difficult to make herbaceous Sauvignon Blanc,' says Kevin Judd, winemaker at Cloudy Bay. '. . . this fresh edge and intense varietal aroma are the

SAUVIGNON BLANC

1992 plantings: 857 hectares

reason for its recent international popularity. [But] the better of the wines have these herbaceous characters in balance with the more tropical-fruit characters associated with riper fruit.'

Many vineyards are producing top-flight Sauvignon Blancs. The variety looks set to enjoy even greater international prominence in the future.

LEADING LABELS:
Babich Marlborough, Cloudy Bay, Collards Marlborough and Rothesay Vineyard, Cooks Winemakers Reserve, Coopers Creek, Corbans Private Bin Fumé Blanc, Delegat's Proprietors Reserve Fumé Blanc, Dry River, Grove Mill, Hunter's Sauvignon Blanc and Sauvignon Blanc Oak Aged, Jackson Estate, Kumeu River, Matua Valley Reserve, Montana Brancott Estate and Marlborough, Neudorf, Nobilo Marlborough, Palliser, Selaks Founders and Sauvignon Blanc/Sémillon, Te Mata Cape Crest and Castle Hill, Vavasour Reserve and Dashwood, Vidal Reserve Fumé Blanc, Villa Maria Reserve, Wairau River

SÉMILLON

1992 plantings: 169 hectares

New Zealand-grown Sémillon can be pungently herbaceous. Overseas, the grape gives rise to a diversity of styles ranging from the fine dry whites of Graves and Australia to the sweet, late-harvested wines of Sauternes and Barsac. Sémillon imparts softness to its blend with Sauvignon Blanc in Graves, and in Sauternes, infection of Sémillon grapes with *Botrytis cinerea* brings a distinctive, 'noble rot' character to the best wines. Although in Europe the variety is invariably blended with other grapes, Sémillon reaches its apogee as an unblended varietal in the smoky, soft, honey-and-toast flavoured 'white burgundies' of the Hunter Valley in New South Wales. Sémillon is also very extensively planted in Chile, where Sémillon/Sauvignon Blanc blends are common, but to date most of the wines have lacked distinction.

The commercial plantings of Sémillon in New Zealand have yielded promising although confusing results. The vines, which display vigorous growth, yield moderately high crops of ten to seventeen tonnes per hectare. Their tough-skinned, greenish-yellow berries ripen slightly later than Sauvignon Blanc, with good weather resistance.

This last quality is very surprising, especially in New Zealand's often humid growing conditions, because the compact grape cluster typical of Sémillon usually renders it highly vulnerable to bunch rot. The clone widely planted in New Zealand — UCD2 — grows much looser bunches.

'Whether UCD2 is true Sémillon or not is open to question,' says viticultural scientist Dr Richard Smart.

Sémillon is New Zealand's seventh most important white-wine variety, with plantings almost doubling between 1986 and 1992. Sémillon's stronghold is Marlborough (where over half of all plantings are concentrated), but the vine also has a significant presence in Poverty Bay and Hawke's Bay.

Sémillon is usually an ideal, softer and milder blending partner for Sauvignon Blanc. In New Zealand's cool climate, however, Sémillon often displays an assertive grassiness that is strikingly reminiscent of under-ripe Sauvignon Blanc. Blended with New Zealand Sauvignon Blanc — as it often is — it contributes a crisp, herbaceous edge and enhances the wine's longevity. Canopy problems and a tendency to over-crop may also be contributing to this variety's typical lack of fruit ripeness.

That Sémillon is a useful, flavour-packed blending variety has been well demonstrated by such wines as Babich Fumé Vert (Sémillon, Sauvignon Blanc and Chardonnay) and Coopers Classic Dry (Chenin Blanc, Sémillon and Chardonnay).

LEADING LABELS:
Collards Barrique Fermented, deRedcliffe Sémillon/Sauvignon Blanc, Neudorf

SYLVANER

1992 plantings: 37 hectares

Sylvaner, in the first half of this century Germany's most important grape variety, is still widely planted in the Rheinhessen, where it produces pleasant, rather unobtrusive wines, most of which are marketed as Liebfraumilch. The vine is also decreasingly common in Alsace, and in Austria and Switzerland, and is established in a small way in New Zealand.

Here it grows with moderate vigour, with an average to high yield and harvest date lying between Müller-Thurgau and Riesling. It is susceptible to wet weather at harvest. Hawke's Bay has the major concentration of vines, followed by Poverty Bay and the Waikato.

Due to its rather neutral flavour, the national crop of Sylvaner is pressed into service for blending.

The range of red-wine grapes grown in New Zealand is far narrower than that of white-wine varieties; of the country's twenty most widely planted grapes, only six are red. Cabernet Sauvignon and Pinot Noir are New Zealand's fourth and fifth most important varieties (with Merlot recently soaring into ninth place), but between 1975 and 1992 the proportion of the national vineyard devoted to red-wine varieties declined from 34.1 percent to 22.5 percent.

The brilliant purple hue of a young red wine comes from the anthocyanins — colouring matter — stored just under the grapes' skins, which are leached out by alcohol during fermentation. The richer the concentration of anthocyanins in the grape — Cabernet Sauvignon and Syrah are very well endowed — the deeper the wine's colour and the greater its potential longevity.

The startling leap in New Zealand's red-wine quality since the mid–late 1980s points to a brilliant future.

RED WINE VARIETIES

Cabernet Franc, a happier vine in cooler regions than Cabernet Sauvignon, is one of New Zealand's more important red wine varieties. It is much valued in Bordeaux, particularly in St Émilion where, under the name of Bouchet, it is the grape primarily responsible for the esteemed Château Cheval Blanc. Cabernet Franc is also widely planted in the middle Loire and in north-eastern Italy.

The vine was established here early this century and 'succeeded well in the northernmost parts of the colony. Unfortunately, it seems to be subject to coulure [failure of the vine flowers to develop] in southern districts' (Bragato, 1906). Its suitability for cooler climates is based on the fact that it buds, and thus ripens, earlier than Cabernet Sauvignon, with slightly heavier crops.

Virtually unknown for several decades, Cabernet Franc has recently surged in popularity and is now our fourth most widely planted red wine variety. From trial vines imported in the late 1960s from the University of California, Davis, and planted at Kumeu — and from which in the 1970s Ross Spence of Matua Valley took cuttings — the

New Zealand industry has derived all its Cabernet Franc. Plantings more than doubled between 1989 and 1992, particularly in Hawke's Bay (where almost half the vines are concentrated), Marlborough and Auckland.

Cabernet Franc's wine is more genial than that of Cabernet Sauvignon, lower in tannin, acids and extract, with an instantly appealing aroma variously described as raspberries, violets and pencil shavings. By coupling a degree of the strength of claret with the suppleness of Beaujolais, Kumeu River's Brajkovich Cabernet Franc first demonstrated this variety's ability to make a delicious varietal red in New Zealand. The production of full-flavoured dry rosés from Cabernet Franc is another recent trend.

More importantly, as part of the encépagement (varietal make-up) of many of New Zealand's top red-wine labels, it is aiding winemakers in their pursuit of Médoc-like complexity.

LEADING LABEL:
Brajkovich

CABERNET FRANC

1992 plantings: 81 hectares

The majority of overseas critics have until recently been lukewarm about New Zealand's reds. 'All in all, New Zealand serves more as a reminder that there are limits to Cabernet's powers rather than as proof of its genius,' English authority Jancis Robinson wrote in 1986.

Jane MacQuitty, wine correspondent of *The Times*, however, more recently sang the praises of Hawke's Bay 'which has produced glorious Bordeaux-inspired wines'. The gap between Robinson's and MacQuitty's opinions largely reflects the startling progress made in New Zealand's red-wine quality since the mid-1980s.

Cabernet Sauvignon — often abbreviated to Cabernet — has a long history in New Zealand. It seems that the vine first arrived with Busby or with the French settlers at Akaroa. Last century the vine was well known in New Zealand and in 1906 Bragato pronounced it to be 'one of the best varieties grown here . . . the wine produced is of an excellent quality.'

Nevertheless, interest in Cabernet Sauvignon slumped during the wasted years of cheap 'plonk' manufacture. The current commercial revival dates from the early 1970s, when Cabernet Sauvignon came to be regarded as the ideal grape to upgrade the overall standard of red wines.

The arrival of Montana Cabernet Sauvignon 1973 ushered in a new era: for the first time a decent quality New Zealand red — unlike the scarce McWilliam's Cabernet Sauvignon and Nobilo's Cabernet Sauvignon — was widely available. By 1983, following a surge of plantings in the late 1970s, Cabernet Sauvignon, the aristocratic grape of the Médoc in Bordeaux, was the second most common variety in the country. The stage was set for the explosion of Cabernet Sauvignon labels over the past few years.

Between 1989 and 1992 Cabernet Sauvignon's plantings rose by more than twenty percent, and although the grape is now New Zealand's fourth most widely planted vine, it is still firmly established as our most popular red wine variety.

In cool climates Cabernet Sauvignon ripens late in the season. Despite some susceptibility to fungous diseases, with proper spray protection the grapes hang well on the vine. Often labelled a shy bearer, Cabernet Sauvignon produces between six and fifteen tonnes per hectare of small, blue-black tough-skinned berries tasting of blackcurrants or blackberries. In New Zealand the grapes are usually picked last, in April and even May, with high levels of acid and tannin.

CABERNET SAUVIGNON

1992 plantings: 522 hectares

Cabernet Sauvignon performs best in the warmer summer temperatures of the North Island. Plantings are most widespread in Hawke's Bay, where the grape yields fragrant, sturdy wine of a richness and complexity only consistently rivalled by the Cabernet-based reds of Waiheke Island. West Auckland (where Cabernet Sauvignon is the most important variety) and Matakana can also produce full-bodied, satisfyingly flavoursome wines.

The Cabernet Sauvignons grown in Gisborne's fertile soils have veered towards blandness. In Marlborough — the second most heavily planted region — they are often rather *too* cool-climate in style, lacking the strength and opulence of optimally ripened fruit. Wherever they are grown, Cabernet Sauvignon vines yield the best wine when grown in dry soils, which avoid late-season vegetative growth.

Many of the early New Zealand Cabernets were thin and marred by green, unripe, excessively herbaceous flavours. A turning point came with the 1983 vintage, which yielded unprecedentedly ripe fruit. The 1985, 1987, 1989, 1990 and 1991 vintages have also yielded fragrant, robust claret-style reds of a standard almost unimaginable at the start of the decade.

The planting of healthier, less-virused vines, the maturing of those vines, the planting of vines on less fertile soils, improved canopy management, much greater appreciation of the way new oak casks enhance quality, and blending with the classic Bordeaux varieties, Merlot and Cabernet Franc, have all played a crucial role in transforming New Zealand's Cabernet-based reds.

Blockbuster reds grown in warmer climates appeal for their forthright flavours and easy-drinking softness, but usually lack the complexity and elegance of the reds grown in cooler climates. The most exciting aspect of the new breed of New Zealand claret-style reds is their impressive flavour delicacy and length.

LEADING LABELS:
Ata Rangi Célèbre, Babich Irongate Cabernet/Merlot, Benfield and Delamare, Brookfields Cabernet/Merlot and Reserve Cabernet Sauvignon, Church Road, C.J. Pask, Cooks Winemakers Reserve, Delegat's Proprietors Reserve, Goldwater Cabernet/Merlot/Franc, Matua Valley Ararimu, Montana Fairhall Estate, Morton Estate Black Label Cabernet/Merlot, St Nesbit, Stonyridge Larose, Te Mata Awatea Cabernet/Merlot and Coleraine Cabernet/Merlot, Vavasour Reserve, Vidal Reserve Cabernet Sauvignon and Reserve Cabernet Sauvignon/Merlot, Villa Maria Reserve Cabernet Sauvignon and Reserve Cabernet Sauvignon/Merlot

MERLOT

1992 plantings: 220 hectares

A vital ingredient in classic Bordeaux is currently stirring up excitement in New Zealand — Merlot.

The potential for Merlot in New Zealand is enormous. Over the lengthy ripening season in our cool climate, Merlot is able to slowly build and concentrate its flavours. And with its early ripening nature — two to three weeks ahead of Cabernet Sauvignon — it can achieve higher sugar levels, lower acidity and riper fruit flavours before late autumn's coolness descends.

Following a 300 percent surge in plantings between 1986 and 1992, Merlot is New Zealand's third most widely planted red-wine variety. Over half of the vines are concentrated in Hawke's Bay; Marlborough and Auckland also have extensive plantings. A proven and increasingly popular variety in the two North Island regions, Merlot's early-ripening ability is also arousing intense interest in Marlborough, where Cabernet Sauvignon frequently fails to reach optimum ripeness. The variety may hold the key to Marlborough's red-wine future.

Merlot produces red wines of alluring richness, plumpness and suppleness. In Bordeaux, Merlot's plantings are almost double those of Cabernet Sauvignon (most wine lovers think the reverse). To the Cabernet Sauvignon-based reds of the Médoc and Graves, Merlot imparts richness and softness. Merlot truly comes into its own in St Émilion and Pomerol, where a typical vineyard is planted two-thirds in Merlot with the rest in Cabernet Franc. Merlot flourishes in clay — precisely where Cabernet Sauvignon vines, which prefer the relative warmth of gravel, struggle to ripen their fruit.

Merlot's early — compared to Cabernet Sauvignon — budding and flowering can be a problem in cooler regions prone to spring frosts. Poor set is a common problem, reducing yield severely. The vine displays moderate vigour, producing eight to ten tonnes per hectare of blue-black, loose-bunched berries, harbouring less tannin than Cabernet Sauvignon.

Merlot plays two major roles in New Zealand red wine-making. In the past it has principally been used as a blending variety, adding its seductive perfume, lush fruit flavours and velvety mouth-feel to the more angular, often leaner, predominant Cabernet Sauvignon; these wines are typically labelled as Cabernet/Merlot. Now that Merlot's status as a premium red-wine variety in its own right has been recognised by winemakers, a cluster of instantly appealing reds, simply labelled Merlot, has also recently reached the shelves. Esk Valley's Merlot-based reds have enjoyed the greatest success.

Corbans, Babich and Collards produced unblended Merlots in the late 1970s and early 1980s, but Kumeu River Merlot 1983 was the first stylish red produced from the variety in New Zealand. Michael Brajkovich, wine-maker at Kumeu River, who has crusaded long and hard on the variety's behalf, sees Merlot's aromas as 'more reminiscent of tobacco and leather than the more aggressive capsicum or blackcurrant characters of Cabernet Sauvignon, and the tannins are softer and more subtle. Where Cabernet may lack weight on the middle palate, Merlot is able to fill out the flavour nicely and add length and persistence to the taste.'

For wine lovers, Merlot's early-drinking appeal is a boon. A high-quality Cabernet Sauvignon needs three to five years before it becomes a pleasure to drink; Merlot can knock your socks off in two. As Cabernet Sauvignon's bridesmaid, or increasingly as the bride, Merlot adds a lush, sensuous appeal to our reds.

LEADING LABELS:
C.J. Pask Reserve, Corbans Private Bin and Private Bin Merlot/Cabernet Sauvignon, Delegat's Proprietors Reserve, Esk Valley Reserve and Reserve Merlot/Cabernet/Franc, Highfield, Kumeu River Merlot/Cabernet, Morton Estate Black Label

---

PINOTAGE

1992 plantings: 77 hectares

New Zealand's fifth most important red wine grape, Pinotage is a South African variety, obtained in 1925 by crossing Pinot Noir with a vine known in South Africa as Hermitage, but which is really the more humble Cinsaut grape of French and Algerian origin. Its breeder, Professor A.I. Perold, was endeavouring to create a softly flavoured red grape of higher yield than the shy-bearing Pinot Noir.

As a commercially grown wine variety Pinotage is unique to South Africa and New Zealand, where it yields soft, rounded reds that are often underrated. In South Africa the grapes ripen easily with high sugars and the top wines, from well-pruned vines, can age well.

Pinotage was established in New Zealand during the late sixties and early seventies, during the rush to replace hybrids with *vinifera* material. The vine grew prolifically, ripening reasonably early with good yields of medium-sized, thick-skinned berries. The variety was especially popular in Auckland, because of its ability to withstand humid conditions. Pinotage's plantings rose by sixty percent between 1986 and 1992, with Marlborough now the most heavily planted region, and smaller pockets in Hawke's Bay, Auckland and Poverty Bay.

Pinotage has had a rather turbulent career in New Zealand. Once heralded as a premium variety capable of producing 'the great New Zealand red', it was later much criticised as 'coarse'. The criticism derived partly from the fact that many so-called Pinotage wines used to include substantial amounts of hybrids, ostensibly to improve the wine's colour. A straight Pinotage is much more worthwhile.

A well-made Pinotage is a soft, medium-bodied, early-maturing, peppery wine, less tannic than Cabernet Sauvignon, with a pleasant berry-like flavour and smooth finish. According to Peter Hubscher, Montana's managing director, 'Pinotage is the most underrated grape variety in New Zealand.'

LEADING LABELS:
Limeburners Bay, Nobilo, Pleasant Valley, Soljans

PINOT NOIR

1992 plantings: 402 hectares

Pinot Noir ranks only behind Cabernet Sauvignon as our second most widely planted red variety, and is our fifth most commonly planted grape overall. In recent years the beguilingly scented, supple and soft Pinot Noirs of the Wairarapa and — to a lesser extent — Central Otago have added an exciting new dimension to New Zealand's premium reds.

Singlehandedly responsible for the majestic, velvety reds of Burgundy, Pinot Noir tastes of strawberries and raspberries in its youth, with fairly high alcohol and often a suggestion of sweetness. Mature Burgundy can be arrestingly complex, with an array of aromas and flavours suggestive of red berry fruits, violets, rotten vegetables, coffee and fruit cake. Pinot Noir is also of pivotal importance in Champagne, where it is prized for its body and longevity; great care is taken to keep the white juice free of tint from the skins. But the vine is notoriously temperamental in its choice of residence and has not readily adapted to regions beyond Europe. In Oregon — where Pinot Noir is by far the most widely planted variety — wines of intriguing scent and elegance have emerged, but elsewhere in the United States and in Australia (with the notable exception of the Yarra Valley in Victoria) its wine has usually lacked real distinction.

Bragato observed in 1906 that Pinot Noir 'ordinarily bears well and yields a nice wine'. The vine is a challenge to viticulturists, however; as an early budder, it is vulnerable to spring frosts and its compact branches are very prone to rot. One advantage is it ripens ahead of Cabernet Sauvignon, typically producing fairly low yields of seven to twelve tonnes per hectare of small berries of varying skin thickness.

Although the first vines in New Zealand were virus-infected, healthier vines have been available since the early 1970s. Several clones are being cultivated, of which the best known are the well-performed AM 10/5, which yields wine with good colour and a soft, fruity palate, and Bachtobel, whose wine lacks colour and weight and is not highly regarded. Clone UCD5 ('Pommard') yields deep-hued, firm-structured wine. The clone previously called 'Gamay Beaujolais' (here and in California) has now been positively identified as Pinot Noir. Over a thousand Pinot Noir clones have been recorded in Burgundy. 'It is a common belief that a single clone will not provide the final answer,' says viticultural scientist Dr David Jordan. 'Other premium clones are sought to blend for added complexity and character.' Low yields are also a key factor in the pursuit of quality.

Pinot Noir thrives in coolish climates, where the grapes are able to hang on the vines for extended periods, picking up the most subtle scents and flavours; with its proneness to rot, this is a risky business. Plantings more than doubled in New Zealand between 1989 and 1992. In Wairarapa, Canterbury and Central Otago, Pinot Noir is now the most important variety. Almost half the vines are found in Marlborough, where the fruit is principally used in the production of bottle-fermented sparkling wines. Hawke's Bay also has extensive plantings.

Nobilo set the early pace with some superb Huapai-grown Pinot Noirs in the 1970s. Later Babich at Henderson and St Helena in Canterbury shared the top show honours. Now clear signs are emerging that the Pinot Noir variety, usually a reluctant traveller beyond its Burgundian stronghold, has found in the Wairarapa a new antipodean home.

LEADING LABELS:
Ata Rangi, Dry River, Gibbston Valley, Larcomb, Martinborough Vineyard, Neudorf, Palliser, Rippon, St Helena, Waipara Springs

SYRAH
(Shiraz)

1992 plantings: 14 hectares

Syrah, a rare variety in New Zealand, is the principal black grape of the upper Rhône Valley of France and is also heavily planted in Australia. In the Rhône it is called Syrah and in Australia, Shiraz or Hermitage. Lauded in France as the foundation of such great reds as Côte Rotie and Hermitage, Syrah nevertheless constitutes less than two percent of France's red-wine vineyard. In Australia, by contrast, Syrah is the most widely planted wine variety. Regardless of the name of the vine or the location of its vineyard, Syrah typically yields robust, richly flavoured reds, peppery in character and with a heady perfume.

Syrah has a long history in New Zealand. Bragato was a fervent supporter, declaring in 1895 that 'The Hermitage will, in your colony, give heavy yields and wine of first quality . . . [it] should compose at least one half of the vineyard . . .' Good wines were made, but in many areas the grapes failed to ripen, lacking sugar and colour, and remaining overly acid. Government Viticulturist S.F. Anderson wrote in 1917 that Syrah was being 'grown in nearly all our vineyards but the trouble with this variety has been an unevenness in ripening its fruit'.

After decades of eclipse, Syrah is currently being re-evaluated. The variety is easy to cultivate, bears well, and new clones and virus-indexed vines show an improved ripening performance. It is vulnerable to excessive rain, however, which causes the berries to swell, split and rot.

Plantings more than doubled between 1989 and 1992, with most vines found in Hawke's Bay and Marlborough. Syrah thrives in a warm growing environment: will it flourish on the sheltered, shingly sites inland from Hastings?

LEADING LABELS:
Matawhero, Rongopai Reserve, Stonecroft, Vavasour

CHAPTER THREE

Wineries, Winemakers and Wines

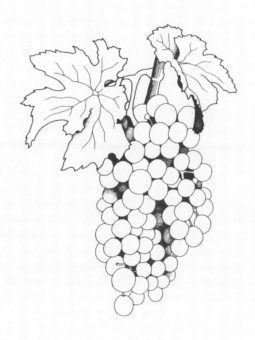

THE PRINCIPAL WINE REGIONS

'I enjoyed your Hawke's Bay last night,' a British wine merchant recently enthused to winemaker Joe Babich. To Babich's ears, this sounded a bit odd — most of us buy New Zealand wine by the name of its predominant grape variety or producer. Overseas consumers, however, often opt for a Bordeaux or Rioja or Chianti — distinctive styles of wine from controlled, strictly delimited regions.

Regions of origin are now also moving boldly into the limelight on New Zealand wine labels. Overseas consumers are increasingly demanding not just New Zealand wines, but our highest profile regional wine styles, like Hawke's Bay Chardonnay and Marlborough Sauvignon Blanc. Wine lovers in New Zealand, too, have recently woken up to the fact that many of our wines display pronounced regional differences.

Why does a Canterbury Riesling taste different from a Gisborne Riesling? Climate is the key influence: our winegrowing regions in northern latitudes are markedly warmer than those in the south. Sunshine hours, rainfall and soil types also vary markedly from region to region. These factors influence the grape varieties suitable for each region and also the styles of wine they yield. Grapes cultivated in the cooler southern regions tend to produce delicate, mouthwateringly crisp wines, whereas fruit grown in the warmer northern regions produces more mouth-filling, softer styles.

You can explore these regional variations for yourself by tasting side-by-side a Sauvignon Blanc from Hawke's Bay and one from Marlborough; and a couple of Cabernet Sauvignons from the same regions. Each pair should be from an identical vintage.

The impact of different regional climates and soils can be obscured by different vinification techniques in the winery, but you will probably find that the Marlborough-grown Sauvignon Blanc is more intensely herbaceous and appetisingly crisp than its riper, rounder Hawke's Bay equivalent. The Cabernet Sauvignon sourced from Hawke's Bay is also likely to be more robust and deep-flavoured than its lighter, 'greener' Marlborough counterpart. Experienced blind-tasters can identify these regional differences fairly easily.

With eighty-five percent of the country's vines concentrated in only three regions, New Zealand wine is overwhelmingly of Marlborough, Hawke's Bay or Poverty Bay origin. Six other regions, however, now boast a sizable cluster of vineyards. This far-flung spread of viticulture has greatly enhanced the variety — and hence fascination — of New Zealand wine.

At various stages of our history, grapevines have been planted, with or without success, over most parts of the country. Bragato in 1895 encountered vines growing in Central Otago, Akaroa, Nelson, the Wairarapa, Hawke's Bay, Bay of Plenty, Wanganui, the Waikato, Auckland and Northland. Such a widespread scattering of early grape-growing and winemaking reflected the isolated, far-flung nature of the first settlements. Although, ideally, considerations of climate and soil should have been uppermost in selecting areas to establish vines, in fact it was the influence of cultural traditions and the availability of cheap land which played leading roles in the early location of the industry in New Zealand.

The early exploitation of the Auckland region was due to the scale of the available market and the presence of Dalmatians and others eager to make wine, rather than to any climatic or physical advantages. Hawke's Bay, with ideal natural conditions for grapegrowing, was sufficiently distant from Auckland to compete for markets in the south. Auckland and Hawke's Bay thus remained the two centres of New Zealand wine for more than half a century. Then in the 1960s, when it became obvious that extensive new plantings would be necessary to cater for the soaring demand for table wines, vineyards spread beyond the traditional grapegrowing zones into Taupaki, Kumeu, Mangatangi and, above all, Poverty Bay.

Poverty Bay, although sharing some of the climatic advantages of Hawke's Bay, is isolated by rugged terrain from the major wine markets. However, Corbans, Cooks and Montana encouraged contract growers there to establish substantial areas in vines. The answer lay in the fertility of Gisborne's soils, which yield bumper crops.

The more recent move into Marlborough by Montana, Corbans, Penfolds and others was in pursuit of another objective. Yields there are relatively light, but land prices were cheap and the region is superbly suited to the production of quality cool-climate table wines.

The 1980s have brought the emergence of the Wairarapa, Canterbury and Central Otago onto the national wine map. By shifting to cooler climate zones, New Zealand winemakers are paralleling a common trend overseas. The search for riper and cleaner fruit has led many newcomers to the industry to avoid the higher rainfall areas of West Auckland, the Waikato and Poverty Bay in favour of the long dry belt extending down the east coast of both islands from Hawke's Bay through the Wairarapa and Marlborough to Canterbury.

With the pronounced variations in soil and climate from Northland to Central Otago, the challenge now facing winemakers is to sort out the best grapes for each region and then to produce the style of wine which fully captures the potential inherent in the fruit. Back in 1896 Whangarei vinegrower Lionel Hanlon stressed this point at a Conference of Australasian Fruitgrowers held in Wellington. 'As has been the case in other countries, so doubtless it will be the case here, that each district will produce one class of wine which will surpass all others in point of excellence . . . The absurdity of every man who has an acre or two of vineyard manufacturing so-called port, sherry, Bordeaux, Burgundy, Chablis, Tokay, etc., need not be discussed. It cannot too forcibly be impressed upon the future wine-growers of New Zealand the great importance of each district producing a class of wine of definite type.'

New Zealand winegrowing is moving in the direction indicated by Hanlon. Sound reasons often exist for blending grapes from a variety of regions, and it is not always economically feasible to produce small batches of regional wines, but in the past decade the production of regional wines has been a powerful trend.

Certain regional strengths are clearly apparent: Auckland in Cabernet Sauvignon-based reds; Poverty Bay in Gewürztraminer and Chardonnay; Hawke's Bay in Cabernet Sauvignon, Merlot, Chardonnay and Sauvignon Blanc; the Wairarapa in Pinot Noir, Riesling and Chardonnay; Marlborough in Sauvignon Blanc, Chardonnay and Riesling; Canterbury in Riesling.

The formation of Hawke's Bay Vintners in 1979 heralded the growing emphasis on regional identification — all wineries in the province agreed to jointly foster Hawke's Bay's image as an area which produces fine quality wines. Recently, other winemakers have banded together to promote their regional identities.

Each of New Zealand's key wine regions has its own distinctive blend of landscapes, individuals and wine styles.

NORTHLAND
7 ha.

AUCKLAND
241 ha.

WAIKATO / BAY OF PLENTY
161 ha.

THE WINEMAKING REGIONS
Areas in Vines 1992

POVERTY BAY / EAST CAPE
1503 ha.

HAWKE'S BAY
1614 ha.

WAIRARAPA
189 ha.

NELSON
81 ha.

MARLBOROUGH
2071 ha.

CANTERBURY
198 ha.

CENTRAL OTAGO
35 ha.

TOTAL AREA 6099 ha.

NORTHLAND/ MATAKANA

The northernmost region of New Zealand stretches out over 440 kilometres of rolling hill country. Its almost sub-tropical climate — warm humid summers, mild winters and abundant rainfall — is less well suited to viticulture than the cooler, drier regions to the south. Northland's main occupation is pastoral farming, yet, from Matakana in the south to Kaitaia in the north, there are six licensed wine-makers, with a cluster of new vineyards on the horizon.

Northland was the cradle of New Zealand wine: here Marsden planted the first vines and here, too, Busby made the first wine. After 1840 and the Treaty of Waitangi, how-ever, the region was exploited mainly for its magnificent kauri forests and later for its gum. Descendants of Dal-matian gumdiggers and the sons and daughters of more recent Dalmatian arrivals until recently almost alone pre-served the winemaking traditions of Busby. Few depend on their vines for a living; with a surveyed total of seven hec-tares under vines, in fact, the region has a mere 0.1 percent of New Zealand's vineyard area. Only a rivulet of wine flows in Northland; the vineyards are the smallest in the country and their wine is mostly sold locally. Château du Brak, Continental and Bryladd are not household names in the rest of New Zealand — or even in Northland.

One of the largest growers in Northland is Continental Wines at Otaika, just south of Whangarei, with more than five hectares under vines. This pretty, terraced vineyard was established in 1964 by Mate Vuletich who, as his widow relates, was born under a grapevine on the family vineyard in Yugoslavia. Vuletich originally planted Baco 22A and Niagara vines, but more recently plots of Cabernet Sauvignon, Pinot Noir, Merlot, Chardonnay, Gewürz-traminer and Müller-Thurgau have been established.

Today the founder's son Mario (41) is one of the few full-time winemakers in the North. In a small, well-equipped winery, about 2000 cases of solid, workmanlike wines are annually produced, all sold at the gate. Although ports and sherries still form part of the range, Mario Vuletich's 'per-sonal goal is to make two top reds, an unblended Cabernet Sauvignon and a Cabernet/Merlot'.

Continental Cabernet Sauvignon is a flavoursome quaf-fing red, spicy and leafy. The Müller-Thurgau is floral, fruity and soft; a bit short of acid 'zing' but clean and easy-drinking. The 'Gum Diggers' Estate Port is a ruby style, full-flavoured, fruity, sweet and smooth.

The traditional wines of Northland — sweet and fortified vin ordinaire — are reminiscent of the national wine style of thirty years ago. Between 1976 and 1990, as the profit-ability of small-scale fortified wine production declined, the number of licensed winemakers in the region nose-dived from nineteen to four. Now a revival is underway. At Mat-akana, forty-five minutes' drive north of Auckland, a new wine district is surfacing. Matakana first entered the spot-light a few years ago when The Antipodean winery spec-tacularly flared and folded. Now there are several quality reds being developed in Matakana. And the launch in early 1992 of Okahu Estate Ninety Mile Red 1989, grown near Kaitaia, heralded the arrival of a new, quality-orientated winery in the Far North.

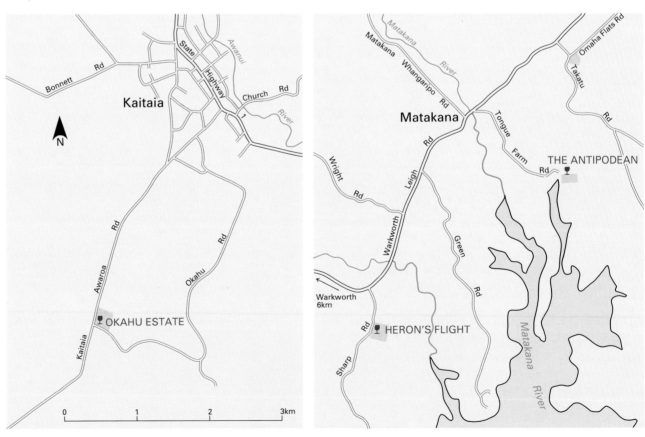

Heron's Flight Vineyard
Sharp Road, Matakana

Owners: David Hoskins and Mary Evans

Key Wines: Cabernet Sauvignon, Chardonnay

On the north-facing slope overlooking Matakana, David Hoskins and Mary Evans have recently produced the first wines at Heron's Flight. 'There are two families of herons living in the pines down by the inlet,' points out Hoskins (45), a Pennsylvania-born science and philosophy graduate who was previously a teacher and community worker but has now moved into full-time winemaking. 'Wine's magical. You're involved in the entire process and are able to exert some control over both the vine and the wine. And wine is so convivial.'

Hoskins and Evans chose Matakana as the site for their vineyard 'not because The Antipodean is here, but because we think the area is clearly suited to viticulture.' Their four-hectare loam-clay vineyard facing Dome Hill is planted in 1.5 hectares of Cabernet Sauvignon ('because we enjoy drinking it'), and smaller plots of Merlot and Cabernet Franc for a blended red; and 1.5 hectares of Chardonnay.

The vines are widely spaced on a unique trellis system designed to maximise the fruit's exposure to the sun. Hoskins and Evans pay tribute in their winery newsletter to Joe Corban for his help in the vineyard, and to Joe's brother, Alex, formerly winemaker at Corbans, who 'walked us through the whole winemaking process: from

planning, purchasing equipment and setting up the winery to actually making the wine . . . '

The first, 1991, red from Heron's Flight is a 100 percent Cabernet Sauvignon. From the 1993 vintage, two Chardonnays are being marketed: one a lower-priced wine fermented in stainless steel and only briefly oak-matured; the other a more complex, barrel fermented style.

The style of Heron's Flight reds will be largely governed by what the site can produce, but Hoskins is adamant he 'doesn't want a massive red; we want sophistication, elegance and longevity'. His Cabernet Sauvignon 1991 has made an extremely auspicious debut: deep ruby hued, with plenty of weight and appealing, fully ripe red berry-fruit and cassis flavours. Heron's Flight should significantly boost Matakana's rise to red-wine prominence.

Dark, deep-scented and packed with brambly flavours, the first 1991 vintage of Heron's Flight announced the arrival of a serious new red-wine producer in Matakana. Chardonnay will be the second string to David Hoskins's and Mary Evans's bow.

Okahu Estate Vineyard and Winery
Okahu Road, Kaitaia

Owners: Monty and Bev Knight

Key Wines: Ninety Mile Red, Ninety Mile White, Clifton Chardonnay

The most northern vineyard and winery in the country, Okahu Estate lies near Kaitaia, on the Pukepoto Road to Ahipara Bay. On a north-west facing slope only a couple of kilometres from the sand dunes of Ninety Mile Beach, Monty and Bev Knight erected their iron-clad timber winery in 1990.

Monty Knight (48), a Kaitaia retailer and commercial property owner, is determined to 'improve the quality of New Zealand's northernmost wines'. In 1984 he planted 250 trial vines of Cabernet Sauvignon, Pinotage and Pinot Noir. After his first crop in 1987 reached high sugar levels (up to 24 degrees brix), in went another 2000 vines.

In the warmth and humidity of the Far North, Knight says Okahu Estate's grapes ripen 'up to three weeks ahead of Gisborne and Hawke's Bay'. The 2.5-hectare loam-clay vineyard is planted mainly in red-wine varieties, which now also include Merlot, Cabernet Franc, Chambourçin

and Syrah. 'To see if the Far North can produce a white', Knight has also trialled Chardonnay and Sémillon. Vineyard expansion over the next few years will concentrate on late-ripening varieties best suited to the region's heat.

Only a few thousand litres of wine, made by Knight with a consultant winemaker, are flowing from Okahu Estate. By 1996, however, the output should reach 3500 cases. The winery is open for sales over the summer, and a vineyard restaurant and chalet-style accommodation are planned for the future, once volumes climb.

Okahu Estate Ninety Mile Red 1989 was an auspicious debut. Labelled as 'a fine claret style wine, resulting from

Monty Knight is determined to elevate the hitherto humble reputation of Northland's wines. Estate-grown, the robust, generously flavoured Ninety Mile Red is Okahu Estate's key achievement.

plenty of Northland sunshine', it was blended from estate-grown Cabernet Sauvignon, Pinotage and Pinot Noir, matured for a year in French oak casks. With its bright ruby hue, reasonable depth of spicy, blackcurrant-like flavours and firm tannin grip, it immediately confirmed the vineyard's ability to produce red wine able to compete with those from further south. The 1991 vintage, which also includes Cabernet Franc and Merlot in the blend, is even better — a warm, generously flavoured red.

Of Okahu Estate's initial white wine releases, made from grapes from Gisborne, Hawke's Bay and Te Hana, near Wellsford, the full, savoury, crisp Te Hana-grown Clifton Chardonnay is the pick.

The Antipodean

The Antipodean Winery
Tongue Farm Road, Matakana

Owner: Petar Vuletic

Key Wine: The Antipodean

The Antipodean's early life, calamitous though it proved for its owners, Auckland lawyers James and Petar Vuletic, effectively launched the new era of Northland wines. As the foundation of their bid to make 'a great wine in the tradition of the Bordeaux first growths', the brothers planted a two-hectare Cabernet Sauvignon, Merlot and Malbec vineyard on a clay slope in Matakana in 1979.

The Vuletic brothers — distant relatives of the Vuletich family at Continental — patiently waited several years for their vines to mature before producing the first vintage of The Antipodean in 1985. Such was their unwavering commitment to quality, to ensure The Antipodean's 1985 crop was fully dry, a helicopter was hired to hover over the vineyard. The wine itself was matured in new, thin-staved French oak barriques for a very long spell by New Zealand standards — up to thirty months.

With the release date of the 1985 vintage approaching, the Vuletics masterminded an extraordinary public relations campaign. For the benefit of the British wine press, the brothers flew to London and lined up the 1985, 1986 and 1987 vintages of The Antipodean in a blind-tasting with several illustrious Bordeaux châteaux. The Antipodean won several glowing endorsements. When a pre-release rehoboam (5L bottle) of The Antipodean 1985 surfaced at auction in Auckland in late 1987, restaurateurs Tony Astle and Mark Bartlett paid a sensational $5100. At this stage hardly anyone in New Zealand had tasted the wine.

James Vuletic, co-founder of The Antipodean, has more recently established a separate, pocket-sized, Merlot-predominant vineyard on a steep slope at Matakana. 'Château Petrus is my model,' says Vuletic.

When the 1985 vintage was finally released in early 1988 at $93 a bottle, a yawning gap was exposed between expectation and reality. The Vuletics' red was impressive, but fell short of the standards already achieved by several other red-wine producers in Hawke's Bay and on Waiheke Island.

Of the three vintages from 1985 to 1987, the 1986 is the best: a fragrant, robust, powerfully oaky red, with hints of green fruit among its mouth-encircling flavours. Although not approaching the class of the finest red Bordeaux, it certainly ranks among the top half-dozen New Zealand reds from 1986.

In 1988, within months of its first release, The Antipodean started to unravel. The 1988 crop was left to rot on the vines. At first this was attributed to poor grape quality, caused by Cyclone Bola; later reports surfaced of a major dispute between the Vuletic brothers.

By mid-1988, following a 'personal falling out', the Vuletic brothers were battling over The Antipodean in the courts. The final act came in March 1990, when the assets of The Antipodean were auctioned; 9000 bottles fetched prices ranging from $20 to $65.

Although the split between the brothers has not been resolved, both remain committed to their dream of making a fine red at Matakana. Petar Vuletic, who still owns the majority of The Antipodean's original vineyard and the right to its striking name, has converted an old haybarn into a winery and produced wine each vintage since 1990.

James Vuletic, who emerged from the break-up with The Antipodean's small, white macrocarpa winery and a minority share of the vineyard, has since sold both to Glen Ireland, who is interested in producing his own red wine. Only 800 metres away, Vuletic is now the owner of a new, steeply sloping 1.5 hectare clay vineyard planted in Merlot (93 percent), Cabernet Franc and Malbec.

Why is he concentrating on Merlot? 'That decision was partly based on Merlot's performance at The Antipodean, where its wine is very full bodied. Château Petrus is my model. But there is also a personal reason: I'd rather avoid direct comparisons between my wine and The Antipodean, which is Cabernet Sauvignon-based.'

The first vintage of James Vuletic's new wine will be after his vines are at least five years old. In the meantime he'll produce 'an excellent rosé in the style of Anjou, which will last for years'.

On the Vuletic brothers' past performance, their reds should be worth waiting for.

Scattered across the softly undulating countryside that cradles the West Auckland townships of Kumeu, Huapai, Waimauku and Hobsonville are numerous quality-orientated wineries. There are no giants like Corbans or Montana here. Familiar names like Coopers Creek and Kumeu River, and less familiar names like Waitakere Road Vineyard and Limeburners Bay, are all small by New Zealand standards. Selaks, the Sauvignon Blanc wizards, are verging on middle size. Only Nobilo — or The House of Nobilo, as the company prefers to be known — and Matua Valley are solidly established among the country's dozen or so medium-sized wineries.

The current rebirth of interest in Auckland viticulture is to a large extent concentrated here. The local climate, according to meteorologists, is one of the best for table winemaking in Auckland, with the district's distance from the Waitakere Ranges giving a lower rainfall than at Henderson. The signs of success are everywhere: in the aesthetically delightful buildings at Matua Valley and Coopers Creek; in the burgeoning of vineyard restaurants at Selaks and Matua Valley; and, most importantly, in the wines themselves. Many are ranked among the country's best.

Kumeu-Huapai accounted for a mere nineteen percent of all vine plantings in the Auckland province in 1960; by 1975 the figure had soared past fifty percent. Henderson winemakers of the 1960s, wishing to expand to meet increasing demand, faced a serious problem in the lack of cheap, reasonably large blocks of land in Henderson. With land at Henderson between 1965 and 1970 valued at $4500 per hectare, and land at Kumeu selling for only $1500 per hectare, expansion soon shifted to the Kumeu-Huapai area. Old established companies were able to plant new vineyards only fifteen kilometres away from their existing wineries.

Many companies — Delegat's, Lincoln, Soljans, Collards, Cooks, Corbans and others — established new vineyards here (some later pulled out). Others, such as Nobilo and Kumeu River, have an extended history in the district. Matua Valley and Coopers Creek are relatively recent arrivals who chose to base their headquarters, although not all of their vineyards, here.

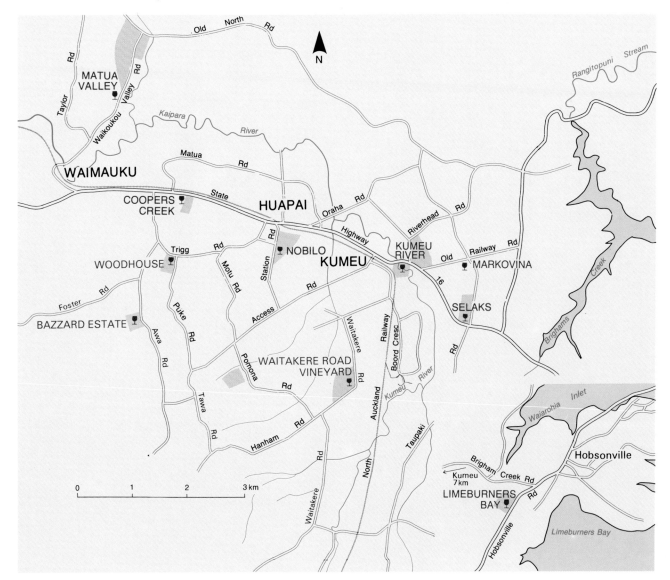

Fledgling producers are still emerging while one or two older ones decline. The latest newcomers are Waitakere Road Vineyard, a red-wine specialist near Kumeu, and Bazzard Estate at Huapai — the source of a promising Pinot Noir. At Woodhouse Wines in Trigg Road, Huapai, George Vitasovich produces a small volume of white and red wines; his Cabernet/Merlot 1989 scored a bronze medal in 1991.

Abel and Co, founded at Kumeu in 1974, flourished as a red-wine specialist until Malcolm Abel's untimely death in 1981. The standard of the wine declined under new ownership and in 1989 production ceased. The winery was later converted into Gracehill Vineyard Restaurant, which, until its recent closure, specialised in 'simple country cooking' washed down with locally produced wines.

It should be emphasised that the majority of the wines produced here are made from fruit trucked in from more southern districts, notably Gisborne, Hawke's Bay and Marlborough. However, confidence in the quality of local grapes is soaring: Kumeu River wines are produced almost exclusively from local fruit.

Rationalisation is underway in the vineyards as the winemakers single out the grape varieties most adaptable to the region's warm, humid summers and heavy clay soils. Cabernet Sauvignon is the leader, followed by Chardonnay, Palomino, Merlot and Sauvignon Blanc. Pinot Noir, Pinotage and Cabernet Franc are also well entrenched.

Further exploring the theme of 'horses for courses', winemakers here are also experimenting with new and healthier clones of existing varieties, use of devigorating rootstocks, 'grassing down' to reduce waterlogging of the soil, improved trellising techniques and other methods to upgrade their fruit quality.

The outcome has been the production of some outstanding wines, both white and red. Kumeu River Chardonnay, Sauvignon and Merlot/Cabernet, some vintages of Matua Valley Reserve Sauvignon Blanc and Collards Rothesay Vineyard Chardonnay have been the most eye-catching successes.

Family-owned enterprises of Dalmatian origin are still a powerful influence here.

Bazzard Estate

Bazzard Estate
Awa Road, Huapai

Owners: Charles and Kay Bazzard

Key Wine: Pinot Noir

A Buckinghamshire solicitor who laboured on the Auckland wharves before moving fulltime into wine, Charles ('Charlie') Bazzard is a complex, forceful personality — one of the characters of the Huapai wine scene. Bazzard Estate is a four-hectare hillside vineyard in the wooded Awa

Valley, two kilometres west of the Huapai-Waimauku highway.

Soon after arriving in New Zealand in 1981 with his Auckland-born wife, Kay, Bazzard purchased the Awa Road vineyard established in hybrids and Palomino seventeen years earlier by Cuthbert Woolcott. 'I had no illusions about making a fortune quickly,' Bazzard recalls. 'But I'd always hankered to be a farmer, and this site reminded me of the vineyards I saw in Germany as a national serviceman.' Today, after much replanting of their east-facing clay slope, the Bazzards grow 2.5 hectares of Pinot Noir and smaller plots of Cabernet Sauvignon, Cabernet Franc, Merlot and Gewürztraminer.

Until the launch of the Bazzard's first vintage in 1991, their grapes were sold to the nearby Kumeu River winery. 'Our plan was to initially concentrate on grape-growing,' says Kay Bazzard, 'and then after we'd mastered that, move into wine production under our own label.' The Bazzards still have no plans to erect their own winery; instead, their wines are made by Kim Crawford at Coopers Creek.

Charlie Bazzard was initially determined to produce a fine quality dry Müller-Thurgau, 'to the standard of the best einzellages [individual sites] of the Rheinpfalz and Nahe'. In 1992, however, Bazzard opted to concentrate solely on the production of red wine.

Bazzard Estate Pinot Noir is a lightly wooded style. 'Our wine is sold in England,' says Bazzard, 'where New Zealand wines are perceived as fresh and fruity. Not using much oak is the best way to show off its fruit qualities.' This is a good wine with vibrant, raspberryish flavours and soft tannins in a fresh and straightforward style.

1991
Bazzard Estate
HUAPAI
PINOT NOIR
11.5% Vol   e 75 cl
Produced & Bottled by: Imported by:
C.J.N. & K.J. Bazzard, Awa Road,   The Organic Wine Co. Ltd, P.O. Box 81,
R.D.1., Kumeu, Auckland, N.Z.   High Wycombe, Bucks, HP13 5QN.

Fresh, buoyantly fruity, briefly oak-aged Pinot Noir flows from Charles ('Charlie') and Kay Bazzard's hill-grown vineyard in the tranquil Awa Valley, near Huapai. Bazzard, a solicitor-turned-waterfront-worker-turned-winemaker, exports much of his wine to his native Buckinghamshire in England.

Coopers Creek Vineyard
State Highway 16, Huapai

Owners: Andrew and Cyndy Hendry and
shareholders

Key Wines: Coopers Classic Dry, Riesling,
Late Harvest Riesling, Sauvignon Blanc, Fumé
Blanc, Hawke's Bay Chardonnay, Swamp Reserve
Chardonnay, Gisborne Chardonnay, Coopers Red,
Cabernet/Merlot

Mouthfilling, succulent and complex, the 1986 vintage of
Swamp Road Chardonnay catapulted Coopers Creek to
prominence. Recently Coopers Creek has demonstrated —
more convincingly than most wineries — that top-flight
Rieslings can also be produced from Hawke's Bay fruit.

Coopers Creek is tucked unobtrusively away alongside
the main road beyond Huapai. The founding partners,
Andrew Hendry and Randy Weaver, first crossed paths at
Penfolds in Henderson during the late 1970s. Hendry,
born in Wanganui and educated in Auckland, is a tall, soft-
spoken, forty-seven-year-old accountant, relaxed in manner
and brimming with wit. A well-respected spokesman for
small-scale winemakers on the Wine Institute's influential
executive committee, Hendry is the financial brain guiding
Coopers Creek and also controller of marketing.

Randy Weaver, an Oregonian armed with a master's
degree specialising in viticulture and oenology, worked for
Montana in Gisborne and with Hendry at Penfolds before
the two decided to set up their own small winery. The
winery, erected in 1981, took its name from an old local
map.

Coopers Creek's path was for a long time blocked by the
local council, which adamantly refused to allow the new
winery — but not the older ones — to process grapes grown
outside the district. The issue was eventually resolved, and
Weaver went on to enjoy gold medal successes with his
1985 Gewürztraminer — still a stunning wine — and 1986
Swamp Road Chardonnay, which peaked early. In mid-
1988, however, Weaver announced his withdrawal from
the partnership and returned to the United States.

Andrew Hendry and his wife, Cyndy, now control
Coopers Creek with a majority shareholding; the balance is
spread over sixteen investors. Cyndy Hendry is also
actively involved in the company's sales and promotional
activities.

Winemaker Kim Crawford has enjoyed eye-catching
competition success since his 1988 arrival at Coopers
Creek. Crawford graduated from Massey University with
a BSc, and Roseworthy College with a postgraduate
diploma in oenology. He then pursued his early career in
the Hunter Valley of New South Wales, at Stag's Leap
Winery in California and Backsberg Estate in South Africa.

Crawford works with fruit drawn from several regions.
The three-hectare vineyard adjacent to the winery,
originally established in Cabernet Sauvignon and Merlot,

*Andrew Hendry (centre) and
his wife, Cyndy, control
Coopers Creek, one of New
Zealand's most consistently
successful exhibitors in show
judgings. Winemaker Kim
Crawford (right) has been
especially acclaimed for his
perfumed, intensely citric-
flavoured Hawke's Bay
Rieslings.*

has recently been partly replanted with Chardonnay for a small-scale bottle-fermented sparkling. The Hendrys themselves also own an eight-hectare Chardonnay block at Havelock North. The majority of the winery's grapes, however, are purchased from growers in Auckland, Gisborne, Hawke's Bay and Marlborough.

Coopers Creek Riesling is an intensely aromatic, vibrantly fruity, penetratingly flavoured wine that in most vintages ranks at the forefront of Hawke's Bay Rieslings. Most of the region's Rieslings struggle to rival the intensity of Marlborough's. Why is Coopers Creek's so successful?

'It's partly the site,' says Hendry. 'The grapes are grown in Jim Scotland's vineyard at Clive. Good air movement close to the sea reduces the risk of disease, allowing us to hang the fruit out late. The grapes don't get wildly ripe, but they build up intense flavours. And we leave part of the block unsprayed to encourage botrytis.'

The equally impressive, slightly freeze-concentrated, sweet Late Harvest Riesling displays a touch of botrytis influence, but places its accent on the lusher fruit flavours of late-picked fruit. Hendry reports that in recent years Coopers Creek's Riesling sales have 'leapt ahead' — so much so that the winery's consistently rewarding Gewürztraminer was discontinued following the 1990 vintage.

Coopers Creek Fumé Blanc is a deep-flavoured dry wine with a penetrating, tropical fruit-like bouquet. Made from Gisborne fruit, with very restrained oak handling, this is a rich, ripe wine with a long finish. In selected years a more complex, partly barrel-fermented version is produced under the Fumé Blanc Oak Cask label.

Coopers Creek's first Marlborough Sauvignon Blanc, from the 1989 vintage, collected a trophy and four gold medals. A vibrant wine with a wealth of pure, gooseberry and passionfruit-like — rather than pungently herbal — flavours, this and subsequent vintages have been classic examples of Marlborough Sauvignon Blanc at its arresting best.

Coopers Classic Dry is one of those rare and delightful wines where both the name and price undersell the merits of the wine itself. In this marriage of Chenin Blanc's fruitiness with Sémillon's crisp, grassy characteristics and Chardonnay's fullness, in its youth the Sémillon holds the upper hand, with the Chenin Blanc asserting itself more

fully as the wine matures.

Three styles of Chardonnay are marketed. The Gisborne Chardonnay, launched from the 1990 vintage, is already the biggest seller of the trio; this is a briefly oak-matured wine with a modest price-tag. The mid-priced Hawke's Bay Chardonnay is tank-fermented and then matured for six months in American oak casks, with no malolactic influence. Using American rather than French oak, says Hendry, gives the wine 'difference and forwardness'.

The high-priced Swamp Reserve (formerly Swamp Road) Chardonnay is one of the jewels in Coopers Creek's range. The label was originally based on fruit grown by Fenton Kelly in Swamp Road, in the Omarunui Valley in Hawke's Bay. Congratulating them on their precise honesty in labelling, the Australian wine writer James Halliday then urged: 'Why not change the name of the road?'

The name proved popular, however — and for a time even survived the winery's swing away from using fruit grown in Swamp Road. Today, Swamp Reserve Chardonnay is based on a selection of Coopers Creek's best fruit — the name is no longer a geographic designation but a brand. Swamp Reserve Chardonnay is French oak-fermented, with about twenty-five percent malolactic influence, and is matured on its lees for up to a year. It is a robust yet stylish wine with an impressive depth of peachy, fig-like and oak flavours.

In the past Coopers Creek's reds have been overshadowed by its whites. The popular Coopers Red, most recently based on a blend of Auckland-grown Cabernet Sauvignon and Pinotage, is a lightly wooded, Beaujolais-style red, fresh and buoyantly fruity. Early vintages of the higher-priced, estate-grown Cabernet-Merlot, however, displayed a lack of optimal ripeness and were assertively leafy-green. The light, elegant 1989 vintage brought a welcome quality lift.

'Looking back,' says Hendry, 'some periods have been very tough. First there were the town-planning restrictions; then Randy Weaver departed. The financial side is always a hassle. But seeing the winery develop physically, becoming an attractive place to visit, is a pleasure and Kim's been a great winemaker, retaining and improving our quality.

'The future holds a good Hawke's Bay Cabernet/Merlot, a Pinot Noir from our Havelock North vineyard, and a bottle-fermented sparkling. But we're not necessarily going to keep getting bigger and bigger.'

Kumeu River

Kumeu River Wines
State Highway 16, Kumeu

Owners: The Brajkovich family

Key Wines: Kumeu River Chardonnay, Sauvignon, Merlot/Cabernet, Brajkovich Chardonnay, Sauvignon, Cabernet Franc, Merlot

'I really love Chardonnay,' says Michael Brajkovich, winemaker at Kumeu River. 'It's a versatile variety and its flavours are so appealing.' Under his premium Kumeu

River label, Brajkovich produces a rich, savoury, multifaceted Chardonnay — one of New Zealand's most talked-about wines.

Kumeu River — under its earlier name, San Marino — was one of the first wineries to establish such classical grape varieties as Chardonnay and Pinotage, and also enjoyed an early reputation for hybrid quaffing wines such as its Kumeu Dry Red. In the past decade the winery has emerged from a flat patch in the 1970s to capture international respect.

The winery is sited on the main highway, one kilometre south of Kumeu. Early Dalmatian settlers tended vines on

the property for several decades before the Brajkovichs' arrival. When nineteen-year-old, Dalmatian-born Mate Brajkovich and his father bought the property in 1944, along with seven hectares of pasture, they acquired a fermenting vat, barrels and a half-hectare of Isabella and Albany Surprise vines.

Shortly after the war Brajkovich began planting hybrids and then, in the 1950s, Pinotage and Chardonnay. The strong impact of this winery on the Auckland wine scene of the 1950s and 1960s owed much to its legendary hospitality: the poets Denis Glover and Rex Fairburn, heart surgeon Douglas Robb and wine and spirit merchant Harold Innes were all frequent visitors.

Kumeu River epitomises the small, family-owned winery. Several family members are involved in the daily operations. The charismatic Mate, until his death in 1992, still worked a full week; in 1985 he was awarded an OBE for his services to the wine industry and the community. Mate's widow, Melba, is the general manager and works in the vineyard shop. Their eldest son Michael (33) controls the winemaking, and is also involved in financial administration and public relations. Another son, Milan (30) who has a degree in chemical engineering, oversees the vineyards, and Paul (25) a commerce graduate, is immersed in marketing.

The Brajkovichs work exclusively with Kumeu fruit. 'It's easy to go out and buy grapes, thus avoiding capital expenditure,' says Michael, 'but quality doesn't start there. Our best wine always comes off the home block.' On the rise across the highway from the winery, eight hectares of Chardonnay, Cabernet Franc and Sauvignon Blanc are cultivated. Recent plantings — including the replanting of the two-hectare original block across the road from the winery — have concentrated on Chardonnay. The one-hectare vineyard next to the winery is planted in Cabernet Sauvignon. The Brajkovichs also own a five-hectare block of Merlot vines planted by Corbans in nearby Waitakere Road in the early mid-1970s.

The winery, with its average annual output of 15,000 cases, in reality is much larger than it looks from the front. Should you step through the rear of the vineyard shop, you will be standing in the original concrete-walled winery erected by Mate in the 1940s. Nearby rests a century-old barrel once owned by Heinrich Breidecker.

Brajkovich's wines are marketed under three labels: Kumeu River (the premium range), Brajkovich (the middle range), and Kumeu (for everyday quaffing). Fortified wines — including port, sherry and an Altar Wine — marketed under the old San Marino brand ceased production in 1990. New Zealand absorbs most of the winery's output. Only ten percent is exported, to the United Kingdom and the United States — almost all Chardonnay.

Kumeu River Sauvignon is usually a splendidly ripe, full, rich-flavoured wine, demonstrating well the Sauvignon Blanc variety's ability when grown in Auckland to yield stylish, rather than strident, wine. Brajkovich is aiming for a 'tropical fruit' rather than an 'herbaceous' style. Fermented in older French oak barrels, so that the

Winemaking at Kumeu River is largely a family affair. From left: Michael Brajkovich, his mother Melba, father Mate (who died in July 1992) and brothers Paul and Milan. The jewel in Kumeu River's range is the distinctive, complex, gloriously full-flavoured Chardonnay.

wood character will be subdued, the Sauvignon is also put through a malolactic fermentation and barrel-matured for nine months. Most vintages yield mouthfilling, moderately herbal, impressively complex wines; in some years botrytis gives a distinct honeyishness.

Kumeu River Chardonnay is outstanding — a powerful, superbly constructed, deep-flavoured wine displaying intense, ripe fruit flavours and a soft, seductive finish. Fermented and matured for about nine months in heavy- and medium-toast Burgundy oak pièces (225L barrels) — about twenty-five percent new each year — it stands out for its distinctive, strongly malolactic-influenced, buttery style and rich, mealy complexity.

The Kumeu River label is also well known for its premium blended red. The wine started its life as a straight, varietal Merlot, but has since evolved — depending on the vintage — into a Cabernet/Merlot or Merlot/Cabernet blend. Brajkovich sees a parallel between the success of Merlot in St Emilion's heavy soils, where it ripens two weeks ahead of Cabernet Sauvignon, and Kumeu's own needs.

'We'll never make blockbuster reds here,' says Brajkovich. 'Our wines are more subdued, but they display elegance and style.' Has he been tempted to draw red wine fruit from Hawke's Bay? 'No. For our top wines we want con-trol of the vineyard.' The wine shows more suppleness and charm than a straight Cabernet Sauvignon, yet retains complexity and flavour depth. The bouquet is plummy and smoky, the colour a bold cherry red. This is a distinctive wine, with the medium body and firm structure of claret, coupled with ripe and berryish flavour.

A quartet of wines are marketed under the mid-priced Brajkovich label. The Sauvignon is fresh and tangy but has tended to lack the individuality of its stablemates. The Chardonnay, compared to the premium Kumeu River wine, is grown in a higher-yielding vineyard, based on heavier-cropping clones, and given restrained or no oak aging. Yet this is still a very fine wine, with much of the style of its illustrious big brother.

Brajkovich Cabernet Franc is delectable in its youth. Fruity, yet with solid extract and an underlying firmness, this wine tastes exactly like a grape should that often is employed to soften and pacify Cabernet Sauvignon. Brajkovich Merlot is deeper and more tannic. Michael Brajkovich delights in Merlot's 'leather, tobacco and coffee' flavours. Matured for a year in seasoned French oak barriques, this is an excellent 'food' wine — the Brajkovichs serve it with pasta.

Michael Brajkovich

More than any other individual, Michael Brajkovich through his outstanding trio of Chardonnay, Sauvignon and Merlot/Cabernet wines has resurrected West Auckland's reputation as a quality wine region.

Brajkovich is a quiet, handsome man with a towering physical presence. After training with distinction at Roseworthy College, South Australia, he returned to his family vineyard ('Nobody asked me to stay in Australia,' he told *Wineglass* magazine, tongue-in-cheek) before spending the 1983 vintage at Chateau Magdelaine, a leading premier grand cru of St Emilion. Here he developed not only a behind-the-scenes appreciation of Bordeaux winemaking techniques, but also, he says, the conviction that a small-scale New Zealand winery had more to learn from the French about handcrafting wines of individuality and finesse than from the Australians or Californians.

From the start he was convinced that Kumeu could produce fruit of outstanding quality. Brajkovich points to how planting on hill sites improves drainage; to the merits of 'grassing down' between the rows as a way to reduce waterlogging of the soil; to the enhanced fruit ripeness he achieves using the U-trellis system; and to the superior method of fruit selection with hand-harvesting. His wines have supported his conviction.

The launch of the first Kumeu River Chardonnay caused a major stir. As Brajkovich has written: 'In 1985 Kumeu River produced a Chardonnay that underwent a total malolactic fermentation, to the surprise of many, and certainly to the disgust of the wine judges who relegated it to the 'no award' level. The style was totally foreign to that of previous New Zealand Chardonnays; it was the style of Burgundy, and unfortunately incorrectly diagnosed as being faulty. Fortunately those people who drink wine for its enjoyment really appreciated the depth, richness and smoothness of this Chardonnay.' The 1987 vintage, which earned accolades in England and the United States, became Kumeu River's first internationally successful wine. The 1989, 1991 and 1992 vintages are equally arresting.

Brajkovich's major ambition is to 'keep making better and better wines, which are distinctive and enjoyable. I'm constantly looking for an improved way of doing things.' He questions why 'so many of New Zealand's top wines are still mechanically harvested, and crushed and drained rather than whole bunch pressed. These are compromises for economic reasons. Yet one can easily recoup the extra costs in the price of a top quality wine.'

Michael Brajkovich in 1989 became the first New Zealander to succeed in the famously rigorous, London-based Master of Wine examination. His right to add the much-coveted letters M.W. after his name generated a surge of publicity. In Brajkovich's eyes: 'I went from being a bit eccentric to being credible.'

Limeburners Bay Vineyards
112 Hobsonville Road, Hobsonville

Owners: Alan and Jetta Laurenson

Key Wines: Cabernet Sauvignon, Cabernet
Sauvignon Dessert Wine, Pinotage, Sauvignon
Blanc, Sémillon/Chardonnay

Limeburners Bay

A dark, sweet, raisiny, fortified Cabernet Sauvignon, in the tradition of the vins doux naturels (fortified dessert wines) of southern France is the most talked-about wine from this small company. Limeburners Bay, owned by forty-nine-year-old Alan Laurenson and his Danish wife Jetta, lies several kilometres east of Kumeu and Huapai in Hobsonville Road. This is one of the newest wineries in the region.

Alan Laurenson is a stocky man, raised in the King Country, who learned aspects of viticulture when employed by Barry Soljan, before settling into his full-time wine-making career. Jetta is a cartographer who has also studied horticulture; she designed the winery and its labels and also oversees the vineyard and the company's administration.

Erected in time for the 1987 vintage, their compact timber winery features a barrel storage area, laboratory and tasting room downstairs, with an upstairs office. The winery's annual output only averages about 4000 cases.

Limeburners Bay's grapes are exclusively Auckland-grown. The 1.5-hectare, gently sloping loam-clay vineyard adjacent to the winery is planted entirely in red grapes: eighty percent Cabernet Sauvignon, the rest Merlot and Cabernet Franc. The Laurensons buy the rest of their fruit from growers in West Auckland.

Limeburners Bay exported fifty percent of its output in 1992, principally to the United Kingdom but also to Germany. The rest is mainly sold at the gate: this is the closest vineyard for people on Auckland's North Shore.

In 1992 Limeburners Bay slid into receivership. 'We were undercapitalised,' admits Laurenson. 'Setting up a winery to make both white and red wines is horrendously expensive. White wines are the problem: you need refrigeration, filters, tanks and lots of other equipment.' An injection of family funds enabled the winery to carry on.

Limeburners Bay's five wines — three reds and two whites — form a tight range. 'The best thing we've done lately is the Pinotage,' says Laurenson. 'Drink it young, with Italian food.' Medium ruby in hue, this is a chewy, savoury, seductively soft red, bargain-priced.

The Cabernet Sauvignon, estate-grown and matured for fifteen months in French oak barrels — fifty percent new each year — is the premium red. This is dark, weighty, slightly leafy, spicy and firm.

Robust, ripe, moderately herbal and soft, Limeburners Bay Sauvignon Blanc is a good example of the style of this variety grown in Auckland's warmer climate. The Laurensons see it as 'suitable for those meals that the big "pongy" Marlborough Sauvignons would completely dominate'.

The Sémillon/Chardonnay marries fresh, crisp Sémillon fruit flavours with the full, savoury character of barrel-fermented Chardonnay. Limeburners Bay also previously produced a Chardonnay, but the 1991 vintage was the last.

The popular Cabernet Sauvignon Dessert Wine is a full-bodied sweet red. Fractionally lower in alcohol (17 percent) than most New Zealand fortified wines, with a liquorice-like intensity and sweet finish, this is a distinctive, delicious mouthful. Laurenson recommends it for the latter stages of a meal, 'when the conversation gets to sex and politics'.

The name of this winery merits an explanation. Limeburners Bay, lying nearby on an upper reach of the Waitemata Harbour, is where a pioneer settler, Rice Owen Clark, finding his land to be overly sour, burned shells for lime.

Alan Laurenson has enjoyed greatest success with red wines. His dark, chunky Cabernet Sauvignon and savoury, soft Pinotage rank among the finest reds grown in West Auckland.

Matua Valley Wines
Waikoukou Valley Road, Waimauku

Owners: The Spence and Margan families

Key Wines: Fumé Blanc, Reserve Sauvignon Blanc, Chardonnay Judd Estate, Gisborne Chardonnay, Ararimu Chardonnay, Gewürztraminer, Chablis, Chenin Blanc/Chardonnay, Shingle Peak Rhine Riesling, Shingle Peak Chardonnay, Shingle Peak Sauvignon Blanc, Cabernet Sauvignon, Merlot, Shingle Peak Cabernet Sauvignon, Ararimu Cabernet Sauvignon

Matua Valley

Matua Valley's handsome, octagonal winery sits on a knoll in the secluded Waikoukou Valley, just a few bends and a climb in the road from Waimauku. Matua Valley — or Matua, as it is increasingly known — is one of the wine industry's greatest success stories. Here the Hunting Lodge Restaurant, in a house adjacent to the winery built in 1868, serves country-style food, emphasising lamb, beef and game.

Matua Valley — named after the original vineyard in Matua Road, Huapai — is controlled by the Spence brothers, Ross (50) and Bill (40). They have winemaking in their blood: a grandfather emigrated from Dalmatia and their father Rod Spence (now in his eighties) founded Spence's Wines in McLeod Road, Henderson, in the 1940s. When

Brothers Ross (left) and Bill Spence are the driving force behind one of New Zealand's most acclaimed medium-sized wineries — Matua Valley. Winemaker Mark Robertson (right) aims to produce 'food styles of wine that are fruit-driven and not aggressive'.

the newly-crowned Queen Elizabeth visited New Zealand in 1953, Spence's Sauterne (sic) and Royal Blend Sherry were at the royal table. The Spence range also featured a 'Pirate Sherry', 'Bedtime Port' and 'Ring a Ding' sparkling wine.

Matua Valley began in a leased tin shed in Swanson, near Henderson, with the brothers holding down full-time jobs elsewhere and producing their wine in the evenings and at weekends. Their first releases won immediate acclaim; the spectacle of the two part-time winemakers rivalling the major established companies in show judgings won Matua Valley an enthusiastic early following.

The tin shed lasted several vintages until an expired lease forced a change of premises. The present vertical-timbered winery, which crushes about 1500 tonnes of fruit annually, was still being built during the first vintage in 1978.

Today members of the Margan family also have a substantial shareholding in Matua Valley, and participate in decision-making at board level. Adrienne, Ross's wife, is involved full-time in the company's administration.

A great deal of Matua Valley's success can be attributed to the partners' acute awareness of the importance of public relations. The brothers themselves are popular personalities, extroverted and informal, with boundless energy. The more and more sophisticated packaging of their wines, their obvious appreciation of architecture, their vineyard restaurant, their sponsorship of the arts: all demonstrate a sure sense of public relations. The outcome is that their wines enjoy an extraordinarily high profile.

In the twenty-seven-hectare, undulating estate vineyard on sandy loam soils are extensive plantings of Cabernet Sauvignon, Pinot Noir, Sauvignon Blanc and Chardonnay. Matua Valley also owns a half share in Ron and Vicki

Smith's vineyard in the Dartmoor Valley, inland from Taradale in Hawke's Bay. Planted in stages since the 1970s, this is a shingly, silty vineyard, established in Cabernet Sauvignon, Merlot and Chardonnay. Growers in Gisborne, Hawke's Bay and Marlborough also supply fruit.

Mark Robertson (31), a tall, fresh-faced Otago University BSc graduate who later spent a year studying at Roseworthy College in South Australia, is in charge of the winemaking at Matua Valley. Robertson worked at Nobilo in 1983, in the London wine trade in the mid 1980s, and as an Auckland wine waiter, before joining Matua Valley in 1987. 'Our goal is to produce "food" wines that are fruit-driven and not aggressive in flavour,' says Robertson.

Matua Valley wines are consistently satisfying and often very high-quality. Among its broad array of white wines, the Judd Estate Chardonnay, Chablis and Fumé Blanc are probably the best known labels.

Judd Estate Chardonnay was for many years the company's white-wine flagship. Based on fruit grown in Maurice Judd's Gisborne vineyard — which Ross Spence says 'regularly produces fruit of special quality' — the wine is barrel-fermented and lees-aged for five months; a proportion also undergoes a malolactic fermentation. Fragrant, peachy-ripe, complex and soft, it is always enjoyable. The 1991 vintage is the most impressive yet.

The rarer, higher-priced Ararimu Chardonnay, launched from the 1991 vintage, is based on Judd Estate fruit given a 100 percent malolactic fermentation and held longer on its lees; Robertson calls it 'our overblown style of Chardonnay'. This is a fat, flavour-packed, creamy-smooth wine of excellent weight and length.

The big-selling Chablis is a pale straw, medium-dry

style, clean and pleasantly fruity. Here Müller-Thurgau, Chasselas, Chenin Blanc and Dr Hogg Muscat deliver straightforward, fresh flavours with a touch of sweetness to broaden the commercial appeal.

The Fumé Blanc, confusingly, is not wood-aged. Produced from Hawke's Bay fruit, this is a crisp, tangy style with a bare hint of sweetness and moderately grassy Sauvignon Blanc flavour. It is overshadowed by the Reserve Sauvignon Blanc, a more subtle and complex wine made from ripe, estate-grown or Hawke's Bay fruit, partly barrel-fermented.

One of the finest value-for-money wines in the Matua Valley range is the strong-flavoured, firm Chenin Blanc/Chardonnay. Other white wines include Müller-Thurgau and Gewürztraminer (both have enjoyed top medal success) and the sweet, freeze-concentrated Late Harvest Muscat and Late Harvest Gewürztraminer.

Matua Valley's range of red wines was until recently dominated by its Cabernet Sauvignon. The voluptuously deep-scented and ripe-tasting 1985 briefly lifted the label into the vanguard of this country's reds. A leafy-green herbaceousness, however, slightly detracted from later vintages of this otherwise impressively rich and robust red from the Smith vineyard. Matua Valley is now tackling this problem by picking the fruit in stages; the earlier ripening vines on the deepest shingle are harvested first, while the grapes planted in siltier sections are left to ripen longer on the vines. Matured for over a year in French oak barriques, this is a mouthfilling, smooth red, hugely enjoyable in its youth. Matua Valley Merlot, grown in the same vineyard, is very similar in style with strong, sweet red berry-fruit flavours.

The dark, brambly Ararimu Cabernet Sauvignon, launched from the 1991 vintage, is of startling quality. Grown 'on a lean shingle escarpment within the Smith-Dartmoor estate', it is packed with very ripe, minty, spicy flavours, firm and lingering. This is clearly Matua Valley's flagship red.

About twenty percent of Matua Valley's output is exported, principally to the United Kingdom, but also to Australia, Japan, Canada, Germany and France.

The recently released Shingle Peak Marlborough wines — Rhine Riesling, Cabernet Sauvignon, Chardonnay and Sauvignon Blanc — have deservedly been winners with their fresh, attractive flavours and striking packaging. Matua Valley's motivation for launching the range is simple: 'In export markets it's obvious Marlborough is the in thing,' says Ross Spence.

Ross Spence

'I'm the thinker, the one who does the background work,' says Ross Spence, leaning back in his chair. Spence is part-owner of Matua Valley and deputy chairman of the Wine Institute. 'Bill, my brother, he's wicked. He's the frontman, the up-front gregarious one. He parties madly.'

Matua Valley, like Morton Estate and Cloudy Bay, is one of the industry's medium-sized 'late-comers'. Ross Spence worked at his father's small Henderson winery until, aged eighteen, he left for two years' winemaking study at California State University, Fresno. He returned briefly to his father's winery, then moved into earthmoving and forestry work, before joining Villa Maria as a winemaker.

I recall leaning against the counter in a run-down tin shed in the mid seventies sipping Matua Valley Burgundy — an unexpectedly smooth and robust, silver medal-winning hybrid red. The Swanson shed swung into action for the 1974 vintage; two years later Bill Spence also joined Matua Valley on a full-time basis. From that humble beginning has unfolded one of the country's most popular and widely admired wineries.

Ross Spence is an influential force in behind-the-scenes policy making at the Wine Institute. Almost every year since 1976 he has sat on the Institute's key executive committee.

'I must have been mad to go into the wine industry,' he reflects. 'What really frustrates me is going to Europe and envisaging New Zealand's potential to be great, and then finding that the work involved in catching up with Europe is so immense. I'd love New Zealand to be selling much more wine overseas than currently. It's amazing we export at all, given the lack of government support. It's not a level playing-field, because European wines are subsidised and Australian wines get bounties.'

In Spence's eyes, New Zealand should concentrate on growing its grapes as cheaply as it can and marketing its wines to the world. 'My outlook on the future is very negative. To survive, you are going to have to produce wines to a price. This could compromise quality, but New Zealand's prices will have to be lower if we are to meet international competition.' Of course, this doesn't preclude the production of smaller volumes of higher-priced, top-flight wines for sale at home and abroad.

Why is Matua Valley so successful? Other wineries produce equally good wines, yet Matua's magic eludes them. 'People say it's because we produce good wines at a reasonable price,' says Spence. 'Also, Bill and I have been accepted as personalities by the trade; people like to be able to sell "Ross's" or "Bill's Chardonnay" — that's very important.'

Nobilo

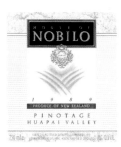

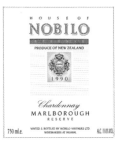

House of Nobilo
Station Road, Huapai

Owners: The Nobilo family

Key Wines: Müller-Thurgau, White Cloud, Poverty Bay Chardonnay, Marlborough Reserve Chardonnay, Dixon Vineyard Chardonnay, Marlborough Sauvignon Blanc, Hawke's Bay Sauvignon Blanc, Pinotage, Huapai Cabernet Sauvignon, Concept

In May 1990 I was lingering in the lobby of the Marriott Hotel in San Francisco. A familiar, snappily dressed figure walked across the other side of the foyer and plunged into the Californian night. His attendance at the influential annual convention of the Wine and Spirit Wholesalers of America completed, Nick Nobilo was off on the next leg of his eye-catchingly successful wine export crusade.

The publicity spotlight usually falls on 49-year-old Nick, the swarthy, good-looking head of this family-owned company, but his brothers, Steve (53) and Mark (44), are also deeply involved. 'You couldn't find three other guys with such diverse interests and strengths,' says Mark. 'Nick is the chief executive and also director of export marketing. Steve is the sales manager, my primary responsibility is the vineyards.'

Their father, Nikola Nobilo, now in his early eighties, is a Dalmatian who abandoned his stonemason's toil on the island of Korcula to found a new vineyard on the far side of the world. Still living on the outskirts of Huapai, where he first planted vines in 1943, Nikola no longer plays an active role in the company, but at board meetings 'Dad is still very much the patriarch,' says Mark.

The Nobilo operation was transformed in the 1960s when Nick Nobilo Jnr joined the company. Gilbey-Nobilo, formed in 1966, saw Gilbey's of England taking up a substantial shareholding, and providing the funds for an ambitious expansion programme based on classical grape varieties. When Gilbey's withdrew in 1974, Nobilo was established with the vineyards and wood-aging facilities necessary to produce fine wines.

The reconstructed company, formed in 1975, brought the Nobilo family into partnership with distribution agents Nathans and the PSIS, and the Development Finance Corporation. One outcome of these financial changes was a loss of overall financial control by the Nobilo family. Yet in the wake of the industry's 1985–1986 price war — in which the company suffered severe financial problems — Nobilo re-emerged as one of the country's largest family-owned vineyards. By selling off surplus land, the family was able to buy out — at 'the right price', it says — all three partners.

Key figures behind the scenes are Malcolm Harre, the long-term winery manager, and assistant winemaker Greg Foster, a Roseworthy College graduate. Each vintage Nobilo crushes about 3000 tonnes of grapes — equal to

The House of Nobilo, owned and run by brothers Steve (left), Nick (centre) and Mark Nobilo, is New Zealand's largest family-owned winery. Over half the company's output is exported — a feat achieved by few other New Zealand wineries.

about 225,000 cases, although a high proportion is exported in bulk.

Nobilo's fruit is almost all drawn from grape growers in the Auckland, Gisborne, Hawke's Bay and Marlborough regions. Its once extensive vineyard at Huapai has shrunk to only eight hectares, planted in a 3:1 ratio of Pinotage and Cabernet Sauvignon. Will this survive?

'West Auckland grows very good grapes in a fine vintage,' says Mark, 'but this happens only two or three times a decade. The one variety that performs consistently well in Huapai is Pinotage. And if you want to have a winery in a rural area around here, the council bylaws demand a vineyard.'

The decline of Nobilo's Huapai-grown reds from their exalted status of the 1970s — Nobilo 1970 Pinotage and 1976 Cabernet Sauvignon were both glamour wines in show judgings — to the mediocrity of the early 1980s, reflected above all else the debilitating effects of viruses in Nobilo's Huapai vineyard. Red wines are no longer the company's flagships, but the old-established Pinotage and Cabernet Sauvignon labels are still in the range. There are two Huapai-grown Cabernets: the cedary, lean, complex, slightly leafy Reserve label is based on a selection of the best fruit, matured for almost two years in the cask.

Although other wineries are releasing fine Cabernet Sauvignon, Nobilo Pinotage has fewer rivals. This is a consistently rewarding red, generously flavoured, peppery and zesty. 'Concept', a blended, predominantly Pinotage-based Huapai red held in wood for twenty-one months, is a distinctive red — robust, meaty and soft.

With five different labels on the market, Chardonnay is a pivotal part of the Nobilo range. The winery style is typically up-front, with strong, toasty, sometimes overwhelming oak, lush, ripe Gisborne fruit flavours, lees-aging complexity, and a soft, buttery finish.

Nobilo Poverty Bay Chardonnay is a fresh, non-wooded style, whereas the Gisborne Chardonnay is oak-matured for five months. Tietjen Vineyard Chardonnay is a partly barrel-fermented Gisborne wine, with about thirty-five percent undergoing a softening malaolactic fermentation. The steely and stylish Marlborough Reserve Chardonnay — my favourite — is fully barrel-fermented, as is the famous Dixon Vineyard Chardonnay, a voluptuous and much softer wine from Mike and Mary Dixon's warm, sheltered vineyard in Gisborne.

The Nobilo range also embraces two Sauvignon Blanc labels: a top-class, penetrating, non-wooded Marlborough style and a barrel-fermented wine from Hawke's Bay. Nobilo Müller-Thurgau, of course, has long been a light, fruity, slightly sweet standard; we all drank it during the summers of the mid-to-late-seventies. Less well known is the fact that Nobilo still produces fortified 'cocktails' like Screwdriver (vodka and orange), Gimlet (vodka and lime), and Stock Cream of Marsala.

Export is a central strand in Nick and Mark Nobilo's conversation. Nobilo exports a far higher proportion of its total output — sixty-five percent — than any other sizable New Zealand winery. Why is Nobilo so export oriented?

'During the industry crash of 1985–86 our output was still largely based on our Müller-Thurgau,' says Mark. 'When our larger competitors' prices went extremely low, our cash flow just dried up. In 1986 we sat down and learned the hard lesson: we were too heavily reliant on the domestic market.'

Scandinavia absorbs most of Nobilo's exports. Nobilo White Cloud — a fruity Müller-Thurgau-based wine boosted by a touch of Sauvignon Blanc — is shipped in 20,000 litre stainless steel tanks and then bottled in Sweden and Denmark. Sold locally and in Asia in tall, frosted bottles, this is a mild, easy-drinking wine, aimed squarely at the slightly sweet market.

In another export coup, in 1991 Nobilo landed a $3 million order from British Airways for 1.4 million mini-bottles of its Poverty Bay Chardonnay; the contract has since been extended up to and including the 1994 vintage. Such glowing success doesn't come entirely without a price. 'There's the loneliness and living out of suitcases,' says Nick, 'but I love the challenge.'

Selaks Wines
Corner Highway 16 and Old North Road, Kumeu

Owners: The Selak family

Key Wines: Sauvignon Blanc/Sémillon, Sauvignon Blanc, Founders Sauvignon Blanc Oak Aged, Chardonnay, Founders Chardonnay, Méthode Champenoise, Cabernet Sauvignon

Mate Selak's death in 1991 brought sorrow to a host of wine lovers, who for decades had basked in the warmth and geniality of his personality. Mate, who as a seventeen-year-old in 1940 journeyed from Dalmatia to New Zealand to join his winemaking uncle Marino, devoted his life to the winery that bore his family name. Mate Selak Blanc de Blancs, a new Chardonnay-based bottle-fermented sparkling, has recently been specially labelled in his honour.

Marino Selak arrived in Auckland in 1906. After many years on the northern gumfields, he planted vines, fruit trees and vegetables at Te Atatu. The year 1934 marked the first Selak vintage in New Zealand.

The vineyard gradually expanded, becoming one of the first to specialise in table wines. But in the early 1960s the north-west motorway sliced through the small vineyard. In 1965 Mate Selak re-established the company at Kumeu, with the first vintage there in 1969.

Today the Selak winery is run by Mate's sons, Ivan and Michael, and his widow, Matija, with winemaker Darryl Woolley the third key figure. Ivan Selak (43) is in charge of the day-to-day administration of the company. After leaving school at fifteen, Ivan plunged straight into

Few New Zealand wineries produce Sauvignon Blanc as fresh, frisky and full-flavoured as Selaks'. Brothers Michael (left) and Ivan (centre) control the company, assisted since 1985 by Australian winemaker Darryl Woolley.

winemaking and made Selaks' wines until 1983, when the company hired an unheralded Australian winemaker, Kevin Judd. Judd's impact was enormous. Selaks' debut Sauvignon Blanc/Sémillon, from the 1983 vintage, earned wide applause; then the 1984 Sauvignon Blanc scooped three gold medals.

From 1987 to 1990, Ivan was usually away from the winery, immersed in Selaks' viticultural operations. In 1990 he returned to head office, but if you ask him whether he or his younger brother, Michael, aged thirty-two, now pulls the strings, he insists: 'We're equals.'

After he left school, Michael worked in the vineyard at Brighams Creek, but recalls: 'I saw that as going nowhere.' After a spell in Australia with Farmer Brothers — the mail-order wine specialists who then imported Selaks — Michael returned in the role of sales representative. In 1987, aged twenty-six, he stepped into the top job.

'It was hard,' he remembers. 'Fortunately, although I had the upfront role, Dad was still there in the background helping me with the key decision-making.' Following Ivan's return to head office, Michael is now focusing on Selaks' marketing.

For many wine lovers, Selaks' name is indivisibly linked with Sauvignon Blanc. The 1984 Sauvignon Blanc, vinted from Hawke's Bay fruit, was a roaring sales success in Australia; most subsequent vintages have been based on Marlborough fruit. Selaks Sauvignon Blanc exhibits a fresh, limey bouquet and tangy, crisp, clearly varietal palate in a straightforward, easy-drinking style.

Selaks Founders Sauvignon Blanc Oak Aged, a higher-priced label launched from the 1991 vintage, clearly out-shines the 'commercial' wine with its richer, lusher flavour. Equally outstanding is the Sauvignon Blanc/

Sémillon, a sixty/forty blend fermented both in stainless steel and oak, then barrel-aged for six months. With its vibrantly fruity, ripely herbal, explosive flavours and mouth-watering acidity, in its youth it can be irresistible.

When is the best time to broach these wines? 'It depends on how you like them,' says winemaker Darryl Woolley. 'At two to three years old they start to develop a touch of complexity; the '83 and '84 are still very much alive. But for my taste, at twelve to eighteen months they're at their peak.'

Woolley, aged forty, is an Australian who worked at Enterprise Wines in the Clare Valley. After arriving in New Zealand in 1981, he spent four years at Corbans' Gisborne winery. Woolley joined Selaks in 1985, in the wake of Kevin Judd's departure for Cloudy Bay, 'because at Corbans I'd been deeply impressed by Marlborough grapes, and Ivan was committed to using an increasing volume of Marlborough fruit.'

The shingly Matador vineyard, near the Cloudy Bay winery, owned by John Webber, now supplies much of Selaks' best fruit. Following the uprooting of the seventeen-hectare Brighams Creek vineyard in 1989, Selaks' company-owned vineyards are a small plot adjacent to the winery and a fourteen-hectare Marlborough vineyard purchased in 1993. Grapes are also drawn from growers in Kumeu, Gisborne and Hawke's Bay.

Selaks' Chardonnays have recently been transformed by the company's shift in emphasis from Gisborne to Marlborough grapes. Up to the 1989 vintage, both of their Chardonnay labels were based on Gisborne fruit. Since 1991, their origin has been 100 percent Marlborough.

In Woolley's eyes, the regional differences in Selaks' Chardonnays are sharply etched. His Gisborne Chardonnays were 'big, early-maturing softies with lush, ripe, melon-like fruit flavours and low acidity.' Selaks' new Marlborough Chardonnays, by contrast, are 'leaner and more austere, with higher acidity and greater longevity'. The 'standard' Chardonnay is both stainless-steel and barrel-fermented; the top-of-the-range Founders label is entirely fermented in new oak barriques and then matured on its lees for nine months.

For $350, you can own the nine-litre bottle of Selaks Méthode Champenoise Brut 1986 which stands tower-like beside the counter in their winery shop. Bottle-fermented sparkling wine has been an integral part of the Selaks range since Mate's pioneering efforts in the 1950s. The wine — then called Champelle — caused many headaches in its development stages, including problems with oxidation and clarification and the perils of exploding bottles. The year 1971 marked the first commercial release.

Today Selaks is renowned for two commercial méthode champenoise styles; the Brut, which harbours a touch of sweetness, and the Extra Dry, which is just what its name says. The pair are produced from an identical base wine — seventy percent Pinot Noir, thirty percent Chardonnay. By using Marlborough, rather than Brighams Creek, Pinot Noir, selecting superior yeasts and working on pH levels to achieve a finer 'mousse', Woolley is confident Selaks will produce better and better sparkling wines.

The Achilles heel of the company has long been its solid but plain reds. 'In the past we tended to make a light, early-drinking style,' says Michael. Since the 1990 vintage, however, Cabernet Sauvignon and Merlot have come on stream from the Dartmoor Valley in Hawke's Bay; this is expected to add weight and flavour richness to Selaks' reds.

Where to from here? Exports are pivotal to the winery's marketing strategies. The sale of Sauvignon Blanc to the United Kingdom is thus far the biggest success. Australia, the Netherlands and Taiwan also hold promise.

Soaring domestic demand for Selaks' wines is also encouraging the company to expand. 'We can't keep up with the orders for our Sauvignon Blanc, Sauvignon Blanc/Sémillon and Chardonnay,' reports Michael Selak.

Waitakere Road Vineyard
Waitakere Road, Kumeu

Owners: Tim and Alix Harris and partners

Key Wines: Harrier Rise Cabernet Sauvignon, Uppercase Red, SBV

'There's definitely a place for West Auckland reds; Brajkovich Cabernet Franc proved that,' enthuses Tim Harris. 'I enjoy the possibilities of that idea.'

Waitakere Road Vineyard lies in gently undulating countryside, three kilometres south-west of Kumeu. Harris (47) is a serious, quiet Canterbury University law graduate who has spent most of his career as an Auckland solicitor, while also writing about wine for 'Metro' magazine and in his 'Friends of Wine' newsletter. For Harris, wine is a 'weekend pursuit, and sometimes in the early morning I spray the vineyard before going to work in the city.'

In 1986 Harris formed a syndicate of seven to buy a six-hectare block at Kumeu. Between 1986 and 1990 the partners close-planted two hectares of Cabernet Franc, two hectares of Merlot and 0.3 hectares of Cabernet Sauvignon. These grapes now form the foundation of the syndicate's Uppercase Red.

The adjacent Harrier Rise vineyard — so named because it 'often has several hawks hovering over it' — is personally owned by Tim and Alix Harris. Established in the late 1970s, the low-yielding, 2.5-hectare vineyard is planted in Cabernet Sauvignon and Seibel 5455. The wines from the two vineyards are kept completely separate.

Harris is specialising in red wines because he doesn't have the time to put into whites. 'For Uppercase Red I'm currently aiming for a delicious, drink-young style of red, along the lines of a cru Beaujolais. We don't get the very high temperature peaks (around 30°C) that may be necessary to produce keeping reds with ripe, long-lived tannins. With Merlot, however, we do get good tannins, and the aging potential of the wine is something we will eventually find out about.'

Harris — whose house sits amidst the vines — makes the wine himself, previously in his garage, now in a small winery. He recalls that for his first vintage in 1989, 'winemaker friends, especially Michael Brajkovich, told me what to do.' Harris expects Waitakere Road Vineyard's output to climb from 150 cases in 1989 to around 5000 by 1995.

Uppercase Red is soft and forward with supple red berry-fruit flavours. The recently launched SBV (Selected Blend Vintage) is a chunky, low-priced, drink-young style. Harrier Rise Cabernet Sauvignon, by contrast, is a weighty wine with the strong, spicy, herbal flavours of unblended Cabernet Sauvignon — a much more tautly tannic, slowly evolving red.

Waitakere Road Vineyard

Lawyer, winemaker and Metro *magazine wine writer Tim Harris is devoted to exploring West Auckland's red-wine potential. His supple, fresh, berryish Uppercase Red is a charming, easy-drinking Beaujolais look-alike.*

HENDERSON

In the shadow of the Waitakere Ranges are grouped several of the oldest wineries in the country. Henderson, twenty kilometres west of Auckland city, has one of the largest clusters of wineries in the country; twenty-five winemakers based their operations here in 1993. Strung out along Lincoln Road and Henderson Valley Road, and nestled in the surrounding hills, all are small or medium-sized. (Corbans retains a head office and shop, but no winery, here.) The oldest, Pleasant Valley, was founded in 1902.

Here a flourishing Dalmatian community has imprinted its energetic, wine-loving way of life on the district. Although winemaking in New Zealand is no longer a Dalmatian preserve, here, numerically at least, they still prevail. Winery names with a distinctively Dalmatian ring abound — Babich, Soljans, Mazuran's, Sapich, Vodanovich.

Generally, these Dalmatian vineyards began life as small mixed holdings of fruit trees, vines and vegetables. Dalmatian settlers who had lived on peasant farms in Dalmatia typically saved funds on the northern gumfields and then looked for self-sufficiency. Cheap parcels of land were available for purchase in the Henderson-Oratia area and the

large Auckland market beckoned. Since 1960 these holdings have shifted towards specialisation in market-gardening, orcharding or winemaking. Also there has been a gradual shift of vine plantings away from Henderson itself. Back in 1960, eighty percent of Auckland's vineyards and orchards were in Henderson and Oratia. Later, the north-west motorway opened West Auckland up to the pressures of urban expansion and reduced the land available for viticulture. The surviving vineyards benefit from differential rating systems that significantly ease their rates burden, but some land values, especially those close to the Great North Road, soared too high to be kept in vines.

Between 1975 and 1992, vineyard areas in the Auckland region slumped from 750 hectares — thirty-two percent of the national area — to only 241 hectares or four percent of the country's entire plantings. Major companies such as Montana and Corbans no longer depend on Auckland as a source of grapes.

Henderson suffers from serious physical and climatic handicaps for viticulture. The rainfall, rising steeply from the city westwards to the Waitakeres, is far from ideal for grape-growing. The plentiful rains, in association with high humidity, create ideal conditions for fungous diseases, especially during the critical February–April ripening period. The heavy clay soils drain poorly and are slow to warm up in spring.

Because of this wet climate, the hardy, weather-and-disease-resistant hybrid varieties were always more popular in Henderson than elsewhere. In 1980 the presence of hybrid varieties (28 percent) was still way ahead of the national average (10 percent).

The 1992 vineyard survey revealed that bulk white-wine grapes, like Müller-Thurgau and Chasselas, are only lightly planted in Auckland. Here is another reason for Auckland's relative demise as a grapegrowing area; most large companies rely heavily on Müller-Thurgau as the basis for a variety of wine styles, and that variety crops relatively poorly in Auckland. Gisborne and Hawke's Bay produce far heavier tonnages.

Henderson may be in eclipse in terms of recent vineyard development, but in other respects the area remains vital to the wine industry. Although Auckland has only four percent of the national vineyard area, much of the wine is made there. The headquarters and bottling plants of many large and medium-sized wine companies are in Auckland, close to the largest market. An enormous amount of the country's output is transported to Auckland as fruit, as partly processed wine, or as finished wine ready for bottling.

The past decade has been notable for the closure of the large Penfolds and Corbans wineries, the demise of smaller companies such as Bellamour and Windy Hill, and the lack of visible progress at such wineries as Mayfair, Mazuran's and Fino Valley.

Penfolds (NZ) was established in 1963 on a small property in Lincoln Road bought from the old firm of Averill Brothers, who had planted vines there in 1922. The new company was founded as a joint venture between Penfolds

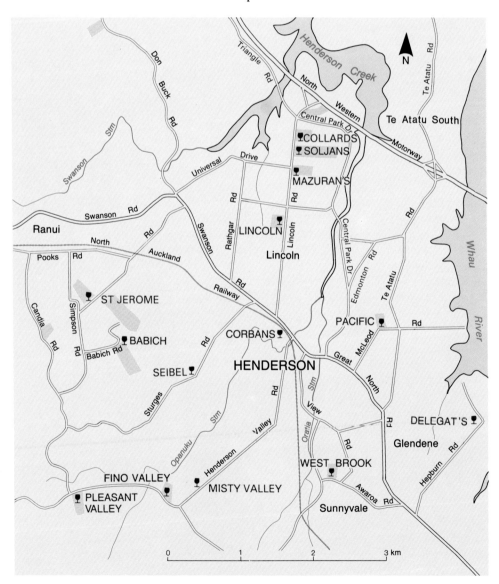

(Australia) and local brewers and merchants. In 1977 Penfolds (Australia) sold most of its shares to Frank Yukich, who in turn sold out to Lion Breweries in 1982. Following Montana's purchase of Penfolds in 1986, the Henderson winery was closed and its winemaking equipment and key staff integrated into Montana's operations.

The launch of the Bellamour label in 1987 signalled that a renaissance was planned for the winery once called Balic Estate, and before that, Golden Sunset. A pioneer firm, Golden Sunset was founded in 1912 by Josip (Joseph) Balich, a Dalmatian who planted his vines at night, by candlelight, after a hard day's labour in the Corban vineyard. In the 1970s and early 1980s, under winemaker Vic Talyancich, Balic produced a great deal of sparkling wine; the most popular was Balic Asti Spumante. But the table wines generally fell one or two steps behind the rest of the industry. In August 1988, despite a belated effort to move upmarket, Bellamour slid into receivership.

Windy Hill was founded by Milan Erceg — who died in 1988 — on the corner of Metcalfe and Simpson Roads in 1934. The winery enjoyed a reputation in the 1970s for good dry sherries and solid reds, but later stagnated.

A couple of decades ago Mayfair Vineyards had a strong reputation for its sherries, cocktails and liqueurs. But today the vogue for such products as Limbo Lime Gin, Rock'n' Roll and Blackberry Nip has vanished, leaving this small

family winery in Sturges Road, run by the Iviceviches, heavily reduced in output.

Fino Valley is another basically undistinguished winery in Henderson Valley Road, run by the Torduch family. Back in the early 1970s, in roaring company at the winery on Saturday afternoons, Fino Valley Medium Dry Red served as my introduction to the world of wine.

Misty Valley — until recently called Old Railway, and before that, Pechar's — is a small winery in Henderson Valley Road, founded by Steve Pechar in 1971. Its wines have typically been mediocre but inexpensive. The medals recently awarded to its Cabernet Sauvignon and Sauvignon Blanc are a positive sign.

Public Vineyards, founded in 1937, is a rare example of the old-time fortified wine producers. Run by the Vitasovich family in Bruce McLaren Road, it specialises in the bulk (200L drum) supply of fill-your-own ports and sherries.

'Purple Death', a bluish-purple port-based drink labelled as 'an unusual 'Rough-as-Guts' wine that has the distinctive bouquet of horse-shit and old tram tickets,' enjoys a high profile at the Sapich Brothers winery in Forest Hill Road. It tastes like cough mixture.

Yet with three high-flying wineries — Babich, Collards and Delegat's — retaining their operations here, backed up by many other reliable producers, the Henderson wine trail still has a multitude of vinous delights for the wine lover.

Babich Wines
Babich Road, Henderson

Owners: The Babich family

Key Wines: Irongate Chardonnay, East Coast Chardonnay, Marlborough Sauvignon Blanc, Fumé Vert, Dry Riesling-Sylvaner, Classic Dry, Pinotage/Cabernet, Cabernet Sauvignon, Irongate Cabernet/Merlot

There is nothing loud or brash about Babich, the country's second largest family-owned winery. As a source of stylish, easy-drinking wines at restrained prices, however, it has long stood out.

Peter and Joe Babich have a cautious approach to expansion. Not for them the frantic pursuit of market share and a headlong rush into grape-growing contracts. What they *are* dedicated to is steady, controlled growth, retaining family control of the winery their father Josip (Joe) established over three-quarters of a century ago.

In 1910, as a boy of fourteen, Josip Babich left Dalmatia to join his four brothers toiling in the gumfields of the Far North. His first wine was produced in 1916. At Kaikino, on the last stretch of land leading to Cape Reinga, he grew grapes, trod them with his feet, and opened a wineshop.

The shift to the Henderson Valley came in 1919. On a twenty-four-hectare wilderness property, Joe milked cows, grew vegetables, established a small orchard — and planted

classical Meunier vines. During the Second World War, winemaking slowly became the family's major business activity. Josip died in 1983, one of the 'grand old men' of the New Zealand wine industry.

Of Josip and Mara Babich's five children, three are involved in the family winery. Peter, a hard-driving 60-year-old, is the general manager; Joe (53) is the talented winemaker and a senior wine judge; Maureen works in the office and behind the winery counter. Neill Culley, a young highly qualified winemaker turned marketing manager, is also a key figure.

Cabernet Sauvignon, Pinot Noir, Pinotage, Chardonnay and Sauvignon Blanc vines grow in the loam-clay soils of the rolling, twenty-three-hectare estate vineyard. The majority of Babich's fruit, however, is drawn from growers in the Gisborne, Hawke's Bay and Marlborough regions.

Babich produces a wide array of bottled white, red and fortified wines, although no wine casks or sparkling wines. Classic Dry, the popular label which has evolved from the older Dry White, is blended from several grape varieties to a crisp, unswervingly bone-dry style. Fumé Vert — a tangy, gently herbaceous, off-dry blend of Sémillon, Sauvignon Blanc and Chardonnay — and the supple Pinotage/Cabernet are also strong sellers in the sub-$10 market.

One of my favourites is the less popular Dry Riesling-Sylvaner, which proves Müller-Thurgau's ability to produce low-priced dryish wines of elegance and spine. The barrel-matured East Coast Chardonnay and full-flavoured Hawke's Bay Cabernet Sauvignon are also good value.

Irongate wines have been a major success story, ever since the first, a 1985 Chardonnay, won a gold medal and the Vintners Trophy as the top-scoring current vintage dry white wine at the 1985 National Wine Competition. This arid vineyard (see page 105) lies in the famous Gimblett Road shingle block west of Hastings.

Fermented and lees-aged for six months in predominantly new French oak barriques, Irongate is a distinctively lean, steely, slowly evolving Chardonnay, subtle rather than up-front in style. The 1987 vintage is currently at the height of its powers. Since 1987, this stony vineyard has also yielded a powerful, concentrated and tautly structured Cabernet/Merlot; the 1989 and 1990 vintages have both been awarded gold medals. Deep ruby in hue, the Irongate red has a full, spicy bouquet and a wealth of blackcurrant-like flavours, impressively firm, complex and lingering.

On even stonier land bordering the Irongate vineyard, Peter and Joe Babich (and their minority partners in Fernhill Holdings) have recently planted another seventeen-hectare vineyard in Syrah, Cabernet Sauvignon, Merlot, Sauvignon Blanc and Chardonnay. 'Syrah is a late-ripener,' observes Peter Babich, 'and late-ripening varieties ripen fully in that terrain.' The vines burst their buds and flower early in the warmth of the deep shingles.

Marlborough wines have also recently started to flow under the Babich label. The first, 1991 and 1992 Marlborough Sauvignon Blancs — both gold medal winners — were deliciously deep-flavoured and a 'steal' at around $12.

Babich also markets a full range of sherries and ports, of which most were re-launched in 1992 under the Henderson's Mill brand. The most distinguished fortified wine is Babich Reserve Port, a deep-scented and mellow old tawny.

Exports account for about twenty percent of Babich's turnover. The United Kingdom, the major market, absorbs several containers per year of Fumé Vert, Sauvignon Blanc and Chardonnay. ('When Prince Edward sat next to me in England recently, enjoying a couple of glasses of Fumé Vert,' Neill Culley has written in the winery newsletter, 'I couldn't help but feel that Babich has really arrived on the United Kingdom wine scene.') Of the other nine countries Babich ships to, Japan and Canada are showing the fastest growth.

Babich's handsome new head office, nestled into a top corner of the estate vineyard, opened in the spring of 1992. The ground floor features plenty of room for browsing and a stunning view. This promises to be a high point, in both senses of the word, of the Henderson wine trail.

Brothers Joe (left) and Peter Babich (right) head one of New Zealand's longest-established wineries, with a reputation for quality wines at sharp prices. Peter's son, David (centre), who is currently studying commerce at university, is a Roseworthy College winemaking graduate.

Peter Babich

While he shuns the limelight, Peter Babich is one of the wine industry's most influential figures: a member of the key executive committee of the Wine Institute almost every year since it was founded; until recently the Institute's deputy chairman; and currently the senior spokesman for the medium-sized, independent, family-owned wineries. Babich entered the wine industry at sixteen — 'I was cheap labour to help around the winery'. By his early twenties he had begun to take over from his father. Today he holds the administrative reins at Babich Wines and oversees the firm's viticultural operations.

Why his life-long immersion in the industry's political affairs? 'I enjoy it,' says Babich. He is proud of his family's achievements and determined to safeguard them. 'Above all, my motive has been to make sure that independent winemakers get a fair go. We always had to battle the large brewery-owned wine companies. Our key concern was the contracting ownership of retail outlets; we've had to fight to safeguard our access to the consumer.'

In a fiercely contested market, the Babich winery is thriving. 'I suppose Joe and I are pretty tenacious and we've always had a large appetite for work. We know all aspects of the industry — we've even dug the post-holes,' says Babich.

'We've always been conscious of the need for good fruit, and value for money. You must always consider the consumer, and not hike prices when the market is in your favour. Financially, we've worked on the basis that our equity must be high, which has eliminated the danger of high interest payments during the industry's difficult periods. And the trade has allowed us to expand — they've appreciated our no-nonsense approach and regular high awards.'

With Peter Babich (and his brother, Joe) at the helm, Babich has been transformed since the 1950s from a small, predominantly fortified-wine producer into a substantial modern winery with a glowing competition record and strong export orientation. In 1989 Peter Babich was awarded the MBE for 'services to viticulture'. 'As a politician of the wine industry, Peter works harder than anyone,' observes Ross Spence of Matua Valley.

Collards

Collard Brothers
303 Lincoln Road, Henderson

Owners: The Collard family

Key Wines: Rothesay Vineyard Chardonnay, Hawke's Bay Chardonnay, Rothesay Vineyard Sauvignon Blanc, Marlborough Sauvignon Blanc, Chenin Blanc Dry, Rhine Riesling, Marlborough Riesling, Private Bin White Burgundy, Cabernet/Merlot, Rothesay Vineyard Cabernet Sauvignon

'I basically don't drink. I'm tasting wine all the time but hardly ever drink it,' says Bruce Collard, winemaker at the small, high-flying winery that bears his family name. Collard's restraint is amazing, for his Rieslings, Chardonnays and Chenin Blancs are consistently among the most delicious in the country.

My respect for the Collards winery, just off the north-western motorway in Henderson's traffic-clogged Lincoln Road, reflects not only the Collard family's meticulous winemaking, but also their reluctance to engage in shrill trumpeting of their own achievements. Lionel Collard (72) can often be found behind the vineyard counter, dispensing their eye-catchingly good wines, but his two sons, Bruce (42) and Geoffrey (40), are much more elusive.

In 1990 Collards celebrated its eightieth anniversary.

Founder John Collard, an English berry-fruit expert, came to New Zealand as an orchard instructor for the Department of Agriculture. He purchased the present Lincoln Road site in 1910, planting it in stone and pip fruits. Although grapes were cultivated at Collard's 'Sutton Baron' property — named after the village in Kent where for generations the Collard family had grown fruit and hops — initially they were for eating, not wine.

John Collard married Dorothy Averill in 1915. Between 1928 and 1963, Dorothy's brothers ran the Averill winery, just along Lincoln Road. At the urging of their uncles — the Averill brothers — in 1946 John Collard's sons, Lionel and Brian, started crushing their own wines. Until 1964, when the present Collards winery was built, Collards wines were always fermented at the Averill cellars ('brewed at the bros', as Bruce Collard puts it).

The Collard family have neatly divided the myriad tasks of a modern winery between them. Geoffrey, who once worked for three years in the Mosel, oversees the grape-growing; Bruce is the talented, self-effacing winemaker; Lionel, in his own words, 'cons the ship', controlling the company's financial administration, sales and public relations. Now the company patriarch, Lionel is an independently minded perfectionist who claims (I'm not entirely convinced) that he'd like to retire, but that 'I wouldn't know what the hell to do with myself.'

What are the key factors underlying the top-flight quality of Collards white wines? 'Good fruit and my sons' pain-

Consistently fine Chardonnays, Rieslings and Chenin Blancs flow from the small Collards winery in Henderson. Geoffrey Collard, pictured in the company's renowned Rothesay Vineyard in the Waikoukou Valley, near Matua Valley, controls the company's viticultural activities.

staking winemaking,' says Lionel. 'Good fruit,' says Bruce. 'We're singleminded in our pursuit of quality grapes, whether we grow them ourselves or buy them from a contract grower.'

The Collard holding in Lincoln Road, originally twenty-five hectares, has been decimated by the construction of the northwestern motorway and other local roads; only three hectares are still in vines. The Collards' major company-owned plantings are now in the Waikoukou Valley, near Matua Valley, where the Rothesay Vineyard has yielded a string of distinguished Chardonnays and Sauvignon Blancs.

Wine judges have for several years showered Collards Chardonnays with high awards. To enhance the individuality of his Chardonnays, Bruce Collard avoids blending the wines of different regions or vineyards, preferring to keep them as single-vineyard bottlings. Each wine is at least partly barrel-fermented, but he is not looking for heavy-handed oak flavours. The softening influence of malolactic fermentation is also present in all his Chardonnays, but Collard shies clear of dominant malolactic characters. The Collards style places its accent on well-judged wood handling and deep, sustained fruit flavours.

Collards Hawke's Bay Chardonnay is a top-class, firm and weighty wine, with intense, lingering fruit flavours and stylish oak; in its youth, locked-in power is the overall impression. Rothesay Vineyard Chardonnay is equally mouthfilling, with powerful fruit and strong oak. The Gisborne Chardonnay is an easy-drinking style — reflecting the lower acidity of Gisborne fruit and more restrained wood handling — with a peachy, fresh, soft palate.

Collards Rieslings are regularly in this country's top drawer. The medium Rhine Riesling, based on fruit drawn from Henderson, Hawke's Bay and Marlborough, displays a wealth of floral/citric Riesling scents, mouth-watering sugar/acid balance and impressive flavour delicacy. The Marlborough Riesling is drier and more penetratingly flavoured.

Collards Private Bin White Burgundy, descended from the old Private Bin Dry White, is strong in body and flavour and cheap to buy. This blend of Chenin Blanc's fruitiness and tangy acidity, Sémillon's herbal characters and Chardonnay's rich flavour and fullness is a happy one.

Collards is also firmly established in the vanguard of New Zealand's Chenin Blanc producers. The wines are produced from obviously very well-ripened fruit, grown in Te Kauwhata and Hawke's Bay, which is fermented to dryness and then partially matured in French oak casks. These are richly scented and deep-flavoured wines, redolent of tropical fruit salad, and bargain-priced.

Rothesay Vineyard Sauvignon Blanc is robust and flinty. Of late it has been partnered by a verdant, vigorously crisp, fresh Marlborough Sauvignon Blanc. Both are non-wooded styles.

The company's Achilles heel has traditionally been its reds. 'Outstanding reds have to be grown on top sites, and we simply haven't had that sort of fruit,' says Bruce Collard. The Cabernet/Merlot has lacked the weight and concentrated flavours of New Zealand's top reds. The recently launched, slightly austere but spicy and firm Rothesay Vineyard Cabernet Sauvignon is a step in the right direction.

With an annual output of about 20,000 cases, Collards is still a small winery, and it plans to stay that way. 'We're in a consolidating rather than expanding mood,' says Lionel. 'We're only too aware of the host of new wineries. They're all small and they're all competing directly with us.' Recent export breakthroughs, to the United Kingdom and Canada, are a signpost to the future.

Corbans Wines
Great North Road, Henderson

Owner: DB Group Limited

Key Wines: Corbans Private Bin Chardonnay, P.B. Fumé Blanc, P.B. Noble Rhine Riesling, P.B. Merlot/Cabernet Sauvignon, P.B. Amadeus Méthode Champenoise, White Label Gisborne Chardonnay, White Label Marlborough Johannisberg Riesling; Liebestraum, St Amand Chablis, Cresta Doré, Velluto Rosso, Marque Vue, Première Cuvée, Italiano Spumante; Cooks Winemakers Reserve Chardonnay, Winemakers Reserve Cabernet Sauvignon, Winemakers Reserve Sauvignon Blanc, Discovery Sauvignon Blanc, Endeavour Chardonnay, Chasseur; Stoneleigh Vineyard Sauvignon Blanc, Rhine Riesling, Chardonnay, Cabernet Sauvignon; Longridge Fumé Blanc, Gewürztraminer, Chardonnay, Cabernet Sauvignon/ Merlot; Robard & Butler Amberley Rhine Riesling, Gisborne Chardonnay

Corbans in 1993 claimed a whopping thirty-eight percent share of all New Zealand wine sold in this country. A leviathan company, Corbans owns vineyards in the Hawke's Bay and Marlborough regions; wineries in Gisborne, Hawke's Bay and Marlborough; and headquarters and a final blending, bottling and warehousing centre in Auckland. Under its key brands — Corbans, Cooks, Robard & Butler, Stoneleigh Vineyard and Longridge — it markets a great diversity of labels — six different Cabernet-based reds, seven different Chardonnays, eight different Sauvignon Blancs.

For forty years, the adroit management of the Corban family ensured their domination of the New Zealand wine industry. From its humble beginnings as a one-and-a-half-hectare vineyard founded by Lebanese immigrant Assid Abraham Corban at Henderson in 1902, the winery flourished through prohibition and depression and early established itself as a household name. But today the company is Corbans only in name, being a wholly owned subsidiary of DB Group Limited, itself owned by Brierley Investments Limited.

A.A. Corban, a stonemason from the village of Shweir on the flanks of Mt Lebanon, inland from Beirut, arrived in New Zealand in 1892. He travelled the goldfields and mining towns of the North Island peddling ornaments and fancy goods, then set up as a dealer in Auckland's Queen Street. Two years later, he sent for his wife Najibie and sons Khaleel and Wadier to join him.

The beginning of Corban's 'Mt Lebanon Vineyards' lies in the 1902 purchase — for 320 pounds — of four hectares of Henderson gumland, complete with a small cottage and a few Isabella vines. His strong ambition to produce wine — a family tradition back in Lebanon — led Assid Abraham Corban to establish a small vineyard: Black Hamburghs for the table and such classic varieties as Chasselas, Hermitage (Syrah) and Cabernet Sauvignon for winemaking; no Albany Surprise — that, said A.A. Corban, was a vine suitable only for lazy winegrowers. At the first Corban vintage in the new country in 1908, the fruit was crushed by hand with a wooden club and an open hogshead used as the fermenting vat.

A small white brick building, still standing at the entrance to Corbans' headquarters, bears testimony to the marketing problems A.A. Corban immediately encountered. West Auckland voted 'dry' in 1908, denying Corban the right to sell his wine directly from his cellar. A railwayman's cottage, standing only a few metres away across the railway tracks in a 'wet' electorate, was pressed into service as a sales depot until 1914, when the surviving white building was erected. This was later superseded by a sales outlet in Auckland city.

By 1916 son Wadier had assumed the duties of winemaker. At the New Zealand and South Seas Exhibition 1925–26, Wadier's Corbans port won first place. Khaleel took charge of sales, travelling the length of the country in an old Dodge van, building up a strong trade in tonic and restorative wines.

Although the arrival of a rotary hoe in 1934, and a caterpillar tractor soon after greatly eased the vineyard toil, by all accounts, until his death from a stroke in 1941, Assid Abraham Corban remained a patriarch in the Old Testament mould, and a strong believer in the virtues of hard work. Najibie, too, until her death in 1957 remained in close touch with all aspects of management.

When the wine boom began in earnest in the 1960s, Corbans' plantings leap-frogged from the Henderson Valley to Riverlea, Kumeu and Taupaki, and later contracts were negotiated with growers in Auckland and Gisborne. Alex Corban, as winemaker, demonstrated a strong flair for technical innovation (see page 15).

To reinforce the company's economic base, A.A. Corban and Sons admitted a nineteen percent shareholding by wine and spirit merchants. But when the challenge from Montana emerged in the late 1960s, the Corban family's own financial resources proved insufficient to pay for the huge expansion necessary if the company was to retain its ascendancy. Rothmans — later renamed Magnum Corporation, and then DB Group Limited — became a shareholder and steadily increased its influence; today the Corban family has altogether lost its financial control.

Alex Corban, Corbans' winemaker from 1949 to 1976, sees no way the company could have stayed family-owned. 'At a time the winery was committed to expansion, our cash flow slowed, due to Montana's impact and the breweries' increasing control of retail outlets. We borrowed to meet our commitments, but then the lenders, including Rothmans, demanded managerial control. Then Rothmans bought out the merchant shareholders, leaving only Rothmans and the family as shareholders. We had the larger

Corbans

Chief winemaker Kerry Hitchcock (left) and general manager Noel Scanlan, pictured at the company's headquarters in the historic Henderson homestead, are the two pivotal figures in the Corbans hierarchy. Some of New Zealand's most arresting wines are marketed under the Corbans Private Bin, Robard & Butler and Cooks Winemakers Reserve labels.

share, but they had the money. At the same time we were being hit with very expensive death duties.' By selling its wine interests, historian Dick Scott has written, the Corban family 'participated in the first share-out in its history. Individual members drew on the first fruits of seventy-five years of denial . . . '

Corbans planted new vineyards at Gisborne in 1968 and three years later sited a second winery there. Subsequently plantings spread up the East Coast to Tolaga Bay, and also to Te Kauwhata.

Through the mid-1970s Corbans struggled to match the quality advances of its rivals. From 1980 onwards, however, there was a marked lift in wine quality, assisted by the arrival of German Norbert Seibel, a Geisenheim graduate with winemaking experience in Germany, France and South Africa. Rothmans, after gaining full ownership of Corbans in 1979, embarked on a multi-million dollar upgrading and expansion programme. New winemaking equipment was purchased and in 1980 the Stoneleigh and Settlement vineyards in Marlborough were planted in Chardonnay, Sauvignon Blanc, Riesling, Sémillon, Gewürztraminer, Cabernet Sauvignon and Merlot. Corbans' climb in prestige in recent years has been strongly assisted by the outstanding fruit coming on stream from these Marlborough vineyards.

In 1987 Corbans bought the assets of the ailing wine company Cooks/McWilliam's Limited for $20 million; both companies had already been subsidiaries of Brierley Investments Limited. According to Paul Treacher, who was appointed general manager of the enlarged company, Corbans' acquisition of Cooks/McWilliam's 'represented an essential major rationalisation within the local winemaking industry . . . it is essential that local large-scale winemakers become and remain as cost efficient as their counterparts abroad . . . '

Long one of Corbans' movers and shakers, Noel Scanlan (41) in 1991 was promoted from marketing manager to the company's top post. Scanlan stresses that Corbans is intensely market-driven: 'The staff in the marketing department write the product specifications for the winemakers and sit in on the tastings.' Brand marketing is a Scanlan specialty: the roaring marketplace success of the Stoneleigh Vineyard range of Marlborough wines even won Corbans the Marketing Institute's top award for 1987.

In Scanlan's eyes, DB Group's 1990 acquisition of a minority shareholding in G. Gramp and Sons, the holding company for Orlando Wines and Wyndham Estate, has given Corbans a priceless link with the Australian wine industry. By drawing some of its wine from across the Tasman, Corbans can supply all market segments, including ones they can't properly address themselves because they lack the right grapes, or grapes at the right price, or the right technology. Quaffing reds such as Bakano, and fortified wines such as Robard & Butler Artillery Port, are now sourced from Australia; so too are cask whites like Liebestraum, when the supply of local grapes runs low.

Scanlan, who works sixty hours a week, says he has 'forgotten what a lunch hour is all about'. To unwind he reads — about marketing.

Shouldering overall responsibility for Corbans' winemaking is Kerry Hitchcock (45), who after leaving school spent six years researching fermentation techniques and new grape varieties at the Te Kauwhata Viticultural Research Station. When David Lucas gave Hitchcock the job of Cooks' founding winemaker in 1972, he started on his lengthy ascent to the top production post at Corbans.

As chief winemaker, Hitchcock controls all the company's viticultural and winemaking operations. His is a multi-faceted role: from negotiating with contract grape-growers, to having the final say on blends, to liaising with the company's finance and marketing arms.

From Hitchcock the division of responsibility extends to senior winemakers installed in each of the company's wineries — all highly qualified professionals in their own right. When Corbans wins the champion Air New Zealand Trophy (as it has twice of late with its opulent Private Bin Noble Rhine Riesling 1986) it is Kerry Hitchcock who is ushered into the media limelight — but in such a diverse organisation winemaking is clearly a team affair.

Corbans is notably reliant on contract growers; only about twelve percent of the fruit which avalanches into its crushers is company-grown. Hitchcock is quick to point out, however, that the company's own 100-hectare Stoneleigh vineyard in Marlborough, and its three Longridge vineyards totalling 100 hectares in Hawke's Bay — at Omarunui, Tukituki and Haumoana — are the key sources of its flagship varietals.

Corbans' trio of wineries in Gisborne, Hawke's Bay and Marlborough all produce wine to a finished state, needing only a final filtering before being bottled in Auckland. An efficiency drive in 1991 generated tidal waves in the production arena: the Cooks winery at Te Kauwhata was

Corbans' 1980 thrust into Marlborough led in the mid-1980s to a dramatic surge in the company's wine quality. Winemaker Alan McCorkindale heads the production team at Marlborough Cellars.

closed, its equipment shifted away and the building sold. The historic Henderson winery was also sold, and its blending, bottling and warehousing operations transferred to Allied Liquor Merchants' much more spacious premises in East Tamaki. The head office and winery shop staff have stayed put at Henderson 'for the time being'.

For decades Corbans has been the second largest wine company in New Zealand. By acquisition and, says Noel Scanlan, concentrating on the key growth segments of the market, they are making ground on Montana. Vertical integration helps — DB Group also owns an extensive liquor wholesale and retail distribution system, including part of the Liquorland, Robbie Burns and Green Bottle chains.

The wines under Corbans' assorted labels range in standard from workmanlike to top-flight. Their price tags are usually an accurate signpost to their quality.

Corbans' yellow Private Bin label is a pointer to some of New Zealand's most distinguished wines. The Private Bin Chardonnay, Marlborough-grown and barrique-aged for one year, is robust, peachy, mealy and deliciously long-flavoured. Its lifted bouquet reveals barrel-ferment complexity; its intense palate is classically poised between power and subtlety. Chardonnay as stylish and concentrated as this is hard to find.

Corbans Private Bin Fumé Blanc, produced from Marlborough fruit fermented and matured in French and German oak barriques, is equally top-flight. The bouquet is packed with rich fruit and toasty oak; the palate is mouthfilling, ripe and soft. Deep-scented, rich, voluptuously ripe and plummy, the Private Bin Merlot/Cabernet Sauvignon and Private Bin Merlot rank among Marl-

borough's few distinguished reds. The ravishingly perfumed and nectareous Private Bin Noble Rhine Riesling — harvested in Marlborough as late as mid-June — is right in the vanguard of New Zealand's sweet white wines.

On a lower plane, Corbans' 'white label' range of sub-ten dollars varietals includes a top value-for-money, oak-matured Gisborne Chardonnay and a tangy, off-dry Marlborough Johannisberg (Rhine) Riesling. Italiano Spumante is a delicious, Dr Hogg Muscat-based sweet sparkling. Corbans Amadeus Méthode Champenoise, based on Hawke's Bay-grown Pinot Noir and Chardonnay, is a satisfyingly lively and full-flavoured bottle-fermented sparkling, competitively priced.

Noel Scanlan has built parallel portfolios around both the Corbans and Cooks brands. At the top end, the Corbans Private Bin range is balanced by the Cooks Winemakers Selection; Stoneleigh Vineyard and Longridge fit into the middle-upper-priced varietal market; Corbans has its 'white label' lower-priced varietals, Cooks its Discovery Collection; Corbans Liebestraum has its match in Cooks Chasseur at the bottom of the market.

Cooks Winemakers Reserve Chardonnay is an upfront style with a richly toasty bouquet. Grown in Hawke's Bay, it is a powerfully wooded wine, ripe and robust. Cooks Winemakers Reserve Sauvignon Blanc is produced from Marlborough grapes, cold fermented in stainless steel tanks and bottled young. Pale lemon-green, it has an arrestingly fresh and penetrating bouquet, mouthfilling body and vibrant, deep-flavoured, mouth-wateringly crisp fruit. Cooks Winemakers Reserve Cabernet Sauvignon is a classic, concentrated Hawke's Bay claret-style red, blackcurrant-like, cedary and tautly structured.

Two consistently outstanding wines are marketed under the Robard & Butler label. Robard & Butler Amberley Rhine Riesling, grown in North Canterbury, is distinctively spicy and complex — these are rare qualities in Riesling. Massive, savoury and soft, Robard & Butler Gisborne Chardonnay has built up a formidable reputation on the show circuit. Partly barrel-fermented, and given a 100 percent malolactic fermentation, it is buttery and toasty in a rich, very upfront style.

Of the quartet of wines marketed under the Stoneleigh Vineyard label, the leafy-green, light Cabernet Sauvignon is the least distinguished. The lemony, slightly savoury Chardonnay is very solid, but the twin peaks of the range are Stoneleigh Vineyard Sauvignon Blanc, which bursts with fresh, tangy, gooseberry and cut-grass flavours, and Stoneleigh Vineyard Rhine Riesling, with its intense, piercing lemon/lime fruit flavours and vigorous acidity.

The Longridge quartet of Hawke's Bay wines features a barrel-fermented, citric-flavoured Chardonnay; a limey, tangy Fumé Blanc; a delicate, well-spiced Gewürztraminer; and a full-bodied, spicy, American oak-perfumed Cabernet Sauvignon/Merlot.

Delegat's

Delegat's Wine Estate
Hepburn Road, Henderson

Owners: Jim and Rose Delegat

Key Wines: Proprietors Reserve Chardonnay, Fumé Blanc, Cabernet Sauvignon, Merlot; Hawke's Bay Chardonnay, Sauvignon Blanc, Cabernet/Merlot; Oyster Bay Chardonnay, Sauvignon Blanc

Throughout the summer of 1992–93, Delegat's pulled in the crowds. The vine-covered courtyard was bustling with visitors attracted by the promise of live music, free wine and food tastings. Food and wine personality Vic Williams was on hand with 'taste sensations' such as green-lipped mussels topped with coriander-flavoured tomato salsa (served, of course, with Delegat's Hawke's Bay Chardonnay). Behind the counter, wine judge Brian Farmer dispensed tastings with his customary wit and passion for wine.

Nikola Delegat, the founder of this winery, purchased land near the Whau River, an arm of the Waitemata Harbour, in 1947. Delegat first arrived in New Zealand in 1923 but later retraced his steps to Yugoslavia, before finally establishing a two-hectare plot of vines at Henderson. If you visit the property, when you step into the concrete block vineyard shop with its exposed beams you are entering Nikola Delegat's original winery.

Delegat's appeared to be content to produce sound, ordinary fortified and table wines until the 1979 arrival of a young Australian winemaker, John Hancock, signalled a change of direction. Almost overnight, Hancock lifted the standard of Delegat's wines: the gold medal won by his 1979 Selected Vintage Riesling-Sylvaner was to be the first of a string of successes. With Nikola's son, Jim, and daughter, Rose, now at the helm, in the early-mid 1980s Delegat's was transformed into a quality-orientated winery with a high reputation for its bold, buttery, richly oaked Chardonnays.

Rocked by the ferocious price war of 1985–86, the family admitted Wilson Neill to a majority shareholding, but in 1991 ownership of the winery reverted entirely to Jim and Rose Delegat.

Jim (44), the company head, oversees its financial administration and is also deeply involved with Delegat's burgeoning export trade. Rose (40), a strong-willed but genial personality, travels the country and the world waving the Delegat's flag. Brent Marris (31), Delegat's winemaker for the past eight vintages, is also an influential figure, with a string of fine wines to his credit.

Delegat's flag has recently been firmly hitched to the Hawke's Bay mast. Wines marketed under the Delegat's — as opposed to the new Oyster Bay — label, are made exclusively from Hawke's Bay fruit. Brent Marris is adamant: 'I've worked with Chardonnay fruit from Gisborne and Hawke's Bay. Gisborne fruit has a much higher pH than Hawke's Bay's. Gisborne Chardonnays have youthful appeal, but they don't seem to go the distance.'

Delegat's has a twin-pronged investment in Hawke's Bay vineyards. The silty twelve-hectare Vicarage vineyard, previously called Swamp Road, in the Omarunui Valley, is planted entirely in Chardonnay. This fruit will in future be earmarked for the Proprietors Reserve Chardonnay.

In 1991 Jim Delegat and Brent Marris in partnership purchased John Kenderdine's twelve-hectare vineyard in shingly Gimblett Road, near Hastings. Planted in Chardonnay, Sauvignon Blanc, Cabernet Sauvignon and Merlot, it sits next to the Babich Irongate and C.J. Pask vineyards and looks sure to supply some of Delegat's finest Hawke's Bay wines.

The high-profile Oyster Bay label was originally designed exclusively for export. Based on Marlborough fruit sourced from several growers, including John Marris, Brent's father, the label burst into the limelight when Oyster Bay Sauvignon Blanc 1990 won the Marquis de Goulaine Trophy for the best Sauvignon Blanc at the 1991 International Wine and Spirit Competition in London. To satisfy demand, the Oyster Bay Chardonnay and Sauvignon Blanc were finally released in New Zealand.

The Delegat's range is narrow: only nine wines, including the pair marketed as Oyster Bay. Riesling, Gewürztraminer and Müller-Thurgau have all recently been dropped in favour of a tighter focus on Chardonnay, Sauvignon Blanc and Cabernet and Merlot-based reds. The seven wines under the Delegat's label are marketed under a two-tier labelling system, with the top end of the range presented as 'Proprietors Reserve'.

The brother-and-sister team of Jim (left) and Rose Delegat produces an array of top-flight wines under Delegat's Proprietors Reserve label. Winemaker Brent Marris, who arrived at Delegat's in 1985, initially won respect with his premium white wines, but has also recently enjoyed acclaim for his Hawke's Bay Merlot and Cabernet-based reds.

My pick of the Delegat's range is the Proprietors Reserve Fumé Blanc. Produced from the winery's ripest Sauvignon Blanc grapes, plus some greener fruit for acid structure and balance, it is partly barrel-fermented but not lees-aged. The bouquet is a lovely amalgam of ripe tropical fruit and oak; the palate is bursting with delicious, ultra-ripe pineappley fruit flavours, fleshed out by oak.

Delegat's Chardonnays of the early 1980s ranked alongside Cooks' and McWilliam's as the best in the land. Fat, with pungent — in retrospect almost overpowering — Nevers oak flavour, they reached a high point with the outstanding 1982 vintage, the champion wine at the 1983 National Wine Competition. Recent vintages of the Proprietors Reserve Chardonnay have been more stylish, with rich, vibrant stone-fruit flavours and a highly appealing, slightly toasty complexity.

Equally impressive was the Proprietors Reserve Auslese Rhine Riesling. The Auslese of vintages between 1980 and 1985 was a lovely, low-alcohol sweet wine, fermented from freeze-concentrated Müller-Thurgau juice. 1986, however, marked a turning point: the freeze-concentration technique was discarded and Marris used only botrytised Riesling fruit for the Auslese. Gisborne-grown, this was a full-bloomed, intensely citrus-flavoured, honey-sweet beauty;

the 1990 vintage, however, was the last.

Vintages in the mid–late 1980s of the Proprietors Reserve Cabernet Sauvignon were always good, without quite achieving real distinction. The 1990 vintage, however, is a distinct advance. Grown in three Hawke's Bay districts, Meeanee, Dartmoor and Ngatarawa, it is a blend of eighty-five percent Cabernet Sauvignon, ten percent Merlot and five percent Cabernet Franc, matured for a year in new and one-year-old French oak casks. Highly concentrated, with strong blackcurrant and spicy flavours, taut and sustained, it looks set to blossom for many years. The Proprietors Reserve Merlot is meaty, plummy and rich.

Delicious in its youth, the mid-priced Hawke's Bay Chardonnay is oak-matured for six months, without any barrel-ferment or malolactic influence. 'We want the fruit to do the talking,' says Brent Marris. Both the Hawke's Bay Cabernet/Merlot and Hawke's Bay Sauvignon Blanc are fresh, straightforward, and easy-drinking in style.

What does the future hold for Delegat's? 'There must be a strong commitment to viticulture,' says Jim Delegat, 'so we'll be aiming to expand our land interests. And we'll need to keep building our presence, both in the domestic and international markets. First and foremost, we're planning on staying a family business.'

Lincoln

Lincoln Vineyards
130 Lincoln Road, Henderson

Owners: The Fredatovich family

Key Wines: Parklands Chardonnay, Hawke's Bay Chardonnay, East Coast Chardonnay, Sauvignon Blanc, Chenin Blanc, Rhine Riesling, The Home Vineyard Cabernet/Merlot, Merlot/Cabernet, Anniversary Show Reserve Port

Lincoln epitomises the long-established West Auckland wineries of Dalmatian origin. From their early years concentrating on fortified wines, in the 1980s they successfully turned to the production of varietal table wines. Brothers John (left) and Peter (centre) Fredatovich control the company, assisted since 1985 by winemaker Nick Chan.

'I never asked myself what I was going to do for a job when I was at school,' says thirty-seven-year-old John Fredatovich. 'I always knew that I was going to be a winemaker. Wine-making is in my blood and I'm in it for life.'

Founded in 1937, Lincoln ranks among the country's half-dozen largest family-owned wine firms. The winery, nestled alongside the strip of American-style takeaway restaurants and unbroken traffic that is Lincoln Road today, initially earned a reputation for its fortified wines, especially dry sherry. In the past decade, however, the Fredatovichs have plunged into the production of fine table wines.

Day-to-day control of the winery rests with John and his brother Peter (42). How do they divide the myriad tasks of a modern winery? 'We overlap,' says Peter, a soft-spoken, slightly shy man who exudes the warmth and sincerity that are hallmarks of the Fredatovich family. 'I tend to focus on administration and sales; John is more involved in viti-culture as well as sales at home and abroad.'

Peter and John, the third generation of the Fredatovich family to produce wine in Lincoln Road, took over the reins from their father, Peter, in the late 1980s. Peter Snr — awarded the MBE in 1989 for his services to viticulture — had controlled Lincoln since the 1955 retirement of his father, also called Peter. 'For the first few years.' recalls

Peter Snr, 'my father worked for the Ministry of Works to earn ready cash so that he could establish our vineyard. He would come home and work on the land, clearing scrub and digging up stumps. It was pretty hard work but he planted one and a half acres (0.6ha) of vines on overhead trellises — and got us started.' To accommodate the harvest, he then coopered his own barrels out of totara.

Today the third key figure at Lincoln is Nick Chan — the son of Gilbert Chan of Totara Vineyards — a Roseworthy College graduate who was appointed the company's wine-maker in 1985. Prior to Chan's arrival the standard of Lincoln's table wines had not shone, but their quality has since surged ahead.

Chan works with fruit drawn from Auckland, Gisborne, Hawke's Bay and Marlborough. The four-hectare block ('The Home Vineyard') adjoining the winery is predomi-nantly planted in Cabernet Sauvignon, with a smaller plot of Palomino, and Muscat vines trained on overhead trellises. Almost encircled by houses, The Home Vineyard survives because for rating purposes the local council zones it rural. At Brighams Creek, near Whenuapai, until recently Lin-coln had a second twelve-hectare vineyard planted in loam-clay soils. The chosen varieties were Merlot, Pinotage, Pinot Noir, Sauvignon Blanc, Sémillon, Chardonnay and Palomino. In the winter of 1992, however, due to the Fredatoviches' increasing preference for Hawke's Bay fruit, and its lower production cost, the Brighams Creek vineyard was uprooted.

Lincoln's most popular wines, Peter Fredatovich reports, are the East Coast Chardonnay, Sauvignon Blanc and Merlot/Cabernet. Soft and buttery, Lincoln East Coast Chardonnay is a distinctive, strongly malolactic-influenced wine, heaven-sent for drinkers who like their Chardonnays in an easy-drinking, upfront style. Partly fermented in tanks, partly in barriques, it delivers great value for money.

Lincoln Sauvignon Blanc, a Hawke's Bay wine, partly fermented in French oak barriques, is a verdant, full-flavoured, tangy mouthful. The Merlot/Cabernet — which has superceded the old Brighams Creek Red — is fleshy, full-flavoured, spicy and soft.

The soft, creamy texture on display in the East Coast Chardonnay is also a feature of the premium Parklands Chardonnay. A barrel-fermented, lees-matured wine grown at Chris Parker's Gisborne vineyard, this is a richly flavoured, forward style, savoury and smooth. Launched from the 1991 vintage, Lincoln's Hawke's Bay Chardonnay is a complex style: 100 percent barrel-fermentation; thirty percent malolactic fermentation; six months aging on yeast lees. Mouthfilling, with rich stone-fruit and toasty oak characters, it is full-flavoured and forward but with obvious cellar potential.

Lincoln Chenin Blanc is a barrel-fermented wine, fruity and tart. With its abundance of citrus-fruit flavours, Lincoln Rhine Riesling is a good Hawke's Bay wine, fresh and vigorous, with an off-dry, appetisingly crisp finish.

Dark and chunky, The Home Vineyard Cabernet/Merlot

is emerging as one of West Auckland's top red-wine labels. A blend of two-thirds Cabernet Sauvignon and one-third Merlot, it is matured for eighteen months in American and French oak barriques. This is a bold, supple red with rich blackcurrant, herbal and plum-like flavours, plenty of weight and a firm finish.

Lincoln has not abandoned fortified wine production, although most of its energies are now poured into table wines. 'For dry sherry we always got silvers, and sometimes golds,' recalls Peter Fredatovich. 'We're no longer pursuing top-level sherries but we're still working hard on ports. There's a market for ports — especially tawny styles.'

Lincoln Anniversary Show Reserve Port won the Fellows Shield for the champion port at the 1991 Air New Zealand Wine Awards — a repeat of its 1988 triumph. Tawny-brown in hue, it is rich, raisiny, creamy-smooth and lingering. 'I don't know how old it is,' admits Peter, 'but a portion of the blend is twenty-five years old.' The mellow, nutty, lighter twelve-year-old Old Tawny Port is equally delicious.

'Making good wine is one thing,' says Peter Fredatovich, 'but you still have to shift it. It used to be hard to find the grapes — now that's no problem. The hardest thing today is selling the wine. Imports have made it so extremely competitive. But we're selling all we make — that's the important thing.'

Distinctive packaging helps. Lincoln's colourful, uncluttered labels, designed by Mark Adams, won a trophy at the Best New Zealand Graphic Design Awards in 1988. After its image was boosted by the labels, the winery enjoyed a marked surge in sales.

What about the future? The Fredatovichs have recently upgraded their vineyard shop and are promoting the winery as a functions venue. Lincoln wines already line retail shelves in the United Kingdom, Canada and Switzerland, but Peter Fredatovich plans to intensify the export drive: 'That's where our future growth will be.'

Mazuran's Vineyards
255 Lincoln Road, Henderson

Owner: Rado Hladilo

Key Wines: Dry Sherry, Gold Medal Dry Sherry, 1954 Tawny Port, Director's Port, Old Tawny Port

West Auckland's sherry and port-producing traditions are still preserved at this tiny winery. In past decades Mazuran's was a real glamour wine in show judgings.

George Mazuran arrived in New Zealand in 1926, a seventeen-year-old intent on avoiding conscription into the Austrian army occupying Dalmatia. After labouring on dairy farms in the Waikato and running an Auckland fish and chip shop — losing his right arm meanwhile in a fishing accident — Mazuran bought the present Lincoln Road property in 1938. The first vines were planted a year later; 1942 marked the first Mazuran vintage in New Zealand.

The manifold difficulties encountered in the post-war years in disposing of this wine led directly to Mazuran's long career as a political lobbyist (see page 9). In 1971, in his twenty-first year as president of the Viticultural Association, George Mazuran was awarded the OBE for his services to the wine industry. He died in 1982, and the company is now run by his son-in-law, Rado Hladilo.

The tall, muscular Hladilo (55) speaks with a thick Dalmatian accent. He arrived in Auckland in 1968. 'Mazuran's needed help because of George's handicap. When I married his daughter, I ended up staying here.' The winery's current output — about 2000 cases each year — is small, allowing Hladilo to do everything himself 'from A to Z'.

Ironically, the wine-growing bonanza that owed so much to George Mazuran's efforts passed his own company by. The Mazuran estate vineyard is small — only 1.5 hectares of Palomino, Muscat Alexandria, Hamburgh, Seibels 5455 and 5437, and Merlot — with no other sources of supply. Hence the wines are hard to buy; they are sold only at the winery. The range of products — concentrated on ports and sherries — has barely changed in the past twenty years.

The small winery shop, built in the 1940s, is crammed with ancient barrels and historic photographs. For $300 you can buy a bottle of Mazuran's Royal Vintage Port 1943; Hladilo recently broached a surviving forty-gallon (182L) cask. Other vintage-dated ports are available from 1945, 1946, 1947 and 1948; 'George would fill a barrel and forget about it,' recalls Hladilo. The Mazuran port style is almost unmistakable: these are dark, almost opaque wines with an almost liqueur-like intensity — rich, treacly and creamy-sweet.

The sherries are as mature as the ports. A madeira, vermouth, liqueurs — Crème de Menthe, Cherry Brandy and Curaçao — and cherry cocktail also line the shelves. Mazuran's table wines carry the labels of another era: Dry

Mazuran's

Dark, almost opaque, creamy-sweet old ports are a specialty of the tiny Mazuran's winery in Henderson. Rado Hladilo, who married Patricia Mazuran, the daughter of founders George and Florence Mazuran, has alone survived to carry on the Mazuran tradition of fortified winemaking.

White (produced from Muscat grapes), Sauternes, Moselle, and Dry Red (made from Seibel 5455).

For thirty years, at home and abroad, judges singled Mazuran's fortified wines out for praise. Today, not only do fortified wines have a low profile in local judgings, the heavy style of dessert wine produced at Mazuran's is no longer in favour. The winery no longer enters New Zealand competitions, but in 1992 its Old Tawny Port was the top port at the annual Ljubljana show in Slovenia.

Is Hladilo tempted to pursue the table wine market? 'No. I'm doing very well. I wouldn't want to get into table wines. I hope to carry on the same for as long as possible.'

Pacific

Pacific Vineyards
90 McLeod Road, Henderson

Owners: The Erceg family

Key Wines: Phoenix Gewürztraminer, Chardonnay, Sauvignon Blanc, Cabernet Sauvignon; Colwyn Park Müller-Thurgau, Pale Dry Sherry, Reserve 20-Year-Old Port

A brew house rose in 1990 at the Pacific winery in West Auckland. 'Our move into beer production will enable us to reposition Pacific as a boutique rather than bulk-wine producing winery,' says Michael Erceg, the founder's son.

Mijo (Mick) Erceg, who died in 1982 aged seventy-five, arrived in New Zealand in 1929. After several years' labour on gumfields, roads and the vineyards of other Dalmatians, he bought a small farm in McLeod Road, Henderson; by 1936 his own vines were in the ground.

Pacific entered into a marketing arrangement with Seppelts of Australia in 1967, and for several years Pacific wines appeared on the shelves under the Seppelt label. A range of 'Saint Stefan' wines followed and, later, Pacific lines were sold carrying the Monlouis, Michael's, Willowbrook and other labels.

Pacific has suffered in recent years from an unfortunate degree of instability, reflected in its changing parade of labels and involved family members. Ivan Erceg, who made the wine in the early 1980s, had studied malolactic fermentations and acidity in wine at graduate level, but his wines failed to stand out. He has since departed from the company. His brother Michael — holder of an American doctorate in mathematics — also temporarily withdrew from the business in 1987, but is once again involved and promotes Pacific wines through his liquor importing, producing and distribution company, Independent Liquor.

Steve Tubic, a Waikato-born winemaker who worked at Corbans between 1979 and early 1986, before working the 1986 vintage at Pacific under the consultant, Norbert Seibel, has been the winemaker since the 1987 vintage.

The concrete-block, fibrolite and timber winery backs onto the five-hectare estate vineyard, which is planted in Cabernet Sauvignon and Merlot. During summer and autumn, to protect the ripening bunches from plundering birds, the entire vineyard is covered in netting. This home vineyard is the sole source of fruit for Pacific's red winemaking. White grapes are all bought in: Chardonnay and Sauvignon Blanc from Hawke's Bay; Gewürztraminer and Müller-Thurgau from Gisborne.

Millie Erceg, Mijo's widow, has worked fulltime at Pacific since 1948. When her sons were both absent from the company, Millie became the managing director; she still oversees the estate vineyard and winery shop.

Michael Erceg predicts that the company's recent move into brewing beer will reshape his family's wine business. 'When I came home [from studying in the United States] in 1981 we had just crushed 100 tonnes of grapes. By 1987 we had moved up to 1500 tonnes [mostly sold as cask wine and coolers]. We are continuing to cut back on the bulk end of the market. We have a decreasing amount going into casks and are stabilising our bottled wines.' Erceg plans to eventually move the brewery away from the vineyard, to 'concentrate on turning the winery into a boutique winery.'

Pacific's top wines are all marketed under the Phoenix brand. The star attraction is the full-bloomed, concentrated, pungently spiced Phoenix Gewürztraminer, grown in the Johnson vineyard in Gisborne. The 1986 vintage — first marketed under the Willowbrook label, then as Pacific Reserve Gewürztraminer — captured gold medals at home and abroad. The 1990 vintage, also a gold medal and trophy winner, is equally arresting.

Phoenix Chardonnay is an oak-fermented style, soft and savoury. Phoenix Sauvignon Blanc — predominantly stainless steel-fermented, but thirty percent barrel-fermented and lees-aged — is vigorously crisp and herbal. The estate-grown, French oak-matured Phoenix Cabernet Sauvignon is dark and gutsy, with strong, spicy, chocolatey flavours.

Fortified wines remain part of the range, with Pacific Reserve 20-Year-Old Port and the amber-brown, fragrant, mellow Pacific Pale Dry Sherry — based on stocks laid down in 1962 — the highlights.

Millie Erceg — whose husband, Mijo (Mick), founded Pacific just before the Second World War — still oversees Pacific's estate vineyard and winery shop at Henderson. Winemaker Steve Tubic produces a superbly aromatic and peppery Gewürztraminer under Pacific's Phoenix label.

Pleasant Valley Wines
322 Henderson Valley Road, Henderson

Owner: Stephan Yelas

Key Wines: Yelas Estate Selection
Chenin/Chardonnay, Gewürztraminer, Chardonnay,
Riesling, Pinotage, Cabernet Sauvignon,
Cabernet/Merlot; Pleasant Valley White Burgundy,
Amontillado Sherry, Amoroso Sherry, Oloroso
Sherry, Founders Port, Tawny Port

Pleasant Valley is far too old to appeal to fashion-conscious
wine drinkers. Yet it is the source of sound, often good
wines offering outstanding value for money. Pleasant
Valley has the dual distinction of being not only the oldest
surviving Dalmatian vineyard in Henderson, but also the
oldest winery in the land under the continuous ownership
of the same family.

Stipan Jelich, the founder, arrived in Auckland in 1890
at the age of twenty-two. After five years' labour on the
northern gumfields he bought, at five pounds an acre,
thirty-two hectares of hill country in the Henderson Valley.
His first crop of grapes, Black Hamburghs, fetched him
less on the Auckland markets than the charges of the carrier
and auctioneer. By 1902, winemaking had emerged as a
better source of income, and from then until Stipan Jelich
(Stephan Yelas) retired in 1939, the vineyard served as a
model of the fruits of peasant stubbornness: the land was
turned over by spade; the vines, tied to manuka stakes,
were hand-hoed, and sprayed from a knapsack mounted on
the back. During Stipan's working life, not even a horse
helped to ease the toil.

Yelas's son Moscow long ago introduced more modern
vineyard techniques, replacing manuka stakes with wire
trellises and knapsack sprayers with machine sprayers.
Until his death in 1984, Moscow, although semi-retired,
still made the wine while his son, Stephan, concentrated on
viticulture.

Around 1980 the cellars were quiet, even sleepy; much
of the grape crop was sold to other companies and the
wines were not widely seen beyond the winery. Now,
under the guidance of Stephan ('Steppie') Yelas (46),
Pleasant Valley has embarked on the comeback trail.
'Taking over was a challenge,' recalls Stephan. 'Back then
we were selling all our grapes. I hired a winemaker and we
turned it around quickly.'

The eight-hectare, hill-grown estate vineyard, entirely
replanted in the past few years, is established in
Chardonnay, Müller-Thurgau, Pinotage, Cabernet Sauvignon and Merlot. Premium white varieties such as
Sauvignon Blanc, Chardonnay and Riesling are also
bought from Hawke's Bay and Gisborne.

The original kauri winery is still used for barrel storage.
The winery is marvellously atmospheric, full of dark nooks
and crannies, old concrete fermentation tanks, old corking
machines and — everywhere you look — aged barrels
housing ports and sherries decades old. Moscow Yelas's
son has inherited a treasure of venerable old fortified wines.

From 1986 until mid-1988 Pleasant Valley wines were
produced under the guidance of Norbert Seibel, formerly
chief winemaker at Corbans. Petter Evans made the 1989
and 1990 vintages, before leaving for St Helena. Since
then, Stephan Yelas himself has made the wine.

Like many wineries, Pleasant Valley markets a two-tier
range of table wines: the premium range is called Yelas
Estate Selection, the lower range Pleasant Valley. Pleasant
Valley White Burgundy, launched in 1988, has swiftly
become one of the winery's most popular labels. Described
by Yelas as 'an easy-drinking style at a good price', it is
blended from several grape varieties, predominantly Chenin
Blanc, and is fresh, off-dry, light and tangy.

The Yelas Estate Selection features uniformly clean,
carefully crafted wines with clear-cut varietal characters,
currently much underrated by wine lovers. The Pinotage,
estate-grown and matured briefly in seasoned oak casks, is
'by far the best red for us, in terms of quality and sales,'
says Yelas. Fruity, peppery and soft, it is very appealing in
its youth. The estate-grown Cabernet Sauvignon and
Cabernet/Merlot, matured in French and American oak
casks, are both fragrant, spicy, full-flavoured and firm.

Yelas Estate Chardonnay is a barrel-fermented, buttery
style, savoury and long-flavoured. The delicate, well-spiced
Gewürztraminer, and crisp, intensely citric-flavoured
Riesling are equally solid.

Of Pleasant Valley's traditional range of fortified wines,
the barrel-matured Amontillado, Amoroso and Oloroso
sherries are all impressively rich, mellow and lingering.
The jewel in the crown, however, is the amber-brown,
nutty-sweet Founder's Port, an almost liqueur-like, fifteen-
year-old tawny style with an illustrious record in show
judgings.

Pleasant Valley's vineyard restaurant, built in 1991 in the
old winery, serves morning and afternoon teas and light
lunches. Bus tours are a specialty. Yelas plants to keep
producing 'sound wines for the average person at reasonable prices. If it's under ten dollars, you can sell it.'

Pleasant Valley

*Stephan ('Steppie') and
Ineke Yelas run the oldest
family-owned winery in the
country. Savoury and
supple, the silver medal and
twice trophy-winning Yelas
Estate Pinotage is their most
consistently successful table
wine.*

Seibel

Seibel Wines
Sturges Road, Henderson

Owners: Norbert and Silvia Seibel

Key Wines: White Riesling Medium-Dry, White Riesling Barrel Fermented, White Riesling Late Harvest, Gewürztraminer Semi-Dry, Sauvignon Blanc Fumé, Chardonnay, Barrel Fermented Chenin Blanc, Cabernet Franc/Merlot/Cabernet Sauvignon

'I didn't want to become a flying winemaker, overseeing wineries around the country,' recalls Norbert Seibel, who was Corbans' chief winemaker in the early–mid 1980s, but now produces an absorbing array of wines under his own label. 'This is much more satisfying, because I can look after each barrel individually.'

Born on the Rhine in Mainz, Germany, Seibel graduated in viticulture and oenology from the acclaimed Geisenheim Institute. He then spent the 1970s in South Africa, firstly as chief winemaker for Nederburg, then as chief winemaker for The Bergkelder. In 1980 he joined Corbans.

'Working in these large firms,' recalls Seibel (51), who speaks with a broad German accent, 'I began to develop the conviction that the true potential of any vineyard or grape variety is best realised on a small scale, with individual and expert attention.' After seven years, Seibel resigned from Corbans, and during the mid–late 1980s worked as a consultant winemaker for Pacific and Pleasant Valley. When in 1987 he launched his first releases under the Seibel label he realized his 'long-held dream to produce a small but

Norbert Seibel produces a distinctive array of wines; rich, tangy, strong-flavoured Rieslings are a highlight. Seibel and his wife, Silvia, are currently rejuvenating the former Balic winery in Henderson.

select range of wines which combine superior taste with healthy composition.'

Seibel's first vintages were produced at Pleasant Valley. Later he based his production at West Brook, where he also assisted Anthony Ivicevich with his production. In 1993, however, Seibel and his wife, Silvia, purchased the old Bellamour (Balic) winery in Sturges Road, Henderson.

Seibel's output is small: he crushes about thirty-five tonnes of grapes annually, equivalent to only 2500 cases of wine. He purchased much of the fruit for his early vintages from Te Kauwhata, but now buys grapes from Hawke's Bay, Gisborne and Marlborough. He is no fan of early bottling; in his pursuit of 'full, round and soft wines', prior to bottling Seibel commonly matures his wines for at least a year in tank or cask.

'I've had a lot of medal success with Gewürztraminer,' replied Seibel, when I asked him what are the key wines in his range. His dry and medium Gewürztraminers have been gently spicy, soft wines with delicate, lingering flavours. In future Seibel plans to produce a single, off-dry style of Gewürztraminer.

If you enjoy Rieslings, watch out for the Seibel range. When vintage conditions permit, three Riesling styles emerge. Fermented and matured in seasoned — or, as Seibel puts it, 'wine-sweetened' — German oak puncheons, the White Riesling Barrel Fermented is distinctively spicy, pungent and dry — a complex style quite unlike other New Zealand Rieslings. The White Riesling Medium-Dry, which is stop-fermented with a touch of sweetness, is packed with lemon/lime fruit flavours and tangy acidity. The botrytised White Riesling Late Harvest — produced about one vintage out of two — is a medium-sweet style, deliciously deep-flavoured and nectareous.

A strapping, mealy barrel-fermented Chardonnay; a mouthfilling Sauvignon Blanc Fumé with rich, fig-like flavours; a dry, yeasty and complex Barrel Fermented Chenin Blanc; and a copper-coloured, strawberry-flavoured Cabernet Sauvignon Blanc de Noir are other highlights of Seibel's intriguingly individual range.

'I left Corbans because our kids were grown up and I was free to do what I wanted to do,' says Siebel. 'In the future I'll increasingly specialise in varieties other winemakers don't produce, like Scheurebe, Morio-Muskat and Bacchus.'

Soljans

Soljans Wines
263 Lincoln Road, Henderson

Owners: Tony and Rex Soljan

Key Wines: Henderson Spumante, Breidecker, Chardonnay, Sauvignon Blanc, Gewürztraminer, Pinotage, Cabernet/Merlot, Pergola Sherry, Reserve Sherry, Founders Port

'A couple of years ago I'd have given winemaking away, it's so competitive,' recalls Tony Soljan. 'But we've recently

done a lot to make our winery more attractive for tourists. And the message is finally getting through that you can make good wines in the Auckland region.'

Soljans, traditionally one of the best small Henderson dessert winemakers, has recently branched out with a sound lineup of white and red varietal table wines. The founder, the late Frank Soljan, journeyed from Dalmatia to New Zealand in 1927 at the age of fifteen. Today, the block of fruit trees and vines he planted in Lincoln Road in 1937 is still a combined orchard and vineyard, run by his sons Tony (48), the winemaker, and Rex (49), who cultivates the impeccable vineyard.

A scene of rare beauty, Black Hamburgh table grapes trained along overhead trellises, greets visitors to the Soljans winery in summer. The six-hectare estate vineyard is also planted in Cabernet Sauvignon, Seibel 5455 (for port), Palomino and Muscats. At Riverlea a second, four-hectare vineyard, planted in Breidecker, Sauvignon Blanc, Muscats and Seibel 5455, was recently sold. Premium varieties such as Chardonnay and Gewürztraminer are bought from growers in Auckland, Gisborne and Hawke's Bay.

Soljans crushes about 100 tonnes of grapes annually — equivalent to about 7000 cases of wine. Much of this is sold directly to the public in the original winery, built before the Second World War, which serves today as the vineyard shop. Here cheerful Tony Soljan, a long-term, influential spokesman for small-scale vineyards, can often be found dispensing the fruits of his labours.

This winery has long been respected as a sherry producer and Soljans' top offering, the Pergola Sherry, emerges after ten to fifteen years in French oak hogsheads — originally used as brandy barrels — seductively full, rich, sweet and raisiny, with plenty of 'rancio' complexity. The eight-year-old Reserve Sherry, predominantly matured in totara casks, is also nutty-brown and creamy-sweet, but less complex than the Pergola Sherry; both are fine examples of this wine style.

Soljans are currently placing less emphasis on everyday-drinking fortified wines — 'You're working for very little', says Tony Soljan — but actively pursuing the premium fortified wine market. 'There's lots of demand for ports — ruby or tawny.' The Founders Port, a ten-year-old tawny

matured in small oak casks, is amber-hued, mellow, sweet and smooth.

Of the table wines, one of the sales success stories is the Henderson Spumante, an unabashedly sweet sparkling wine with the heady fragrance of Muscat grapes and a fresh, very fruity flavour. For no-fuss occasions Soljans produce a popular, light, slightly sweet Breidecker, sold as 'a good alternative to Müller-Thurgau'.

The premium range of table wines includes a bone-dry Auckland Sauvignon Blanc; a cask-matured, Gisborne-grown Chardonnay; a medium Hawke's Bay Gewürztraminer; a rewardingly savoury and supple, meaty, lightly wooded Pinotage; and a fragrant, French oak-matured, red berry-fruit-flavoured Cabernet/Merlot.

In the barbecue garden, a tranquil resting place in the midst of lawns, apple trees and overhead-trellised vines, each year thousands of visitors devour succulent spit-roasted lamb — washed down with Soljans' rock-solid wines.

The chance to chat to affable winemaker Tony Soljan in the winery shop, and eye-catching overhead-trellised vines, are just two of the attractions at Soljans. Old sweet sherries and the flavour-packed Pinotage are highlights of the range.

St Jerome Wines
219 Metcalfe Road

Owners: The Ozich family

Key Wines: Cabernet Sauvignon/Merlot, Dry Red, Chardonnay, Chablis, Physicians Port

One of West Auckland's most compelling reds flows from this tiny winery. Sited at the top of Metcalfe Road, it was founded by Mate Ozich, now in his seventies, who brought his family to New Zealand in 1954. Since the first vintage in the 1960s, the vineyard, on a warm, north-facing clay slope behind the winery, has grown to eight hectares, predominantly planted in Cabernet Sauvignon and Merlot.

Mate's sons, Davorin (42) and Miro (39), who 'grew up in the nearby Babich vineyard', now run the winery, Davorin concentrating on winemaking and marketing, Miro on viticulture. 'A life-long feel for grapes and wines' led Davorin, who holds an MSc in biochemistry and spent his early career in the pharmaceuticals industry, to join St Jerome on a full-time basis in 1987. After working the 1987 vintage at Château Cos d'Estournel in St Estephe, Davorin came back with a 'serious, uncompromising approach to

red winemaking, which endeavours to get the very best out of the grapes'.

The wines — originally sold as Nova and later Ozich — are now labelled St Jerome, in honour of the patron saint of the Ozich family. In the Dalmatian village of Rascane, by tradition each new vintage was first tasted on 30 September — St Jerome's Day.

The highlight of St Jerome's range is the strapping, dark-hued Cabernet Sauvignon/Merlot. Fermented with natural yeasts, and matured for eighteen months in French Nevers oak barriques, it displays strong, herbaceous, blackcurrant-like flavours, underpinned with plentiful oak and taut

St Jerome

St JEROME
Cabernet Sauvignon Merlot
1989
75 cl Alc 13.2% v/v
Produced and bottled in the Cellars of:
ST. JEROME WINES,
219 Metcalf Road, Henderson,
Auckland, New Zealand.

The pride and joy of brothers Davorin (left) and Miro Ozich is their estate-grown, chunky, rich-flavoured, tautly structured Cabernet Sauvignon/Merlot. The robust, ripe-tasting, soft Chardonnay is their white-wine flagship.

tannins. The silver medal-winning 1986 and 1987 vintages displayed a distinct green-leafiness, but the riper-tasting 1988, 1989 and 1990 vintages are distinctly superior. Robust and chewy, St Jerome Cabernet Sauvignon/Merlot cries out for at least five years in the cellar.

For earlier consumption the Ozichs produce two reds: a Cabernet Sauvignon with fresh red berry-fruit flavours, restrained oak and easy tannins, and a fruity, supple, non-wooded Dry Red.

St Jerome's white-wine flagship is undoubtedly the Hawke's Bay Chardonnay. Fermented in French oak casks,

given a 100 percent malolactic fermentation, and lees-aged for several months, it is mouthfilling, peachy, buttery and soft — deliciously upfront and forward in style. The lower-priced Chablis is light, lemony and dry. A solid Sauvignon Blanc, Gewürztraminer, Rhine Riesling and Müller-Thurgau — in ascending levels of sweetness from bone-dry to medium — round out the white wine lineup.

The St Jerome range also features two ports: a predominantly Cabernet Sauvignon-based Vintage Port, and the popular Physicians Port, a cask-matured tawny style blended from stocks up to twelve years old.

West Brook

West Brook Winery
34 Awaroa Road, Henderson

Owners: Anthony and Sue Ivicevich

Key Wines: Chenin Blanc, Sémillon, Sauvignon Blanc, Chardonnay, Blue Ridge Rhine Riesling, Cabernet Rosé, Merlot, Cabernet Sauvignon, Vintage Port

'The quiet achievers' is how West Brook recently described itself on its winery brochure. West Brook — until a few years ago called Panorama — is a small, family-operated winery in the built-up Awaroa Road area of Henderson.

Tony Ivicevich and his father Mick arrived in New Zealand from their native Dalmatia in 1934. After only one year, the present property had been purchased and planted in trees and grapevines. By 1937 the first port and sherry were on the market, sold for one shilling and sixpence in beer bottles without labels.

Today the winery is run by Tony's rangy son Anthony (44) and his wife, Sue, who, she says, 'basically does the paperwork and handles the winery sales'. Their son, Michael, and Anthony's mother, Mary, also help, making West Brook very much a family affair.

The three-hectare home vineyard behind the concrete-block winery is planted principally in Cabernet Sauvignon and Merlot; Chardonnay was added in 1992. This small, house-encircled vineyard survives, says Ivicevich, 'because

it keeps my hand in on the viticulture side, and also for nostalgic reasons — it's nice to have vineyards still in Henderson. If the vineyard wasn't there, we might as well uproot the winery and move out.' Premium white varieties are bought from growers, principally in Hawke's Bay.

The level of production is not high — about 100 tonnes of grapes are crushed annually, equivalent to 7000 cases of wine — and the wines are only lightly scattered through North Island retail outlets. Although West Brook still makes sherries, 'it's only a matter of time before they're phased out,' says Ivicecich. To symbolise the winery's switch in emphasis from fortified to table wines, Panorama, its long-standing name, was dropped in 1987 in favour of West Brook. In future the company's top wines will be marketed under the Blue Ridge label.

Chenin Blanc and Merlot are the fastest-selling wines in the West Brook range. 'We're pushing Chenin Blanc,' says Ivicevich. 'I know it's usually hard to sell, but it's catching on because Müller-Thurgau is increasingly perceived as "too fruity". People want clean, pleasant fruit flavours and a crisp finish.' This is a fresh Hawke's Bay wine, its lemony, slightly tart flavour balanced by a hint of sweetness.

Anthony Ivicevich is also 'keen to pursue Sémillon; I like its crispness and herbaceousness'. Full-flavoured yet easy-drinking, West Brook Sémillon is a tasty amalgam of grassy, tropical-fruit and — with age — slightly honeyish flavours. It flourishes in the bottle for several years.

The Merlot is made 'for early consumption, and as a good first step into red wine-drinking'. Estate-grown, to underline its fruitiness it is not barrel-matured, and its soft, easy tannins enhance its youthful appeal. This is a dark, chunky red with plenty of blackcurrant and plum-like flavours and a smooth, undemanding finish.

A fat, peachy, buttery-soft, strongly 'malo'-influenced Chardonnay; a coral-pink, fresh and fruity, slightly sweet Cabernet Rosé; and a dark-hued, full-flavoured Vintage Port are also features of West Brook's sound range.

Where to from here? 'The quality of our wines is getting stronger,' observes Anthony Ivicevich — correctly. 'We're sourcing better fruit but we haven't really pushed marketing hard. In future I want to make some of the best wines in Henderson — and sell them in the mid-price range.'

Anthony Ivicevich has steadily shifted the focus of his small West Brook winery from fortified wines to table wines. The estate-grown, gutsy, plummy West Brook Merlot delivers soft, enjoyable drinking in its youth.

Montana Wines
Head Office: 171 Pilkington Road, Glen Innes,
Auckland

Owner: Corporate Investments Limited

Key Wines: Montana Marlborough Sauvignon
Blanc, Marlborough Rhine Riesling, Marlborough
Cabernet Sauvignon, Gisborne Chardonnay,
Marlborough Chardonnay, Ormond Estate
Chardonnay, Renwick Estate Chardonnay,
Brancott Estate Sauvignon Blanc, Fairhall
Estate Cabernet Sauvignon, Wohnsiedler
Müller-Thurgau, Wohnsiedler Sauternes,
Wohnsiedler Rosé, Blenheimer, Chablisse,
Fairhall River Claret, Timara Chardonnay/
Sémillon, Timara Cabernet/Merlot, Bernadino
Spumante, Fricanté, Lindauer (Special Reserve
Brut de Brut, Brut, Sec and Rosé); Deutz
Marlborough Cuvée (Brut and Blanc de Blancs);
Penfolds Chardon (white and pink), Hyland,
Cottle Bush Sauvignon Blanc, Buchanan
Point Chardonnay, Lightfoot Hill Rhine
Riesling, Jackman Ridge Cabernet Sauvignon,
Winemakers Reserve Chardonnay, Winemakers
Reserve Gewürztraminer, Winemakers Reserve
Sauvignon Blanc, Winemakers Reserve Cabernet
Sauvignon; Woodhill's Country Dry, Country
Medium

Montana is the colossus of New Zealand wine. The company's headquarters are in Auckland, but it also owns vineyards in Hawke's Bay, Gisborne and Marlborough, and a quartet of wineries located in those four regions. Its range of labels, rock solid in quality and unrivalled for their value for money, have a commanding forty percent share of the domestic market for New Zealand wine.

Ivan Yukich, founder of this giant company, arrived in New Zealand from Dalmatia as a youth of fifteen. After returning to his homeland, he came back to New Zealand in 1934, this time with a wife and two sons. After years devoted to market gardening, Yukich later planted a fifth-hectare vineyard high in the bush-clad folds of the Waitakere Ranges west of Auckland, calling it Montana, the Croatian word for mountain. 1944 saw the first Montana wine on the market.

What would Ivan Yukich feel about Montana's industry dominance today? Fierce pride, no doubt, tempered with deep disappointment that, in the end, Montana and the Yukich family parted company.

Under the direction of sons Mate — the viticulturist — and Frank — winemaker and salesman — the vineyard grew to ten hectares by the end of the 1950s. The company then embarked on a whirlwind period of expansion unparalleled in New Zealand wine history. To build up its financial and

distribution clout, Montana joined forces with Campbell and Ehrenfried, the liquor wholesaling giant, and Auckland financier Rolf Porter. A new 120-hectare vineyard was established at Mangatangi in the Waikato and in the late 1960s Gisborne farmers plunged into grape-growing at the Yukiches' urgings. A gleaming new winery rose on the outskirts of Gisborne in 1972 and a year later Montana absorbed the old family firm of Waihirere.

Although production was booming the company at this stage earned a reputation for placing sales volume goals ahead of quality. The launch-pad for Montana's spectacular growth was a series of sparkling 'pop' wines — Pearl, Cold Duck and Poulet Poulet — which briefly won a following. For those with a finer appreciation of wine the company somehow managed to produce an array of classic labels.

The real force behind Montana's early rise was Frank Yukich (see page 12). He early perceived the trend away from sherry to white table wine and was the first to adopt aggressive marketing strategies. The year 1973 was a momentous one. The giant multinational distilling and winemaking company Seagram obtained a forty percent shareholding in Montana, contributing money, technical resources and marketing expertise. Seagram's investment was originally trumpeted as 'basically an export deal . . . [to] export three million gallons [13.6 million L] a year within five years.' The anticipated river of wine from New Zealand to the United States never flowed, however, because just as Montana's output again surged, American domestic wine prices dropped.

The same year, Montana made an issue of 2.4 million public shares. Seagram's investments, shareholders' funds and independent loans together provided $8 million over the next three years for development purposes.

Next — also in 1973 — came the pivotal move into Marlborough. As Wayne Thomas, then a scientist in the Plant Diseases Division of the DSIR, has related: 'In March 1973, Montana under the guidance of its founder and managing director, Mr F.I. Yukich, intended to undertake a major vineyard planting programme in New Zealand . . . Although plenty of suitable land was available in both the Poverty Bay and Hawke's Bay regions, my own impression was that it was too highly priced for vineyards . . .

'I gave the subject of alternative vineyard areas in New Zealand considerable thought and . . . then phoned Mr Frank Yukich and suggested that . . . he should consider the possibility of establishing vineyards in the Marlborough region as it had all the necessary criteria on the surface to make it successful . . . [Later] Mr Yukich rang, requesting that I have suitable authorities in the Viticulture Department at the University of California, Davis, confirm that the Marlborough region would be suitable for growing wine grapes . . . Confirmation was duly obtained from Professors Winkler, Lider, Berg and Cook . . .'

The first vine was planted in Marlborough on 24 August 1973: a silver coin, the traditional token of good fortune, was dropped in the hole and Sir David Beattie, then chairman of the company, with a sprinkling of sparkling

Montana

wine dedicated the historic vine. The first grapes were harvested on 15 and 16 March 1976; fifteen tonnes of Müller-Thurgau were trucked aboard the inter-island ferry at Picton and driven through the night by Mate Yukich to Montana's Gisborne winery. A 'token' picking of Cabernet Sauvignon followed in April.

Montana was moving swiftly to rectify its quality problems. The standard of the 1974 and subsequent vintages soon lifted the company into the ranks of the industry's leaders. Still pursuing the mass market, the company now shifted its emphasis to non-sparkling table wines. Bernkaizler Riesling (later called Benmorven) began to open up a huge market for slightly sweet white wines later developed with Blenheimer, one of New Zealand's most popular wines.

A year after the pivotal moves of 1973, Frank Yukich, the key visionary behind Montana's rapid rise, was gone — the loser when his relationship with Seagram turned sour. Soon after, the company also severed its link with the old Yukich vineyard at Titirangi. The twenty-hectare vineyard site and substantial winery was unsuited to further development and the company chose instead to expand elsewhere. The old winery was dismantled and most of the equipment sent to Blenheim.

Montana's costly move into Marlborough contributed to the company's depressed financial condition from 1974 to 1976. But the subsequent recovery represents a major business success story. After two years of losses, Montana showed a small profit in 1975/76 and by 1978 had paid its maiden dividend. Profits in the year ending 30 June 1983 totalled $6.4 million.

In late 1985 Corporate Investments Limited took control of Montana, by adding Seagram's 43.8 percent stake to its own already substantial shareholding. Seagram pulled out when the industry's fortunes turned sour: in the year to 30 June 1986 the company recorded a loss of almost $1.6 million. Corporate Investments, a company listed on the stock exchange, is principally owned by its chairman, Peter Masfen, who has served as a director of Montana for twenty years.

Masfen, one of New Zealand's wealthiest men, has a personal fortune estimated to be in the region of $40 million. His wife, Joanna, is the daughter of Rolf Porter, Montana's early financial backer. An accountant, Masfen owns a string of private businesses in addition to his Corporate Investments holding. 'Our company tends to buy into out-of-favour sectors and does well out of them,' observes Masfen.

Following its acquisition of Penfolds Wines (NZ) Limited in 1986 from Lion Corporation Limited, Montana moved back into the black, posting a $5.14 million profit for the year ending 30 June 1987. Later that year, Corporate Investments secured a 100 percent shareholding in Montana and then de-listed the company from the stock exchange.

Under managing director Peter Hubscher, Montana has geared itself to repel the onslaught of Australian wines under CER. 'Our whole strategy is planned to prevent us being swamped by the Australians,' says Hubscher.

In Gisborne, the Montana and former Penfolds wineries have been linked by pipelines. At the Riverlands winery a few kilometres on the seaward side of Blenheim, towering 550,000-litre insulated tanks have been installed to store reserves of top-selling wines in optimum condition. The 'tank farm' here has the capacity to store up to 20 million litres of wine. During vintage up to 500 tonnes of fruit avalanches in each day from contract vineyards and the company's own plantings covering 500 hectares at Brancott, Renwick, Fairhall and Woodbourne. The complex also features a cask-filling plant (producing 20,000 casks daily), a barrel hall, cooperage, offices and a retail shop.

From Blenheim and Gisborne much wine then rolls north in bulk rail tankers to Auckland for bottling. All finishing and maturing of bottled wines is carried out at the Glen Innes complex in Auckland; since 1989 an overflow of bulk grapes from Hawke's Bay has also been crushed and fermented there. If you stroll around this expansive complex with Peter Hubscher, through a labyrinth of storage tanks and pulsating bottling lines, finally to a large wall-mounted photograph of Montana's sweeping Marlborough vineyards, your inevitable lingering impression is of the company's enormous scale.

What is Montana's strategy to keep its market dominance? 'The company has been erected on the premise of good value,' former managing director, Bryan Mogridge, observed in 1991. 'You are never cheated when you buy a bottle of Montana; usually you get a slightly better wine for your dollar. From that cornerstone, we look at every market segment.'

Montana's cost structure directly reflects its economies of scale. The company draws about forty percent of its grape intake from its own vineyards — a much higher percentage than its major rivals. Montana's strategy is to control its means of production via a heavy commitment to vineyard and winery assets, and price aggressively.

'With the wines that count, we want to control the vineyards', says Hubscher. Since 1987 Corporate Investments Limited has poured over $30 million into Montana, expanding its vineyard holdings and purchasing and refurbishing The McDonald Winery in Hawke's Bay (see page 123). Montana now owns 500 hectares of vineyards in Marlborough, eighty hectares in Gisborne and 250 hectares in Hawke's Bay — fourteen percent of the country's total vineyards.

In 1988 Montana made a crucial decision — to expand its share of the premium (over $15 per bottle) market, where traditionally it has not been a major force. Previously, geared to crush huge tonnages, it simply wasn't able to handle small, superior batches of wine. Now it has two locations reserved for small-scale, premium wine production: a separate flow system at the Blenheim winery for hand-picked, bunch-pressed fruit, and The McDonald Winery at Taradale.

And can Montana compete with Australian bulk wines at the bottom end of the market? The company is adamant it can, on the grounds that its grape costs are now com-

Montana, which earlier concentrated its vineyards in Marlborough and Gisborne, made a deep thrust into Hawke's Bay in the late 1980s. Among the first fruits of the company's heavy investment in vineyards and The McDonald Winery have been the impressive Cabernet Sauvignon and Chardonnay under the Church Road label.

petitive with Australian prices; its bulk wines have better flavour; and its bottling lines go just as fast.

Montana is certainly footing it on the international stage. The company's 1992 exports were worth more than $10 million — a fourfold increase on 1989. Montana is focusing its export effort on the United Kingdom and Europe, where its sales are showing rapid growth.

Accorded praise by many observers a decade ago as New Zealand's largest and best winery, Montana is facing more formidable competition today. Yet several of its labels — notably the Marlborough Sauvignon Blanc and Rhine Riesling, Deutz Marlborough Cuvée and the new Montana Estates quartet — must be rated among New Zealand's finest wines. Montana's white wines often match the boutiques on quality and outperform them in the value-for-money stakes.

Montana's popular Blenheimer, launched in 1977, is a household name. Made predominantly from Marlborough-grown Müller-Thurgau, it is a fruity, light, mild wine, soft and gently sweet. Although lacking the fragrance and fruit intensity of a top Müller-Thurgau, this is a perfectly acceptable, undemanding wine for occasions when wine is the backdrop to, rather than the focus of, conversation.

Montana Wohnsiedler Müller-Thurgau, launched in 1981, has also established itself as one of the country's fastest-selling white-wine brands. Blended from Gisborne and Marlborough fruit, it is slightly sweet, delicately flavoured, flowery and fresh — a tangy, refreshing mouthful. Montana Chablisse, another big seller, is a dryish, but not bone-dry, blend displaying Müller-Thurgau's typical lightness and fruitiness and a hint of Sauvignon Blanc-derived herbaceousness.

The popular Gisborne Chardonnay — with an annual output of about 60,000 cases, one of the country's biggest-selling Chardonnays — has traditionally been a light, fresh style, placing its accent squarely on its soft, peachy fruit flavours. Until a few years ago it had no wood treatment, but now a 'portion' is aged in French and American oak puncheons. This is a beguilingly easy-to-drink Chardonnay.

Montana Marlborough Chardonnay displays much greater wood character, from its twelve months' maturation in French oak puncheons. It is peachy-ripe and extremely easy-drinking.

One of Montana's Achilles heels until recently was its lack of a Chardonnay capable of competing at the very top end of the market. However, its recently launched Ormond Estate (Gisborne) and Renwick Estate (Marlborough) Chardonnays are both arrestingly stylish and flavourful, their quality consistent with their high prices. (See also Church Road Chardonnay, page 124.)

Montana's range of Marlborough white wines has fully justified the company's faith in the region. Since the first 1979 vintage, Montana's Marlborough Rhine Riesling has stood out — a fragrant, flowery, polished wine with abundant fruit flavour and crisp acidity. At three to five — even ten — years old most vintages are toasty and honeyish and awash with delectable, apple and pineapple fruit flavours. In *Cuisine* magazine's 1992 tasting of New Zealand Rieslings, the 1989 and 1983 vintages both scored five stars, backed up by four stars for the 1991 and 1985 vintages. The miracle is that a wine of such indisputably high quality is so widely available and so bargain-priced.

Montana Marlborough Sauvignon Blanc is full of distinctive herbal varietal character. Unmistakably Sauvignon Blanc in its youth, with its assertive capsicum-like bouquet and flavour, after a couple of years' bottle age this dry wine develops a less pungent, more gooseberryish character, softer and lusher.

It is beyond doubt that this label — and Cloudy Bay's

Montana, by planting the first commercial vineyards in Marlborough in 1973, laid the seed for the flowering of one of the New World's most acclaimed wine regions. Montana's Marlborough Sauvignon Blanc and Rhine Riesling are consistently top-flight and rate among the country's greatest wine bargains.

Sauvignon Blanc — have focused more international attention on the soaring standard of New Zealand wine than any others. It is that rare combination: a world-class wine that is nonetheless freely available and affordable. Montana Marlborough Sauvignon Blanc has proved its top-flight quality over fourteen years; the 1991 vintage was crowned the champion white wine of the 1992 Sydney International Winemakers' Competition.

Montana's much rarer and more subtle Brancott Estate Sauvignon Blanc, partly fermented and lees-aged in French oak barriques, is equally arresting with its sustained, complex herbal flavours.

Turning to the red wines, Montana's Cabernet Sauvignon led the range from 1973 to 1977, and then was superseded by the Marlborough Cabernet Sauvignon. Those early reds, made from Gisborne fruit, pioneered in New Zealand the production of quality red wine in large volumes. Generally fruity, straightforward Cabernet Sauvignons, they ranged in standard from the last, nondescript 1977, to the memorable red label 1973 reserve bin.

Montana's Marlborough Cabernet Sauvignon, matured for a year in American oak puncheons, has been the pick of the company's commercial range of Marlborough reds. A uniform style has emerged: medium-bodied, with plenty of Cabernet Sauvignon fruit and adequate tannin. However, their typically slightly vegetative bouquets and flavours, coupled with a lightness of mouthfeel, are *too* cool-climate in style to rival the premium reds of Hawke's Bay and Auckland. The much rarer, perfumed Fairhall Estate Cabernet

Sauvignon is vastly superior; it is fleshy, ripe and supple.

Fairhall River Claret is an everyday-drinking, rosy-hued blend of Cabernet Sauvignon, Pinotage and Pinot Noir grown in Montana's Fairhall vineyard in Marlborough. This is a non-wooded, light, raspberryish red, simple, soft and low-priced. Fortunately for red-wine lovers, the exciting quality of the Church Road Cabernet Sauvignon (see page 124) has clearly signalled Montana's devotion to producing fine quality reds in Hawke's Bay.

Several sparkling wines are of interest. Fricanté, Muscat-based and bottle-fermented, is fresh, slightly sweet and very fruity. Bernadino Spumante is a popular, low-priced blend of Gisborne grapes, including Dr Hogg Muscat. At a recent meeting of the Rangitoto Beefsteak and Burgundy Club in Auckland, a highly aromatic, light and unabashedly sweet sparkling wine was served 'blind' with dessert. The consensus around the tables was that this delicious wine was a good Italian Asti Spumante — until out of the paper bags came Montana Bernadino Spumante.

Lindauer, launched in 1981, and now available in Brut (dry) Sec (medium), Rosé and Special Reserve Brut de Brut versions, was this country's first widely released bottle-fermented sparkling. The Brut, the best-known label of the quartet, is based on Pinot Noir and Chardonnay fruit grown in Gisborne and Marlborough. Matured on its lees for eighteen months, Lindauer Brut displays a restrained rather than pungent yeastiness and a stylishly crisp, delicate, lemony flavour. The Special Reserve Brut de Brut — my favourite — is markedly fuller, richer and creamier.

Montana's flagship sparkling wine is clearly its outstanding Deutz Marlborough Cuvée, produced since 1988 under a joint agreement between Montana and the Champagne house of Deutz. The partners' ambition, says Peter Hubscher, is 'to produce the best sparkling wine outside Champagne itself.'

The hand-picked, Marlborough-grown Chardonnay and Pinot Noir grapes on which Deutz Marlborough Cuvée is based, are pressed in a computer-controlled French Coquard Champagne press which yields juice of great delicacy. After each vintage, the wine is blended in Blenheim under the guidance of André Lallier, chairman of Champagne Deutz. Then it is shipped to Auckland, bottled and matured on its yeast lees for two years in a specially built, $800,000 climate-controlled cellar.

The result is a rich, toasty, yeasty wine with a sustained 'mousse' and impressive complexity, slightly riper-tasting than the true Champagnes with which it is inevitably compared. A 100 percent Chardonnay Blanc de Blancs version was also launched in 1993.

If there was one Penfolds label Montana was sure to preserve, it was Chardon, long one of New Zealand's favourite low-priced sparklings, which has a name equally reminiscent of Moët and Chandon, the Champagne house, and the Chardonnay variety, but in fact not the vaguest connection to either. Chardon is a carbonated bubbly, predominantly produced from Müller-Thurgau grapes, backblended with Muscat juice. The palate is fresh, low alcohol, and sweet; its lightness and fruitiness have won it countless fans.

Other releases under the Penfolds brand include the quartet of quality varietals under the Winemakers Reserve banner, of which the weighty, buttery-soft Chardonnay and fresh, piercing Sauvignon Blanc are the highlights. Penfolds Hyland is a stylish but inexpensive bottle-fermented sparkling with crisp, delicate fruit flavours.

Of Penfolds' range of sharply priced varietal wines bearing the names of pioneer winemakers — Cottle Bush Sauvignon Blanc, Buchanan Point Chardonnay, Lightfoot Hill Rhine Riesling and Jackman Ridge Cabernet Sauvignon — the best buy is the Cottle Bush Sauvignon Blanc, a typical Marlborough style with its sustained herbal flavours and appetising crisp finish.

Peter Hubscher

'Hubscher is the custodian of the long term,' says Peter Scutts, Montana's former marketing director. 'He arrived at Montana in 1973, the year they planted the first vines in Marlborough.'

Auckland-born Hubscher (50) is a Massey University graduate in food technology. He worked the 1964 and 1965 vintages at McWilliam's in Hawke's Bay while still a student, and after graduating plunged straight into the wine industry: the first graduate New Zealander to join a non-family-owned winery. He stayed at McWilliam's until 1973, climbing to the posts of plant manager and technical assistant to Tom McDonald.

After spending his first years at Montana as national wineries manager, in the early 1980s Hubscher also took over the viticultural reins. Following the transfer of Bryan Mogridge, the company's managing director, to a key role in Corporate Investments' head office, in 1991 Hubscher stepped into the top job at Montana.

'Teamwork' is an important word in the Hubscher vocabulary. Montana, he stresses, cannot be a company of individualists; there is no room here for prima donna winemakers. 'The duty of a Montana winemaker is to carry out the team decision.'

For a technical expert, Hubscher has an impressive grasp of marketing. 'I'm very consumer orientated,' he says. 'Our basic goal is to produce wines that satisfy consumers; they must taste good and always taste the same. Montana's customers want certainty.'

Hubscher points to Montana's market leader, Gisborne Chardonnay, as a prime example of his dedication to consumer needs. Montana Gisborne Chardonnay has always been a non-oaked or very lightly oaked style. 'Past marketing managers used to urge us to produce a strongly wooded style, because oaky wines often scored well in competitions. We stayed with the accent on the fruit and the wine rockets out of the winery.'

Hubscher has also been deeply involved in pursuing Montana's close technical links with overseas producers. The joint venture with the Champagne house of Deutz to make high-class bottle-fermented sparkling wine in Marlborough is a triumphant example. Cordier, the owner of a half-dozen well-known châteaux in Bordeaux, has recently started making a technical contribution to Montana's red wine production. The valuable impact of Montana's past technical links with Seagram taught Hubscher a powerful lesson: 'We mustn't waste time reinventing the wheel.'

Hubscher is passionately devoted to New Zealand. 'My greatest satisfaction after over twenty-five years in the industry is knowing Montana makes good wine, and that in the future we'll make even better.'

In Bryan Mogridge's eyes, Peter Hubscher is 'the finest winemaker New Zealand's ever seen. He's made far more wine of high quality than anyone else. When wine from a 550,000-litre tank at Blenheim wins a trophy as the best Sauvignon Blanc in the world [the Marquis de Goulaine Trophy at the 1990 International Wine and Spirit Competition in London], it has to tell you something.'

WAIHEKE ISLAND

Rumours circulating fifteen years ago about the establishment of a new 'boutique' winery on Waiheke Island met in most quarters with disbelief. Was not viticulture rapidly shifting away from the long-established Auckland region to Gisborne, Hawke's Bay and Marlborough? Few believed that outstanding red wine would soon be flowing from this lovely, sprawling island in Auckland's Hauraki Gulf.

The key to the island's wine magic is its warm, dry climate. As Stephen White, the owner of Stonyridge Vineyard, has written: 'We believe that wine quality is primarily determined by the vineyard and the climate it enjoys. Like France, New Zealand has a marginal climate for growing

Cabernet Sauvignon, so I was looking for an area with higher summer temperatures (26–32°C), a maritime influence, low rainfall in January, February and March, and a poor but free-draining soil . . . I chose Waiheke.'

Although the Goldwaters at Putiki Bay pioneered the new era of Waiheke wine, they were not the first vintners the island had attracted. Old, rambling hybrid vines are tangible reminders of the Gradiska family winery, which produced both fortified and a trickle of table wines around the 1950s, until a series of personal tragedies overtook the family and winemaking ceased.

Goldwater Estate and Stonyridge Vineyard, the island's longest-established and highest-profile wineries, have recently been joined by a third red-wine specialist, Peninsula Estate. Other ventures are also getting off the ground. Nick Delamore, whose vineyard lies at Owhanake Bay, near Oneroa, picked his first, tiny Chardonnay crop in 1990; since 1991 his Chardonnay grapes have been sold to Goldwater.

Barry Fenton picked his first crop of Cabernet Sauvignon and Merlot at Oneroa in 1992, which was vinified at Stonyridge Vineyard. 1993 brought the first vintage at the Dunleavy and Buffalora families' three-hectare vineyard. Planted in Cabernet Sauvignon, Merlot and Cabernet Franc, this vineyard is separated from Stonyridge Vineyard only by the width of a narrow airstrip.

The viticultural potential of Great Barrier Island, further out in the Hauraki Gulf, is also being explored. Ex-Aucklander John Mellars has given up a twenty-year career in computer consultancy to plant Cabernet Sauvignon, Merlot, Pinot Noir and Chardonnay on a steep hill overlooking Okupu Bay. Mount St Paul, a nearby landmark, will feature on Mellars' labels.

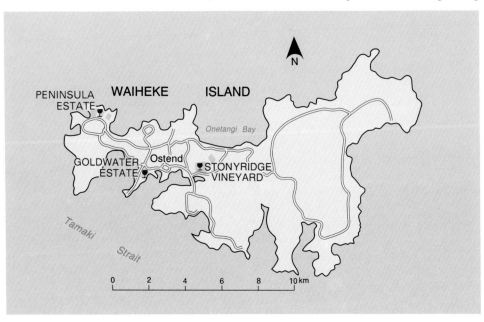

Goldwater Estate

Goldwater Estate
Causeway Road, Putiki Bay, Waiheke Island

Owners: Kim and Jeanette Goldwater

Key Wines: Cabernet/Merlot/Franc, Delamore Chardonnay, Marlborough Chardonnay

The modern flush of viticultural enthusiasm on Waiheke Island was triggered in 1978 when Kim Goldwater, now aged fifty-seven — a former engineer and fashion photographer — and his wife Jeanette planted their first experimental vines in poor, sandy clay soils on the hillside overlooking Putiki Bay.

The Goldwaters, who were fuelled as students by Babich and San Marino dry reds, recall how they were later 'seduced by the Mediterranean lifestyle and especially the idea of serving wine with food every day'. The Goldwaters are a warm and hospitable couple who take obvious delight in one another's company. Jeanette manages the vineyard and expertly handles the company's public relations ('I enjoy the people'); Kim's winemaking ambition is 'to win

international recognition for making one of the world's great wines'.

Under the initial guidance of viticultural expert Dr Richard Smart, the Goldwaters have established a three-hectare vineyard of classic Bordeaux varieties: Cabernet Sauvignon, Merlot and Cabernet Franc. Sauvignon Blanc was uprooted in 1992. When the fruit from recent vine plantings comes on stream in the mid-1990s, Goldwater's output will double to about 3500 cases.

'When you pass fifty, the physical side gets harder,' says Kim Goldwater, who is moving to a less 'hands-on' role and increasingly immersed in paperwork. 'But to employ someone, you have to increase your income.' Hence the Goldwaters' decision to raise their output. Allan McWilliams, an Auckland-born Roseworthy College graduate with winemaking experience in France, Italy, Australia and Oregon, assisted with the production of the Goldwaters' wine in 1991 and 1992, but has since departed.

A quartet of Cabernet Sauvignons was released from the 1982 to 1985 vintages, all labelled as straight varietal wines. Each shared the Goldwater stamp of deep, near-opaque colour and a rich, although slightly leafy-green, palate. For their mouthfilling body and sheer intensity of flavour, these wines quickly staked out a position among New Zealand's foremost reds.

Commencing with the 1985 vintage — and climaxing with the glorious 1990 — the Goldwaters have marketed a Cabernet/Merlot/Franc blend whose quality has far out-stripped its predecessors. Matured for twelve to eighteen months in fifty percent new French oak barriques, this is a strapping yet elegant red, brambly, spicy, richly flavoured and taut.

In Kim Goldwater's eyes, how has his wine's style evolved since 1982? 'The rough edges have been knocked off. To begin with we were preoccupied with enormity; now we aim for elegance.'

Goldwater likes to drink his reds young, 'while they're exuberant and tannic and gutsy and brilliantly hued.' To broach them at less than five years old, however, is to ignore their obvious cellaring potential. Both the 1989 and 1990 vintages are of arresting quality.

Goldwater's wood-fermented, lees-matured Sauvignon Blanc Fumé typically displayed ripe, tropical-fruit flavours but very restrained varietal characteristics; the last vintage of this label was 1992. Goldwater's two white wines in future will be a fat, peachy-ripe, soft Chardonnay made from Nick Delamore's fruit grown near Oneroa and — primarily for export — a crisp, citric-flavoured Marlborough Chardonnay.

The Goldwaters relish their island, wine-enveloped lifestyle. 'I've been amazed,' says Jeanette. 'When we planted those first vines, we had no idea our lives would become so wondrous and so good.'

Kim and Jeanette Goldwater are firmly established in the vanguard of New Zealand's red-wine producers.

Peninsula Estate
52 Korora Road, Oneroa, Waiheke Island

Owners: Doug and Anne Hamilton

Key Wine: Cabernet/Merlot

On their breathtakingly beautiful, wind-buffeted site on Hakaimango Point, overlooking Big Oneroa Beach, Doug and Anne Hamilton produce a muscular, deep-flavoured red in the classic Waiheke mould.

Doug Hamilton (50), a distinctive figure with his broad-brimmed, floppy hat, was raised in Pukekohe and spent his early career as a mechanic. Anne, his wife, is a life-long Waiheke Island resident. When the Hamiltons purchased their Korora Road, Oneroa property in 1984, they intended only to graze horses on it.

Doug Hamilton's interest in wine was kindled by working the harvest at Goldwater Estate. After studying horti-culture, he concluded 'the only damn thing that you could grow up here was grapes'. The Hamiltons planted their first vines in 1986; three years later Peninsula Estate's first vintage was ensconced in barrels.

So exposed is the Hamiltons' elevated coastal vineyard, their young vines had to be irrigated — an extraordinary requirement in Auckland. The 2.5 hectare, sloping vine-yard of friable clays over broken rock is planted in Cabernet Sauvignon, Merlot, Cabernet Franc and Malbec.

In the winery Doug Hamilton built in 1988, the annual grape crush is only about fourteen tonnes — equivalent to a thousand cases of wine. Hamilton and his full-time assistant winemaker, Christopher Lush, hand-pick their fruit; macerate the skins in the wine for a lengthy (up to two weeks) period following the fermentation; and then mature their red for twelve to fifteen months in

Peninsula Estate

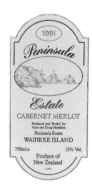

Doug and Anne Hamilton are true specialists, producing only a single, blended, claret-style red. Their Peninsula Estate Cabernet/Merlot is dark-hued and weighty, with the rich, ripe flavours typical of Waiheke Island wines.

French and American oak casks. Only a single wine is currently produced, but in future, says Hamilton, 'we may select.'

Looking back, Anne Hamilton admits 'it's been a lot harder than we expected. It's the physical labour; Jeanette [Goldwater] tried to warn us how hard it is physically. Netting the entire vineyard in summer is back-breaking —

there are heaps of birds and not enough vineyards.'

Peninsula Estate is a big, generous red, not yet in the same league as those of Goldwater and Stonyridge, but impressively savoury, spicy and full-flavoured. 'In the future I want to make a wine as good as [Stonyridge] Larose '87,' says Hamilton. 'From what we've seen of our vines, they're capable of that.'

Stonyridge Vineyard

Stonyridge Vineyard
Onetangi Road, Waiheke Island

Owner: Stephen White

Key Wines: Larose Cabernets, Airfield Cabernets

Stephen White, the driving force behind the high-flying Stonyridge Vineyard, near Onetangi, is a thirty-five-year-old Aucklander who graduated with a Diploma of Horticulture from Lincoln College. After working in vineyards in Tuscany and California, and at Chateaux d'Angludet and Palmer in Bordeaux, White returned to New Zealand to pursue his high ambition: 'to make one of the best Médoc-style reds in the world.'

Stonyridge, named after a nearby hundred-metre-high rocky outcrop, lies only one kilometre from the sea. White and his early partner, Dr John McLeod — who withdrew from the venture in 1987 — planted their first vines here in 1982 on poor, free-draining clay soils saturated with manganese nodules. Kiwifruit failed here but an olive grove has flourished.

The Stonyridge vineyard runs to three hectares of north-facing, predominantly Cabernet Sauvignon and Merlot vines, with smaller amounts of Cabernet Franc and Malbec. The vines are all trained in classic Bordeaux fashion, and the berries purposely kept small — giving proportionately more skins to their juice — in a bid to build deeper flavour into the finished wine. The vines only yield about five

tonnes of grapes per hectare — well below the New Zealand average for the chosen varieties. White is adamant: 'With heavier-cropping systems, you can't achieve the flavour concentration.' During 1992 and 1993 the vineyard was extended, allowing Stonyridge to grow, in White's words, from 'an extremely small winegrowing operation to a very small winegrowing operation'.

In his elegant terracotta concrete and timber winery, with its three processing levels and underground barrel cellar, White specialises in the production of claret-style reds. 'From the start I deliberately sought to produce a Bordeaux-style blend using low cropping, long skin maceration, a second label to allow "selection", barrique-aging and selling "en primeur". Many of these were firsts for New Zealand.'

Inspired initially by Te Mata's 1982 and 1983 reds, and more recently by Villa Maria's reserve reds, Stephen White produced his top wine, Larose — named after the many roses growing in the vineyard at Stonyridge — which now rivals those big names at the top of the New Zealand red wine ladder. The critical applause for this power-packed, minty, excitingly concentrated red has been loud, and Larose has an equally illustrious record in blind tastings. Why is it so successful?

'The vineyard is great,' says White. 'It's north-facing, gets lots of sun, is well-drained and of low fertility. We're very careful about our viticulture. And we have a high-quality philosophy throughout the winemaking process.'

Matured in seventy percent new French oak barriques, Larose is a consistently powerful yet stylish red, dark hued and massive in superior vintages, and even in lesser years attractively scented and ripe. After his legendary 1987 Larose, which thrilled White with its 'huge extract yet cool-climate intensity', he rates the 1990 and 1991 vintages as the finest (with the 1993 potentially 'the best ever').

The second-string label, Airfield, named after a nearby landing strip, is used 'as a back-up for poorer-performing rows, or for a poor year'. In top vintages like 1990 and 1991, no Airfield is marketed. Stonyridge's total annual output is only about 1200 cases, predominantly sold on an 'en primeur' basis whereby customers pay about nine months in advance of delivery in return for a hefty discount.

Where to from here? 'The wines should get better and better,' says White. 'The new plantings should be amazing, reflecting what I've learned running the existing vineyard over the last decade. I'm very bullish about the future.'

The massive, minty 1987 Larose catapulted Stonyridge into the ranks of the country's leading red-wine labels — and subsequent vintages have enhanced the early reputation. Stephen White has simultaneously built Stonyridge into a wine lover's mecca, with vineyard and winery tours, accommodation and a weekend verandah café.

SOUTH
AUCKLAND

[Map showing South Auckland region with locations: Devonport, Waitemata Harbour, Tamaki, Balmoral Rd, Mt Albert Rd, MONTANA, Panmure, Howick, Pakuranga Rd, Manukau Harbour, Otahuhu, Great South Rd, East Tamaki Rd, VILLA MARIA, Mangere, Massey Rd, Papatoetoe, AUCKLAND INTERNATIONAL AIRPORT, Manukau Harbour, Pahurehure Inlet, Papakura, Papakura Interchange, ST NESBIT, Drury. Scale 0-10km. North arrow marked N.]

St Nesbit Winery
Hingaia Road, Karaka

Owner: Dr Tony Molloy, QC

Key Wines: St Nesbit, Rosé

How many of New Zealand's Cabernet Sauvignon-based reds produced outside Hawke's Bay or Waiheke Island can foot it in the quality stakes with the top wines of those acclaimed red-wine regions? St Nesbit is a rare example. The release of the first 1984 vintage, with a gold medal under its belt, heralded the arrival of tax lawyer Dr Tony Molloy, QC, as a force to be reckoned with among this country's red-winemaking fraternity.

St Nesbit is a small winery tucked away on a southern arm of the Manukau Harbour in Hingaia Road, Karaka,

only a couple of kilometres from Auckland's southern motorway. Molloy bought a fourteen-hectare dairy farm here in 1980, and two years later planted his first vines in Karaka's fertile, loamy soils. His passion is 'to set new standards of excellence in making New Zealand wine'.

Tony Molloy (49) is an outgoing, razor-sharp man, the owner of a study lined with a thousand and one books on the law and winemaking. In naming his winery, Molloy has chosen to canonise his late grandfather, a former New Zealand cricket captain, Nesbit Snedden.

With the demands of Molloy's high-flying legal career, what triggered his decision to plunge into winemaking? 'My uncle, who was the Bishop of Wellington, and my local parish priest, both acquired a taste for wine in Italy and passed it on to me. But in the late 1970s there seemed to be so few decent reds in New Zealand. And when an overseas friend said New Zealand's top reds "didn't taste like wine",

St Nesbit

Tax lawyer and textbook author Dr Tony Molloy, QC, fashions a stylish and subtle claret-style red at his tranquil property in South Auckland. Molloy markets St Nesbit after lengthy ageing in barrel and bottle; the 1988 vintage was on sale in 1993.

it rankled. I thought: we must be able to do better.'

Molloy handles the workload of being a winery owner with apparent ease. Having a fulltime vineyard worker and cellarhand is essential. 'I take time off for vintage. We pick with friends — there are lots of judges and QCs — and I also take the next few days off. The rest is done in the evenings, around midnight, or at dawn.'

For the 1984 vintage, Molloy was totally reliant upon bought-in fruit. Later, both home-vineyard and contract-grown grapes were processed. Since 1989, however, the crush has been exclusively of estate-grown fruit.

The St Nesbit vineyard is planted on a relatively cool site by Auckland standards; Molloy sometimes — as in 1990 — picks his Cabernet Sauvignon at the end of May. By reducing the vines' yields and leaf-plucking, he is moving to eliminate the slight vegetative characters noticeable in some vintages. The 4.5-hectare vineyard is predominantly planted in Cabernet Sauvignon, with smaller plots of Merlot, Cabernet Franc, Petit Verdot — a dark, peppery but late-ripening traditional Bordeaux red-wine variety — and Malbec.

In pursuit of Molloy's goal of 'greater flavour concentration', since 1989 at least a quarter of the bunches have been cut from the vines before flowering. With their average yield of not much over four tonnes per hectare, St Nesbit's vines are cropping at about half the New Zealand average for the planted varieties.

The assaults of rabbits and birds (which destroyed over half of the 1989 Cabernet Sauvignon crop), and losses of vines to hormone sprays, have also combined to keep St Nesbit's output small. After an allocation has been exported

to Germany and the United Kingdom ('The legal world supports my wines well, both here and in the United Kingdom,' reports Molloy), usually only about 250 cases are left for release in New Zealand.

Molloy's roomy (600 square metres) winery started off, he recalls, as 'a new floor put in an old cowshed. Then we put a new cowshed over the floor.' Despite this modesty, the St Nesbit winery is in reality highly sophisticated. Molloy makes the wine himself.

To enhance St Nesbit's individuality, Molloy ferments his wine with natural 'wild' yeasts. New oak plays a vital role; since 1986 all the wine has been matured in brand-new or freshly shaven barrels, typically for two years. In future Molloy plans to market a lower-priced red, based on bought-in fruit, to find a use for his one-year-old casks. A sturdy, wood-matured rosé has also been launched from the 1992 vintage.

St Nesbit is an impressively deep-scented wine, with a complex, strongly oak-influenced bouquet that is strikingly Bordeaux-like. It has a slightly lighter mouthfeel than the top reds of Hawke's Bay and Waiheke Island, but is consistently elegant and full-flavoured.

In St Nesbit's extraordinarily comprehensive newsletter, Molloy reveals the depth of his love affair with wine. 'By wine, food is transmuted from mere fuel into an occasion which enhances our humanity and our wellbeing. When the wine is New Zealand wine, and the food is fresh New Zealand food, the occasion becomes an epiphany; a celebration; a revelation of our fortune in living in a country providing such space, sunshine, and food and wine to gladden the human spirit.'

Villa Maria Estate
5 Kirkbride Road, Mangere

Owners: George Fistonich and Ian Montgomerie

Key Wines: Reserve Barrique Fermented
Chardonnay, Res. Marlborough Chardonnay, Res.
Gewürztraminer, Res. Sauvignon Blanc, Res. Noble
Riesling, Res. Cabernet Sauvignon, Res. Cabernet
Sauvignon/Merlot; Cellar Selection Chardonnay,
Sauvignon Blanc, Cabernet Sauvignon/Merlot,
Shiraz; Private Bin Chardonnay, P.B. Sauvignon
Blanc, P.B. Gewürztraminer, P.B. Rhine Riesling,
P.B. Chenin Blanc/Chardonnay, P.B. Müller-
Thurgau, P.B. Cabernet Sauvignon, P.B. Pinot
Noir

What thoughts does the name 'Villa Maria' conjure up in
your mind? I think instantly of two things: top-flight
Reserve Chardonnays and Cabernet-based reds; and the
triumphal thirty-year expansion of the wine interests of co-
owner George V. Fistonich.

How big is Villa Maria? It claims a ten percent share of
the domestic market for New Zealand wine. In the eyes of
Montana or Corbans, Villa Maria is relatively small fry.
Yet Villa Maria is clearly the third-largest winery in the
country, much bigger than any of the medium-sized producers like Nobilo, Babich or Matua Valley.

The origins of Villa Maria lie in a tiny operation called
Mountain Vineyards, which was run as a hobby by Dalmatian immigrant Andrew Fistonich. Fistonich worked on
the gumfields, then later made a few bottles of wine for
himself and friends before becoming a licensed winemaker
in 1949. When illness slowed him down, his son George
abandoned his career plans in carpentry, leased his father's
vineyard, formed a new company, and bought a press,
barrels and pumps from Maungatapu Vineyards at
Tauranga. In 1961, Villa Maria Hock nosed out into the
market.

The winery initially made its presence felt at the bottom
end of the market. The slogan 'Let Villa Maria introduce
you to wine' associated with the sale of sherries and
quaffing table wines, created an image the company for
years struggled to overcome. But in recent years Villa
Maria has established an illustrious track record in wine
competitions.

'I started off selling at the gate,' recalls Fistonich, 'but
then we branched out to supply half-a-dozen wine shops.
In the early days, when we were not known, a lot of
retailers took the attitude they had never heard of Villa
Maria. I travelled virtually the whole of the North Island,
and we made contact with strong and independent people
who supported us right from the start. Eventually, as our
popularity grew, we put on a full-time sales manager in
1968. We kept concentrating on getting outlets throughout
New Zealand, and then we started an advertising campaign

in 1970. We were pretty much doubling our sales every
year at that stage.'

Villa Maria expanded rapidly through the 1970s —
absorbing Vidal in 1976 — and early 1980s, emerging as a
fierce rival of the largest wineries such as Montana and
Corbans. John Spencer of the Caxton group of companies
was then a silent but substantial shareholder.

At the height of the wine industry's price war late in
1985, Villa Maria slid into a much-publicised receivership.
With its limited capital reserves, the winery was simply
unable to survive in the heavy loss-making trading environment created by its larger rivals. It was rescued by a capital
injection from its new part-owner, Grant Adams, then
deputy chairman of the investment company Equiticorp.
(In July 1991 Adams sold his fifty percent share in Villa
Maria to Mangere grapegrower Ian Montgomerie, a long-
time Villa Maria supplier.) Barely a year later, Villa Maria
astounded observers by absorbing the Bird family's ailing
Glenvale (now Esk Valley) winery. Villa Maria was on the
comeback trail.

The team at Villa Maria is young and vigorous. Ian
Clark, the indefatigable public relations manager, generates
a wave of publicity every time Villa Maria picks up an
award at home or abroad. Steve Smith, who trained under
viticultural guru Dr Richard Smart, liaises closely with the
company's grapegrowers in a bid to constantly raise the
quality of their fruit.

Villa Maria owns an eight-hectare vineyard in Gisborne,
which customarily supplies the fruit for the company's
prestigious Barrique Fermented Chardonnay. It also owns
the estate vineyard at Esk Valley, and in 1992 planting
began at a new company vineyard at Flaxmere, near
Gimblett Road, inland from Hastings. But it relies on contract grapegrowers for the vast majority of its fruit.

Villa Maria's grape purchasing scheme, organised by
Steve Smith, is designed to give growers 'a real incentive
to produce quality'. If a grower's Chardonnay grapes are
selected by Villa Maria for one of its Reserve labels, the
winery might pay $1500 per tonne; if the grapes are down-
graded to a 'commercial' range like Forest Flower, they
would fetch about $600 per tonne.

Villa Maria's top wines, marketed under a Reserve label,
have of late enjoyed a phenomenal run of gold-medal and
trophy-winning successes. Its Reserve Chardonnays and
Cabernet Sauvignon-based reds are indisputably among
New Zealand's most outstanding wines. Their quality is
matched by the top-flight Reserve wines from the two other
wineries in the Villa Maria group's stable, Esk Valley and
Vidal.

At the 1993 Liquorland Royal Easter Wine Show, for
instance, the group scooped the trophies for Champion
Wine of the Show (Vidal Reserve Cabernet Sauvignon/
Merlot 1990), Best Red Wine — High Price (Vidal Reserve
Cabernet Sauvignon/Merlot 1990), Best Red Wine —
Average Price (Villa Maria Cellar Selection Cabernet
Sauvignon/Merlot 1991), and Winemaker of the Show
(George Fistonich).

Villa Maria

Grant Edmonds, winemaker at Esk Valley between 1989 and 1993, was recently appointed winemaker of Villa Maria.

The winemakers in the three wineries are encouraged to strive to outperform each other. For marketing reasons, however, it is important that each label has a consistent style. Take Sauvignon Blanc. The Villa Maria Private Bin Sauvignon Blanc is designed to be 'light and racy', with early-picked fruit giving it a tangy herbaceousness; the Vidal wine is a medley of herbaceous and riper fruit flavours; the Esk Valley version places its accent firmly on ripeness and richness.

Grant Edmonds (37), formerly winemaker at Esk Valley, was appointed chief winemaker of the Villa Maria/Vidal/ Esk Valley empire in 1993. Directly beneath the chief winemaker in the Villa Maria production pyramid are a trio of winemakers based at Mangere, Vidal and Esk Valley. The Vidal and Esk Valley winemakers prepare their wines to a 'ready for bottling' state, before they are tankered by road to Villa Maria for their final filtering and bottling.

Villa Maria's top end of the range Chardonnays, the Reserve Barrique Fermented and the Reserve Marlborough, afford an absorbing style contrast. 'In the Barrique Fermented,' Kym Milne, chief winemaker from 1987 to 1992, observed, 'we are using a lot of new oak, not quite 100 percent, but a lot. Big and rich, based on Gisborne fruit, this is a high extract, weighty sort of wine. With Marlborough we are aiming at a much more elegant, fruit-driven style, only about fifty percent new oak, and about four to five months in oak.' Intensely aromatic, fat and savoury, the Reserve Barrique Fermented Chardonnay is consistently arresting in its youth and has a formidable track record in show judgings. Leaner, invigoratingly crisp and stylish, the Reserve Marlborough Chardonnay is more slowly evolving.

Three other white wines carry Villa Maria's prestigious Reserve label. The Reserve Gewürztraminer, usually produced from fruit grown at Ihumatao near Auckland International Airport, in the mid-1980s was pungently aromatic and richly seasoned — the finest in the country. Vintages since 1987, however, have displayed reduced flavour concentration; this is now a good, but not great, wine.

Villa Maria's Reserve Sauvignon Blanc is made from its ripest fruit, fermented in new German oak barriques. This wine exhibits substantial body, a hint of lees-aging complexity and deep, rich tropical-fruit flavours. The Reserve Noble Riesling, launched from the 1991 vintage, is a succulent, thrillingly perfumed, honey-sweet beauty.

The highlights of the second-tier Cellar Selection range, launched in 1992, are the Marlborough Chardonnay and Sauvignon Blanc and the Hawke's Bay Cabernet Sauvignon/Merlot. These excellent wines place their accent on fruit quality — the grapes are sourced from some of Villa Maria's top vineyards — with markedly less oak influence than the Reserve wines.

The quality of Villa Maria's third-tier Private Bin range has improved sharply in recent vintages. At its best, the low-priced Private Bin Müller-Thurgau is a delightfully floral, light wine with a perfect balance of slight sweetness, fruitiness and acidity. As an aperitif, this is hard to beat. With its enjoyable melon-like flavours, freshness and crisp finish, the Private Bin Chenin Blanc/Chardonnay (a 70/30 blend) is a good everyday-drinking dry wine.

The Private Bin Gewürztraminer is a Gisborne-grown wine, fruity and lightly spiced with a touch of sweetness. The garden-fresh, herbal and tangy Private Bin Sauvignon Blanc is a Marlborough-grown, tank-fermented style.

With its light citrus-fruit flavours and restrained oak, the Private Bin Chardonnay, grown in Gisborne, is a no-fuss style ideal for early consumption. The latest addition to the ranks of Villa Maria's Private Bin white wines is the Rhine Riesling, launched in 1991. This is a deep-scented, impressively flavour-packed Marlborough wine, its hint of sweetness balanced by vigorous acidity.

The sextet of varietal wines marketed by Villa Maria under the Forest Flower label offer enjoyable, flavoursome drinking at a bargain price. Under the St Aubyns label, Villa Maria produces an even lower-priced range of bottled wines for no-fuss quaffing.

The distinctive black labels of Villa Maria's reserve range are a key force among New Zealand's top reds. The Reserve Cabernet Sauvignon has captured a formidable string of gold medals. Formerly Kumeu-grown, but now based on Hawke's Bay fruit, it is matured for eighteen months in a mix of French and American oak barriques. This bold, dark hued, full-flavoured red exhibits the sweet perfume of American oak, great depth of herbal, cassis-like flavours and a tight tannin grip. Its voluptuous Reserve Cabernet Sauvignon/Merlot stablemate is equally rich and complex, with mouth-encircling, spicy, plummy flavours and the firmness of structure to flourish long-term in the bottle. These are very distinguished wines — a match for any reds in the land.

Villa Maria Private Bin Cabernet Sauvignon — in 1991 labelled as a Cabernet/Merlot/Franc — is a chunky, full-flavoured red that in riper vintages like 1991 and 1989 delivers fine value for money. This claret-style red has recently been partnered by a briefly wood-aged Private Bin Pinot Noir — light, strawberryish, appealingly supple and round.

George Fistonich

With his quiet manner and slow drawl, it would be easy to underestimate George Fistonich. He also avoids the limelight, happily stepping back in favour of his winemaking team. Yet Fistonich, Villa Maria's managing director, ranks among the wine industry's most powerful figures.

George's father, Andrew Fistonich, arrived in New Zealand from Dalmatia just before the Depression. George (53) spent his childhood in the little brick house still standing next to the winery; today it's used by visiting winemakers. After leaving school at sixteen, Fistonich initially headed for the building trade — 'In many Dalmatian families the oldest son was sent to university and the second son was expected to be a tradesman' — before he plunged into the wine industry.

Today, George Fistonich likes being involved in all aspects of the company. 'Some GMs are accountancy-orientated; I'm not. I find people much more fascinating. I love the element of psychology in business; it surfaces in marketing, negotiating and motivating. I'd be bored as a small-scale winemaker. There's much more scope for all that interests me in a large company.'

Villa Maria's output peaked in the early to mid-1980s, when it crushed around 6000 tonnes of fruit; today the scaled-down company processes about 5000 tonnes each vintage. Its former strong presence in the price-sensitive, bulk cask-wine market has been deliberately reduced, leaving only the Maison Vin range.

In its pursuit of the more profitable, bottled wine market, Villa Maria has evolved a three-tier range of 'Reserve' (low output), 'Cellar Selection' and 'Private Bin' (in fact readily available) wines, with its lesser wines released under the St Aubyns, Forest Flower and Old Masters labels. The overall marketing goal, says Fistonich, is 'to provide a comprehensive range of wines for consumers at different price-points for different occasions. With our several brand levels, we're able to rigorously select the best of each year's crop for our top labels.'

Villa Maria was a conspicuous absentee from the large group of New Zealand wineries which swung into export in the mid-1980s. 'Before launching into export as a top-quality producer, we needed to standardise our wine styles and develop new labels,' says Fistonich. 'Long term, that's paid off. We're now shipping to the UK (by far our largest export market), Finland, Sweden, Norway, Canada, Hong Kong, Singapore, Tonga, Samoa and Fiji.'

In the eyes of Kym Milne, his former chief winemaker, Fistonich is 'an entrepreneur, an ideas man, rather than someone who gets bogged down in paperwork. When he was thinking about buying the old Glenvale winery, it looked to me like a concrete jumble, but he could see its potential. And he loves getting out in the market and doing deals. Above all, he's a great "people" person.'

Looking back over three decades at 'the Villa', what gives Fistonich the deepest satisfaction? 'Improving our quality with the right people and vineyards. It's all come together for us in the last five years. We now have an overriding desire to be the best.'

WAIKATO/BAY OF PLENTY

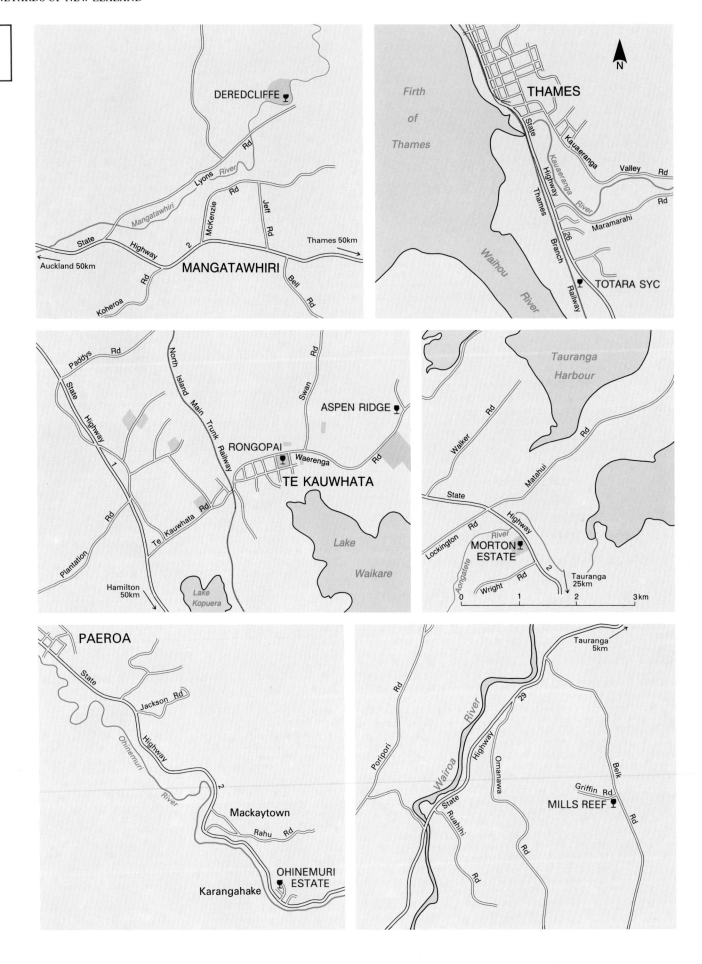

The Waikato, bounded to the east by the Coromandel Ranges and to the west by the Tasman, is languishing as a wine region. The foundation of the government Viticultural Research Station at Te Kauwhata in 1897 gave an early boost to grapegrowing and by 1982 growers had 336 hectares, or 5.7 percent of the country's total plantings, under vines. With the southwards drift of viticulture, however, the area in vines has since plummeted; in 1992 only 161 hectares of vines (2.6 percent of the national vineyard) were planted there.

At Te Kauwhata, the 'space age' Cooks winery erected in the early 1970s and the old Viticultural Research Station have recently closed. Rongopai at Te Kauwhata, deRedcliffe in the Mangatawhiri Valley and — to a lesser extent — Totara near Thames are still producing fine wines, but new ventures are rare in the Waikato.

Aspen Ridge is a very small winery with a six-hectare vineyard near Lake Waikare, east of Te Kauwhata township. The company was established in 1963 by Alister McKissock — who also directed the Te Kauwhata research station from 1963 until his resignation in 1966 — with the assistance of Nathan's liquor interests. Today the wines, which are 'bottle-aged [sometimes for over a decade] and sold ex-vineyard to a clientele of aficionados', are only plain. Aspen Ridge also markets a range of grapejuices and highly regarded grape jellies. The winery was for sale in 1993.

Another tiny winery with a local following is Vilagrad, founded by Ivan Milicich Snr at Ngahinepouri, just south of Hamilton, in 1922. Nelda and Pieter Nooyen, the third generation of the family, now produce a range of table and fortified wines, including a Müller-Thurgau/Breidecker, Chardonnay, Sauvignon Blanc and Cabernet Sauvignon. Most of the wine produced from Vilagrad's five-hectare vineyard and bought-in fruit is consumed at social functions held at the winery.

Awaiti Vineyard is a relatively new label. The Pett family, orchardists and grapegrowers, cultivate Cabernet Sauvignon and Chenin Blanc at their three-hectare vineyard in Awaiti South Road, near Paeroa. The Cabernet Sauvignon and Chenin Blanc recently marketed under their own brand were made by Nick Chan of Lincoln Vineyards.

Grape yields are lower in the Waikato's clay soils than in the fertile Gisborne plains. The region also shares West Auckland's climatic disadvantages: although temperatures and sunshine hours are as high as in classic European table wine regions, the average rainfall and humidity are much higher. The outlook is bleak for the Waikato's viticultural future. As Tom van Dam of Rongopai Wines at Te Kauwhata put it: 'We're in the middle of everywhere and the middle of nowhere.' In the Bay of Plenty to the east, the area under vines is tiny. Yet the 1989 emergence of Mills Reef, near Tauranga, to join Morton Estate near Katikati, has doubled the number of wineries in the region.

---

deRedcliffe Estates
Lyons Road, Mangatawhiri Valley

Owner: Otaka Holdings (NZ) Limited

Key Wines: Sémillon/Chardonnay, Sémillon/Sauvignon Blanc, Sauvignon Blanc, Hawke's Bay Chardonnay, Mangatawhiri Chardonnay, Rhine Riesling, White Lady, Coral Reef, Cabernet Sauvignon/Merlot

'Where in the world,' asks the advertisement, 'can you find a stone-walled winery as grand as a château, in a river valley carpeted with vines and roses, where visitors are welcome every day to tour the barrel hall and watch the winemakers at work, or join a study group to enhance their knowledge and appreciation? And where in the world can you stay on for lunch, dinner or a weekend of fine food and relaxation in an atmosphere that inspires the senses?' The answer is Hotel du Vin, at the beautiful, isolated deRedcliffe vineyard in the Mangatawhiri Valley.

The founder of deRedcliffe, Wellington-born Chris Canning, returned to New Zealand in 1975 after many years running an advertising agency in London and New York, having also tended vines in France and been part-owner of a vineyard in Italy. In a natural basin among the Hunua's bush-clad hills, Canning began in 1976 to plant his own vineyard: 'I wouldn't say it was a case of a man of destiny who had to make wine. The reason I bought the place was to escape foreign capital gains tax.'

Canning's entrepreneurial talents — and those of his wife Pamela — were again demonstrated in 1987 by the rise at deRedcliffe of the $8 million Hotel du Vin, an accommodation and conference complex set amidst native trees, which boasts a top class restaurant. The hotel is principally aimed at Aucklanders who are 'forty-five minutes away by road, much less by helicopter'. To fund this major development, deRedcliffe Group Limited, incorporating the vineyard, winery and hotel, was floated on the stock exchange in 1987, with Canning the majority shareholder.

Following early financial difficulties — including a $485,000 trading loss in the year to 30 June 1989 — the deRedcliffe Group was purchased by a Japanese-owned company, Otaka Holdings (NZ) Limited. This was the first Japanese investment in the New Zealand wine industry. Mr Michio Otaka, the principal shareholder, is 'a genial and very observant politician who made his fortune in the construction industry,' says winemaker Mark Compton. Chris Canning soon after resigned as managing director.

The Hotel du Vin was enlarged and upgraded. The main building had grown out of the original concrete block and timber house on the property. According to Chris Canning, for the new owners the hotel represented 'a wholly European experience. Significantly, they saw nothing "South Pacific" or "Oceanic" in the concept of a country hotel and

deRedcliffe

The lovely hill-ringed deRedcliffe Estates vineyard is principally planted in Chardonnay. Winemaker Mark Compton, who joined deRedcliffe in 1987, produces distinctive Sémillon-based white wines, recently partnered by a fresh, penetrating Marlborough Sauvignon Blanc and Rhine Riesling.

winery . . . Plaster erased the Kiwiana blocks and the roofline was punctuated with an Etruscan tower.'

Chardonnay is the principal variety in the silty, six-hectare estate vineyard skirted by the Mangatawhiri River, with smaller plots of Cabernet Sauvignon, Sémillon and Pinot Noir. Most of deRedcliffe's fruit, however, is bought from growers in Marlborough, Hawke's Bay and Auckland.

Until the 1986 vintage deRedcliffe wines were crushed and fermented elsewhere but barrel-aged at the vineyard. Now the wines are crushed and fermented at the new on-site winery. 'Of all the [original] buildings which failed to fulfil [Japanese] expectations, the winery was pre-eminent,' says Canning. 'Practical yet charmless, it committed the ultimate sin of being clad in corrugated iron!' In late 1990 rose a handsome, spacious stone-walled winery with exposed redwood and cedar timbers, high arches and vaulted ceilings. Viewing gantries and informative story boards permit self-conducted tours along the mezzanine floor. This stylish, extremely visitor-orientated winery is unlike any other in New Zealand.

Responsible for the 15,000-case annual production is forty-year-old Mark Compton, a Wellington-born BSc, who followed his three-year winemaking stint in Australia with a long period at Montana. Compton concedes that before his 1987 arrival deRedcliffe wines were 'good solid commercial wines', but rarely special. 'Today the pressure is on the winemaker to match the culinary efforts of the hotel'.

'I've learned to really like Sémillon,' says Compton. 'We're becoming a Sémillon specialist.'

DeRedcliffe Sémillon/Chardonnay is a medium-bodied wine with lemony/herbal flavours in a crisp, easy-drinking style. DeRedcliffe Sémillon/Sauvignon Blanc is a softly mouth-filling wine, blended from two-thirds Sémillon and one-third Sauvignon Blanc, and matured in French and German oak casks. With its lush, ripe, herbal fruit fleshed out with toasty wood, this is a rich-flavoured dry wine with loads of character.

Two styles of Chardonnay are produced. The Hawke's Bay Chardonnay is the most distinguished — a barrel-fermented, lees-aged wine with rich, complex, buttery-soft flavours. The estate-grown Mangatawhiri Chardonnay is 'more of a Chablis style' — light, delicate and lemony.

Two other highlights of deRedcliffe's white-wine range are the Sauvignon Blanc and Rhine Riesling. The Marlborough Sauvignon Blanc is a medley of riper and early-picked fruit with penetrating herbal flavours and bracing acidity. The medium-dry Rhine Riesling, also from Marlborough, is packed with classic lemon/lime varietal aromas and flavours, in a mouthwateringly crisp style.

The silky Cabernet/Merlot — of which the delightful 1980 vintage first established deRedcliffe's reputation — is a stylish light red. Blended from Hawke's Bay and Mangatawhiri fruit, it is matured for twelve months in predominantly American oak casks, emerging ripe, scented and soft.

Mills Reef Winery
Belk Road, Tauranga

Owners: The Preston family

Key Wines: Chardonnay, Elspeth Chardonnay,
Sauvignon Blanc, Gewürztraminer, Dry Riesling,
Mills Reef

'If you can make good wine out of kiwifruit, you can make good wine out of anything,' a top winemaker once told me. Warren 'Paddy' Preston, of Preston's kiwifruit wine renown, has recently turned his hand to grape winemaking under the Mills Reef label — with instant success.

Winemaking is a rare pursuit in the Bay of Plenty: Morton Estate at Aongatete, near Katikati, is the only other significant winery in the region. The Mills Reef winery lies in Belk Road, in the lower Kaimais west of Tauranga.

Paddy Preston plunged into grape winemaking in 1989 because, he says, 'it's a bit of a challenge, and the whole family were keen to make something well that's more widely recognised than fruit wines.' Winemaking at Preston's/Mills Reef is truly a family affair. Paddy, a gentlemanly, modest fifty-seven-year-old, now concentrates on making the Mills Reef grape wines. Three of his children are also immersed in the business. Warren (34) handles the marketing of both the Preston's and Mills Reef ranges. Tim (31) has recently assumed responsibility for the fruit-wine production, and Melissa (27) runs the office.

Mills Reef's objectives are 'to make the highest quality wine and maintain mid-range prices'. The wines are labelled in honour of Charles Mills, Paddy's great-grandfather, who spent several years at sea and as a miner — hence the rather imaginative reference to the reef.

Mills Reef owns no vineyards. The company is not committed to any particular growing region, but so far has bought all its grapes from Hawke's Bay. With its annual crush of about 130 tonnes of grapes, equivalent to 9000 cases of wine, Mills Reef is a small winery by New Zealand standards.

The winery building started life as a cowshed, to which a cool-store, tank-room and bottling-room have been added. When the Mills Reef grape-wine venture was launched, 'we already had most of the gear, except for barrels,' recalls Paddy.

The Preston family is moving to keep its grape- and fruit-winemaking activities clearly apart in the public's eye. A new, specialist grape winery will open at Bethlehem, north of Tauranga, in 1994.

The Mills Reef range is very tight: 'We thought we'd master white wines first, and later make reds,' says Paddy Preston. Only six wines have been marketed, all white (the first Cabernet-based red will flow from the 1993 vintage). These are uniformly well-crafted and deep-scented wines with delicate, lingering varietal flavours.

Mills Reef

'Chardonnay is definitely my favourite wine,' enthuses Preston, 'the one I'm especially determined to crack.' Mouthfilling, with peachy-ripe, lush fruit flavours, subtle lees-aging complexity and a buttery-soft finish, Mills Reef Chardonnay is typically forward and deliciously easy-drinking in style. Elspeth Chardonnay, based on a selection of the winery's best Chardonnay grapes and cask-matured longer than the 'standard' Chardonnay, is named in honour of Paddy Preston's mother.

By harvesting the grapes at two ripeness levels — with herbaceous and fully ripe, tropical-fruit flavours — and fermenting the lot in French oak barrels, Paddy Preston produces a very stylish Sauvignon Blanc. With its lovely, vibrant fruit flavour underpinned by a hint of toasty oak, this is a well-balanced and skilfully made wine.

Mills Reef Dry Riesling is a high-scented, off-dry wine with crisp, delicate citrus-fruit flavours. The Gewürztraminer is fruity and dryish, with delicate but clearly defined varietal spiciness and lemony hints.

Labelled simply 'Mills Reef', Paddy Preston's sparkling is deliciously rich and creamy — a celebration of Chardonnay fruit flavours. By barrel-fermenting the base wine, followed by a secondary bottle-fermentation, Preston produces a distinctively full-flavoured, complex and stylish sparkling.

Do the Prestons feel out on a limb, with only one other winery based in the Bay of Plenty? 'No. It's an advantage being here. We're very close to the Auckland market and there's a big population in the Bay of Plenty.' Mills Reef wines are sold in retail outlets around the country and in the United Kingdom, at the winery, and by mail-order. 'We're very pleased with progress,' grins Paddy Preston.

Paddy Preston, previously a champion fruit-wine maker, has turned his talents to grape winemaking under the Mills Reef label — with instant success. Mills Reef's lush, creamy-soft Chardonnays and Chardonnay-based Méthode Champenoise reflect Preston's fascination with the variety.

Morton Estate

Morton Estate Winery
State Highway 2, Aongatete, via Katikati

Owner: Appellation Vineyards Limited

Key Wines: Yellow Label Chardonnay, Cabernet Sauvignon; White Label Chardonnay, Sauvignon Blanc, Fumé Blanc, Riesling, Cabernet Sauvignon; Black Label Chardonnay, Cabernet/Merlot, Merlot; Méthode Champenoise, Morton

For several years after its first vintage in 1983 the Morton Estate winery in the Bay of Plenty sold all its wine with impressive ease. The gold medal success of its 1983 'White Label' Chardonnay spurred interest, the striking Cape Dutch-style winery on the highway near Katikati, attracted widespread attention, and winemaker John Hancock had already built a cult following during four spectacularly successful vintages at Delegat's. Today, the market is tighter and the going is a bit tougher at Morton Estate. 'When we released our 1985 'Black Label' Chardonnay it rocketed out of the winery,' recalls Hancock. 'Now the stocks last twelve months.' Yet in other ways little has changed at Morton Estate. Hancock is still in the driver's seat and the wine quality is as good as, indeed better than, ever before.

John Hancock is an affable, irrepressible character, but beneath the popular 'bonhomie' image, there's an ambitious, much more serious side to his personality. His swift climb through the technical ranks of the Australian wine industry over a decade ago bears ample witness to his talents.

Born in South Australia, Hancock (42) spent much of his childhood 'out in the bush' on his family's sheep farm near Coonawarra. The teenage Hancock began experimenting with fruit wine-making, working with mulberries, peaches and apricots. After leaving school he wanted to study agricultural science, but he was also eager to learn winemaking. At Roseworthy College he studied both, and in 1973 he graduated with a Diploma in Oenology.

Hancock now plunged into the wine industry as assistant winemaker at the Lindeman's-owned Leo Buring winery in the Barossa Valley. Later he was appointed one of two winemakers at the giant Berri co-operative winery in the Riverland. 'Then Mr Delegat advertised for a winemaker.'

Hancock signed up for two vintages at Delegat's but stayed for four (1979–1982), making an immediate impact by scoring a gold medal with Delegat's 1979 Riesling-Sylvaner. When he heard someone was planting vines at Katikati, he thought it was 'madness'. Morton Brown, the entrepreneurial founder of Morton Estate, offered him a partnership. Hancock refused. Brown persisted; Hancock grew to admire his attitude. 'Morton wasn't a wine man at all, but he was very progressive, very marketing orientated, and prepared to do whatever was necessary.'

John Hancock of Morton Estate produces an exceptional bottle-fermented sparkling wine and dark, ripe-tasting Hawke's Bay reds. It is on Chardonnay, however — especially the increasingly restrained and refined, top-end-of-the-range Black Label — that Morton Estate's reputation rests.

Like his earlier shift to Delegat's, Hancock's move to Morton Estate met with immediate show success. 'We hit the Chardonnay market at the right time and grew with it.'

In 1988 Morton Brown sold his shares in Morton Estate to Mildara Wines of Australia and Hancock sold his own shares in the winery to Mildara, in return for shares in the Australian company. The Australians had little detailed input into Morton Estate. In mid-1993 the winery was purchased by Appellation Vineyards Limited, returning Morton Estate to New Zealand ownership.

As Morton Estate's general manager, these days Hancock is deeply immersed in paperwork and public relations. The winemaking here is now largely the domain of Steve Bird, another Roseworthy graduate who arrived in 1986.

Hancock has enormous faith in Hawke's Bay — most Morton Estate wines are grown in that region. The small, two-hectare vineyard at the Bay of Plenty winery produces Pinot Noir fruit used in the bottle-fermented sparkling. However, the company-owned Riverview vineyard at Mangatahi — inland from Ngatarawa in Hawke's Bay — is far more important.

Here Morton Estate has planted Chardonnay, Sauvignon Blanc, Pinot Noir, Cabernet Sauvignon, Merlot and Syrah vines. This is a late-ripening vineyard by Hawke's Bay standards.

My favourite wine in the Morton Estate range is its eye-catchingly deep-flavoured Méthode Champenoise. Says Hancock: 'We went into bottle-fermented sparklings because I was bored and needed something to do. We saw there was a gap for a premium bubbly, so decided to have a crack at it.' Lees-matured for at least two years, this is an impressively delicate, complex wine with a stylish bouquet and persistent 'bead'. Its flavours are fine and dry, the mature yeast autolysis characters giving subtlety and depth. This is one of the country's most outstanding sparkling wines. It has recently been partnered by Morton, a lower-priced, non-vintage, bottle-fermented sparkling, which is light and fruity in style, with much more restrained yeast characters.

Chardonnay pours out of the Morton Estate winery; of its annual 40,000-case output, over 15,000 cases are Char-donnay. The style is evolving. 'From the 1991 vintage I'm placing more emphasis on fruit flavours, with less augmentation with winemaking techniques,' says Hancock. 'This means less malolactic fermentation influence, and no overuse of wood.'

Morton Estate's rich and peachy-ripe White Label Chardonnay is by far its most popular label. Made from Hawke's Bay fruit, fermented and matured for nine months in thirty percent new French oak casks, it is more restrained than its lush Black Label big brother, but a proven performer in the cellar and a fine wine in its own right. The non-wooded Yellow Label Chardonnay, launched from the 1990 vintage, is low-priced and designed for early consumption.

Mouthfilling and complex, with layers of intense stone-fruit and wood flavours, the outstanding Black Label Chardonnay is the consummation of John Hancock's life-long love affair with Chardonnay. Based on Riverview vineyard fruit, it is fermented and lees-aged for twelve months in new French oak barriques. This is a justly celebrated wine. Once golden, very substantial and huge in flavour, its style has changed in recent vintages, with greater emphasis on flavour delicacy, tightness and longevity.

Of Morton Estate's three Sauvignon Blancs, the White Label Sauvignon Blanc is a non-wooded Hawke's Bay wine, fresh, tangy and gently herbaceous. It is partnered by a ripe, briefly oak-aged White Label Fumé Blanc; Hancock sees this as 'a better match with food'. The pick of the trio is the barrel-fermented Black Label Fumé Blanc, a delicious wine brimming with ripe tropical-fruit and toasty French oak flavours. The 1991 vintage, however, was the last of this label.

In the past the reds failed to stand out. Since the emergence of its top-flight Black Label Hawke's Bay reds from the 1989 vintage, however, Morton Estate finally has red wines it can be proud of. The Black Label Cabernet/Merlot and Merlot are both voluptuous reds — bold, very ripe and full-flavoured.

Morton Estate's exports hold the key to its future growth. Already fifteen to twenty percent of its production is sold overseas, in the United Kingdom and the United States, Australia, Germany and Holland.

Ohinemuri Estate Wines
Moresby Street, Karangahake

Owners: Horst and Wendy Hillerich

Key Wines: Riesling, Gewürztraminer, Chardonnay, Sauvignon Blanc Oak Aged, Classique, Pinotage

The rearing, bush-tangled slopes of the Karangahake Gorge, between Paeroa and Waihi, are an extremely unlikely site for a winery. The terrain is too precipitous for grape-growing. Yet amidst these rugged hills Horst and Wendy Hillerich have recently established their distinctive, charming chalet-style winery — Ohinemuri Estate.

Horst Hillerich (34) was born near Frankfurt in Germany and served his winemaking apprenticeship between 1976 and 1978 at Weingut Hulbert, a six-hectare family estate in the Rheingau. After two years' study at Veitshochheim University in Franconia, Hillerich graduated with the equivalent of a BSc in viticulture and winemaking. He arrived in this country in June 1987 at the invitation of James and Annie Millton, owners of The Millton Vineyard near Gisborne.

'We had met when I worked next door to him in the Rheinhessen,' recalls James Millton. 'I liked him because he wasn't always as straight or correct as the others — he was more socially relaxed. But his winemaking is very clean and tight.'

Ohinemuri Estate

Horst Hillerich, who served his winemaking apprenticeship in the Rheingau, wants Ohinemuri Estate's wine bar to be known 'for good wine and as a good place to go. And it's important to us to keep contact with our customers.'

Hillerich spent his first few months in New Zealand working for the Milltons. 'I'll always remember the day he told me off about our winery hose,' says Millton. 'In Germany he'd been taught that after using a hose, you should always neatly roll it up. I told him that when I'm busy there isn't time. He replied that if you're flat out, you should be able to unroll a hose knowing it won't be in a tangle. He's a very precise winemaker.'

Later that year Hillerich met his Nelson-born wife, Wendy. From 1988 until early 1990 he worked under winemaker Gilbert Chan at Totara Vineyards. In 1989, while still working at Totara, he crushed and fermented the first wines under his own Ohinemuri Estate label.

The Ohinemuri Estate winery is delightful. A replica of a Latvian chalet, built on the banks of the Ohinemuri River in 1985, the buildings originally served as a house, stables and hay-loft. The Hillerichs fell in love with the complex

at first sight and have lovingly preserved its atmosphere. Customers in the winery shop call for service with the aid of an Austrian cow-bell.

Ohinemuri Estate is a winery, not a vineyard. Hillerich draws his grapes from Hawke's Bay, Gisborne and the Waikato. A few demonstration rows of vines will be cultivated near the winery, but only for 'image' reasons, says Hillerich. 'It's too wet.'

Horst and Wendy Hillerich are pouring extraordinary energy into Ohinemuri Estate. Their first wines were crushed and fermented in Auckland but matured in the Karangahake Gorge. After 'seemingly endless bureaucratic and contractor delays', their winery shop finally opened to the public on 20 December 1990.

For the 1992 vintage, the Hilleriches converted their garage into a temporary winery — concreting the floor, insulating the walls, moving tanks inside. The harvest was crushed in Auckland, then the juice was transported to the Karangahake Gorge for fermentation, storage and bottling.

Ohinemuri Estate will always be a very small company; each year Hillerich only intends to produce about 2000 cases. A new and larger winery, using gravity-feed systems, was built for the 1993 vintage; with its steep-pitched roof, locals call it 'the church'. The Hillerichs also plan to open a wine bar and sell two-thirds of their output on the site.

Hillerich's wines are of rock-solid quality. Ohinemuri Estate Riesling, which — unusually for New Zealand — is briefly oak-aged, is deep-scented and fruity with strong lemon/lime flavours and a fresh, off-dry finish. The Sauvignon Blanc Oak Aged (formerly labelled Fumé Blanc) is a robust Hawke's Bay wine, its ripe gooseberry and herbal fruit flavours enhanced by restrained oak handling.

My favourite is the Pinotage, a delicious 'nouveau' style in which strong, fresh raspberryish fruit flavours hold sway. About thirty percent of the fruit is 'whole bunch fermented' — using the traditional Beaujolais technique — giving the wine its very appealing suppleness and softness.

Rongopai

Rongopai Wines
71 Waerenga Road, Te Kauwhata

Owners: Tom and Faith van Dam

Key Wines: Riesling, Botrytised Selection, Botrytised Riesling, Botrytised Chardonnay, Sauvignon Blanc, Te Kauwhata Chardonnay, Chardonnay, Cabernet Sauvignon/Merlot, Merlot, Pinot Noir, Syrah

Rongopai burst unheralded onto the wine scene in 1986 with a 300-case release of three botrytised wines at a standard — and at prices — that compelled widespread interest. Today sweet wines are still the highlight of the range, but Rongopai also produces an interesting array of dry whites and reds.

Rongopai — meaning 'good taste' or 'good feeling' — winery in Waerenga Road, Te Kauwhata, was founded by Dr Rainer Eschenbruch (54) and Tom van Dam (38), who worked together at the Te Kauwhata viticultural research station. German-born Eschenbruch is a tall and angular, vastly experienced winemaker, holder of a PhD from the Geisenheim Institute. After several years at Stellenbosch Farmers' Wineries in South Africa, he arrived in this country in 1973 to take charge of the programme of wine-making research at Te Kauwhata.

Van Dam is an MSc graduate who after starting work at Ruakura in 1978 ended up as the officer-in-charge at Te Kauwhata. He and Eschenbruch formed their partnership in 1982, but in early 1993 Eschenbruch withdrew from the company. Van Dam's wife, Faith, is also involved in marketing their small annual output of 6000 cases in New Zealand, the United Kingdom and Germany.

A derelict winery built in 1919 by Thomas Hutchinson,

who also planted a vineyard on the Rongopai block of land at Te Kauwhata, still stands on the property, its original concrete fermentation vats long empty. Another recently renovated winery now used by van Dam was built between the First and Second World Wars by Lou Gordon, who also operated under the name Rongopai. 1988 was the first vintage processed by van Dam at his own winery; earlier crops were crushed at St Nesbit.

On loam-clay soils sloping gently to the north — earlier covered in Albany Surprise vines, which he promptly bulldozed — van Dam has established a two-hectare vineyard of Riesling, Chardonnay, Cabernet Sauvignon, Bacchus and Scheurebe (the latter two for late harvest styles), densely planted to lower the vines' vigour. He also owns another two-hectare vineyard not far from the winery and purchases grapes from Te Kauwhata and Hawke's Bay.

Rongopai's fame hinges on its late-harvested, botrytised sweet whites, a style clearly traceable back to Rainer Eschenbruch's German origins. These are not produced every vintage: Cyclone Bola intervened in 1988; 1989 was too hot and dry for 'noble rot' to spread; 1990 was too wet. The delectable botrytised wines of 1991 and 1992 were the first since 1987.

Rongopai Botrytised Selection (previously labelled Late Harvest) is blended from a quartet of German grape varieties — Bacchus, Scheurebe, Würzer and Müller-Thurgau — which ripen early, by the end of February, but are held on the vines to shrivel until late March or early April. With its heady scent, concentrated fruit flavours and slender body (the alcohol level can dip below eight percent) this is a highly seductive wine, all sweetness and delicacy.

The summit of the Rongopai range is undoubtedly the opulent Botrytised Riesling, previously labelled Riesling Auslese. Tom van Dam reports that 'the ladies who hand-pick our botrytised grapes never believe we could possibly fashion a decent wine out of such "nobly rotten" fruit — until they taste the finished product!' This is a ravishingly perfumed sweet Riesling enhanced by its thrillingly intense botrytis character. The rarer Botrytised Chardonnay, launched from the 1991 vintage, is equally honeyish and succulent.

Rongopai Riesling is an off-dry style with delicate citrus-fruit flavours and an emerging hint of honey. Two Chardonnays are marketed. Made from very ripe fruit, which is barrel-fermented and lees-matured for nine months, the Te Kauwhata Chardonnay is the pick of the two — this is a delicious, peach and apricot-flavoured wine. The lower-priced 'standard' Chardonnay — two-thirds stainless steel-fermented and one-third barrel-fermented — is an easy-drinking style with fresh, buoyant stone-fruit flavours and a soft finish.

Fat and creamy-soft, with ripe tropical-fruit rather than strongly herbal flavours, Rongopai Sauvignon Blanc is a very distinctive style — a world apart from the rapier-like Sauvignons of Marlborough. 'I don't like herbaceous Sauvignon Blancs,' says van Dam. 'It's also climatic; the further north you go they tend to lose that anyway. We pick our fruit as ripe as possible, avoid skin contact, and use partial barrel fermentation, lees-aging and malolactic fermentation. All this adds to the wine's complexity and tones down its herbaceousness.'

Rongopai also produces a solid lineup of Cabernet Sauvignon, Merlot, Pinot Noir and Syrah-based reds from Te Kauwhata fruit. The Rongopai range is not sold directly from the winery, but in 1992 the company opened a shop near the main highway in Wayside Road, Te Kauwhata.

Totara

Totara Vineyards SYC
Main Road, Thames

Owners: Gilbert, David and Ken Chan and
Kevin Honiss

Key Wines: Totara Café, Fu Gai, Kiwifruit Liqueur,
Winemakers Reserve Sauvignon Blanc, Winemakers
Reserve Chardonnay, Chardonnay, Cabernet
Sauvignon

Ah Chan of Canton — 'Kumara Joe' to most people — planted in 1925 a small plot of Albany Surprise table grapes, together with kumara, to supply the Thames market. Stanley Chan, no relation of Ah Chan although from the same village in Canton, bought the tiny vineyard in 1950, and started winemaking. So began Totara Vineyards SYC.

The name Totara SYC was derived from the Totara Valley — near the winery site just outside Thames — and Stanley Young Chan's initials. Stanley Chan's mother's family brewed and sold rice wine in their Canton shop; his father was a distiller before emigrating to Dargaville at the turn of the century. Today, this low-profile vineyard is run by one of Stanley Chan's sons, Gilbert (52) and Kevin Honiss, who joined Totara in 1990.

Until a decade ago, Totara's reputation hinged primarily on its fortified wines and sweet white table wines. Yet although the company is not widely considered a producer of quality wine, it still enjoys success in competitions: its Winemakers Reserve Chardonnay 1990 won the top Chardonnay trophy at the 1992 Air New Zealand Wine Awards.

Totara has struggled in recent years. Its lengthy involvement in the cask wine market, which only ended in 1990, was 'disastrous', admits Gilbert Chan. The company uprooted all its vineyards in 1986, except for a hectare of Cabernet Sauvignon adjoining the winery, which is retained for image reasons. Almost all the winery's grapes are now purchased from growers in Hawke's Bay.

Totara's current strategy is to 'concentrate as a boutique winery on smaller volumes'. Liqueurs like Totara Café are still 'the mainstay', says Chan, and sherry and port are sold in bulk to retailers. The table and fortified wine ranges have been severely rationalised.

A two-tier labelling system has recently been evolved, with Totara's top wines now being marketed under the Winemakers Reserve label. The quality of his two recent gold and silver medal winners proves Chan's ability to produce wines of high quality. The Winemakers Reserve Sauvignon Blanc 1990 — one-half barrel and one-half tank-fermented — displays ripe, gently herbaceous varietal characters, a touch of complexity and good flavour persistence. The Winemakers Reserve Chardonnay 1990 is powerfully wooded but impressively robust, lush and soft.

Totara Fu Gai has been listed by many Chinese restaurants, although its sales are now slowing. Fu Gai is a Sauvignon Blanc-based, fresh, moderately herbal wine with a whisker of sweetness.

Another feature of this winery is its liqueurs, including Kiwifruit, which is more popular with Japanese tourists than New Zealanders, and the coffee-based Totara Café, reminiscent of the Mexican product Kahlua but sold at a much lower price.

Gilbert Chan is part-owner of one of New Zealand's most distinctive wineries. Out on a limb at the base of the Coromandel Peninsula, Totara is the country's only predominantly Chinese-owned wine company.

Gisborne is the wine industry's bread basket, the prime source of bulk grapes for its hungry cask production lines. Yet according to local grapegrowers, who in 1991 erected bold signs alongside the two major roads into the city, Gisborne is also the 'Chardonnay capital of New Zealand'.

Gisborne's traditional non-glamorous image partly stems from the relative paucity of fully self-contained wineries here; the Gisborne wine trail currently extends only to Matawhero, The Millton Vineyard and Landfall (supplemented by winemaker Phil Parker's 'Smash Palace' wine bar in the city). The upmarket image of these 'boutique' vineyards has tended to be overwhelmed by the broader picture of Gisborne as a 'plain Jane' region. Thus the tag 'carafe country' has been easy to apply.

Yet numerous top-flight wines are produced in Gisborne. Villa Maria Barrique Fermented Chardonnay, Robard & Butler Gisborne Chardonnay and Revington Vineyard Chardonnay have all enjoyed eye-catching success in show judgings. The Millton Vineyard's Riesling Opou Vineyard and Chenin Blanc are consistently outstanding; at its best Matawhero Gewürztraminer is arrestingly aromatic and deep-flavoured. Gisborne has had an undeservedly bad press.

The East Cape, dominated by the Raukumara Range, has only limited lowland areas suitable for viticulture. Grapegrowing is confined to the Poverty Bay flats around Gisborne, which form the largest of the coastal alluvial plains, and to smaller ones further north at Tolaga Bay and Tikitiki, and to the south near Wairoa.

Friedrich Wohnsiedler pioneered winemaking in Gisborne after a false start by Marist missionaries, who landed by mistake at Turanganui (Gisborne) in 1850 and there planted vines before departing for their original destination, Hawke's Bay.

Wohnsiedler, born on a tributary of the Rhine, arrived in New Zealand around the turn of the century. When patriots laid waste his Gisborne smallgoods business during the 1914–18 war, Wohnsiedler moved out and onto the land, planting vines at Ormond in 1921. His first vintage, a sweet red, was labelled simply as 'Wine'.

When Wohnsiedler died in 1956, his Waihirere vineyard covered only four hectares. (His name lives on, of course,

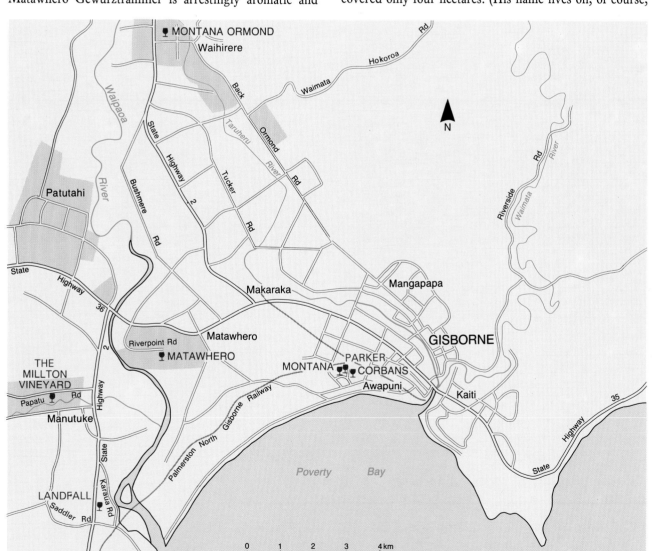

on the label of Montana's Wohnsiedler Müller-Thurgau.) In 1961, a rapid expansion programme began which, after a series of financial restructurings, saw the Wohnsiedler family eventually lose control. By 1973 Montana had completely absorbed Waihirere.

From a paltry acreage of vines supplying the old Waihirere winery, since 1965 viticulture has swept the Gisborne plains. Corbans and Montana between them have three large wineries in the area (not open to visitors), and by 1992 Gisborne had 24.6 percent of New Zealand's total area in vines.

Gisborne is white wine country. Müller-Thurgau, Chardonnay and Muscat Dr Hogg are — in that order — the region's three foremost varieties. Run your eye down a list of the ten leading grape varieties planted in Gisborne and you will not find a single red. 'The easterly sea breezes which come in at about eleven o'clock in the morning and drop at about three in the afternoon cool things down,' says James Millton, Gisborne's most accomplished winemaker. Here the later-ripening red varieties tend to become swollen, at the cost of flavour and colour intensity: with the odd exception, Cabernet Sauvignon has not performed well in Gisborne.

The doubts over grape quality centre principally on the fact that although the vines get ample amounts of sunshine and heat, the highly fertile soils and plentiful autumn rains combine to produce both excessively dense vine-foliage growth and bumper crops. Professor H.W. Berg, a Californian viticultural expert hired as a consultant by Montana in 1973, found New Zealand 'an extremely fertile country. Crops of thirteen tons to the acre [33 tonnes per hectare] on three-year-old vines at Gisborne are absolutely amazing, as are seventeen and eighteen tons per acre [43–45 tonnes per hectare] on four-year-old vines . . . Forget about the deep, fertile soils . . .'

Millton agrees that the region's soils have 'the ability to produce a lot of grapes. The soils have good water-holding capacity, but in dry years the wines have plenty of flavour.' And the soil types vary from silt to sand to clay: 'every kilometre it's a different situation. Those who control the vegetative nature of the vines can do better.'

The rainfall during the critical February-April harvest period averages seventy percent higher than in Marlborough, and thirty-three percent higher than in Hawke's Bay. In the past, this has encouraged growers to pick their crops as soon as an acceptable sugar level was achieved, in order to avoid bunch rot.

Selected vineyards, however, are now employing a variety of viticultural techniques to achieve fruit quality far above the norm. By planting on less fertile clay soils; selecting devigorating rootstocks; planting more phylloxera-resistant vines (the bug has made rapid inroads into Gisborne's vineyards since its discovery there in 1970); planting new, improved clones; planting virus-free vines; plucking leaves to reduce fruit shading; later harvesting to advance ripening; and a range of other approaches, Gisborne viticulturists have of late been exploring more fully their region's wine potential.

Millton believes the region's finest wines will eventually come from the surrounding hill country. 'The best sites are in the foothills on the eastern side of the Ormond Valley. Here you'd get better drainage and more air movement to reduce humidity. Unfortunately, people don't like growing things on the hills.'

Landfall

Landfall Wines
State Highway 2, Manutuke

Owner: John Thorpe

Key Wines: Chardonnay, Gewürztraminer, Pinot Noir, Méthode Champenoise

Landfall, a relatively recent arrival on the Gisborne wine scene, received a major shot in the arm when its Revington Vineyard Chardonnay 1989 won the trophy for champion Chardonnay at the 1990 Air New Zealand Wine Awards. In 1993, however, the winery's ownership structure and range were significantly altered.

The name of the winery, sited on the main highway at Manutuke, south of Gisborne, commemorates the nearby landfalls of early Maori canoes and Captain Cook. Founding partners John Thorpe (38) and Ross Revington (39) met while playing in the Gisborne Civic Orchestra (Revington on the double bass, Thorpe on the flute). Born in Gisborne, in his twenties Thorpe helped to set up his family's horticulture blocks, and has previously produced fruit wines and mead.

Revington, a lawyer, came to Gisborne in 1982 ('partly influenced by Matawhero's success') and in 1987 he and his wife, Mary Jane, purchased an established four-hectare vineyard in the Ormond Valley, now called Revington Vineyard, which previously had supplied the grapes for several gold medal-winning Cooks Chardonnays. 'Kerry Hitchcock, the winemaker, used to say "the fairies live there",' recalls Revington.

The Landfall partnership — originally called White Cliffs — was formed in 1989. Thorpe was the 'practical, hands-on person' who made the wine; Revington, who still pursues a full-time law career, saw his role as keeping an 'overview'. Recently, however, Revington withdrew from the partnership, leaving Thorpe as the sole owner of the winery (although not of the Revington Vineyard).

The silty eight-hectare vineyard adjoining the winery is planted entirely in Pinot Noir. Landfall also buys fruit from other growers in the region. The roomy, corrugated iron winery started life as a packing shed, to which the partners added a pergola and vineyard bar serving fresh luncheons seven days per week.

The dual highlights of the early Landfall range were the Revington Vineyard Chardonnay and Gewürztraminer.

Barrel-fermented and lees-matured, the Chardonnay is savoury and lemony; subtle rather than sledgehammer in style. The Gewürztraminer is a powerful dry wine with impressive length of citric-fruit and spicy flavours. With Revington's departure from the partnership, however, these wines will no longer be part of the Landfall range.

Landfall's lower-priced initial releases featured a dry Sauvignon Blanc, Chardonnay, Gisborne Red and Pinot Noir, and a slightly sweet Müller-Thurgau and Pinot Noir Blush. These were all light, pleasant and easy-drinking in style, although lacking the distinction of the Revington Vineyard pair. A promising new range of Longbush wines (the label is owned by Thorpe's brother, Bill) was launched from the 1992 vintage, with the full-flavoured, buttery-soft Chardonnay the pick of the bunch.

In future Thorpe would like Landfall's annual output to climb to about 8000 cases, concentrating on Chardonnay, Gewürztraminer, Pinot Noir and a bottle-fermented sparkling.

John Thorpe (left) and Ross Revington established the Landfall partnership in 1989, but Revington withdrew in 1993. Thorpe has tightened the Landfall range and also produces wine for his brother Bill's Longbush label.

Matawhero Wines
Riverpoint Road, Matawhero

Owner: Denis Irwin

Key Wines: Gewürztraminer, Chardonnay, Sauvignon Blanc/Sémillon, Chenin Blanc/ Chardonnay, Cabernet Sauvignon/Merlot, Pinot Noir, Syrah, Bridge Estate

Matawhero a decade ago enjoyed a reputation second to none in New Zealand for its handling of the Gewürztraminer variety. Today Matawhero's star has dimmed, but the wines flowing from the end of Riverpoint Road can still be absorbing.

Company head Denis Irwin, in his mid-forties, is one of the great individualists of the New Zealand wine industry, a man whose rollercoaster career has reflected his unique personal blend of innovative winemaking, entrepreneurial business style — and love of letting his hair down. Dubbed the 'Matawhero Maniac' in some quarters, he is reputedly proud that Gisborne, the first city in the world to greet the sun each day, enjoys 'fresh sunlight'. According to an Australian report Irwin once rang Robert Mondavi, the famous Californian winemaker, and told him that California wasn't any good for wine because by the time the sunlight got there it was pretty used up. 'After telling him that, I knew I had his full attention.'

Matawhero's thirty-hectare vineyard surrounding the winery is planted predominantly in Gewürztraminer and Chardonnay, with smaller plots of Sauvignon Blanc, Chenin Blanc, Cabernet Sauvignon, Merlot, Malbec, Syrah and Pinot Noir. The nearby two-hectare Bridge Estate vineyard is entirely planted in red-wine varieties, principally Merlot. Much of the crop is sold to larger wine companies, but

Irwin has first choice of the available fruit.

A restless spirit, Irwin spent most of the 1980s across the Tasman, unsuccessfully attempting to establish a second winery in Victoria. In his absence, Hatsch Kalberer, a tall, gentle Swiss who, Irwin says, 'came wandering down Riverpoint Road escaping the nuclear holocaust,' assumed the day-to-day responsibility for running Matawhero. Kalberer, who has recently departed after nine vintages at Matawhero, fitted Irwin's winemaking approach: 'I'd loathe to have anyone on my place who'd spent more than two months at a wine college.'

Matawhero's white wines of the 1980s — unlike the delightfully perfumed, incisively flavoured wines on which the company's fame was built in the mid–late 1970s — often displayed blurred varietal characteristics. 'We're still training and returning to a philosophy that goes back to the basic roots of the art of winemaking; using our own yeasts and not releasing any wine before it's aged,' says Irwin. By fermenting with 'wild' rather than cultured yeasts, making extensive use of malolactic fermentation, and lengthy wood handling, Matawhero has sought to produce, as Kalberer put it, 'wines to enjoy drinking, rather than impress'. But the standard of the wines — judged as 'style' rather than 'varietal' wines — has been uneven: sometimes impressive, sometimes poor.

So far, Matawhero's success has hinged primarily on Gewürztraminer. The Gewürztraminer at its best is everything that wine of this variety should be: pungent, very aromatic, unmistakably spicy in taste. The first vintage, 1976, was made in a converted chicken shed and scored a silver medal at that year's National Wine Competition. The 1978 scored a gold medal. Recent vintages, often matured in seasoned oak casks, have been more subdued in spiciness than their ebullient predecessors, but still enjoyable wines — weighty, dry and soft.

Matawhero Chardonnay, although barrel-fermented,

Matawhero

Denis Irwin, after a long sojourn in Victoria, is back down the end of Riverpoint Road, where the Matawhero legend was born. Whether Irwin can restore Matawhero's former prestige will be intriguing to trace.

given a full malolactic fermentation, lees-aged, held up to eighteen months in wood and high-priced, has been of unpredictable quality. The Sauvignon Blanc/Sémillon, an eighty/twenty blend, displays restrained herbal flavours; the Chenin Blanc/Chardonnay is mouthfilling, ripe and soft.

Matawhero's Cabernet Sauvignon/Merlot is Irwin's flagship red. At its best this is a rewarding wine with deep colour, generous flavour and a strong tannin backbone. The Pinot Noir is unusually dark hued, robust and full-flavoured — more reminiscent of the Rhône than Burgundy. Matawhero Syrah is also a gutsy red with warm, ripe, chocolatey fruit flavours. Bridge Estate, a single vineyard blended red with a strong Merlot influence, is attractively fleshy and supple. Matawhero's four reds are among the finest produced in the region.

'One certain thing about winemaking,' says Denis Irwin, 'is that you only go out feet-first. Withdrawal is impossible: you've always got young wines, old wines and buds in spring demanding your attention.'

Parker

Parker Méthode Champenoise
24 Banks Street, Gisborne

Owner: Phil Parker

Key Wines: Classical Brut, Dry Flint, Rosé Brut, First Light Red, Juddy's Ruby Port

'Smash Palace', squatting in the midst of Gisborne's factories, according to its roadside sign is 'the oasis in the industrial desert'. In this pink and blue corrugated iron wine bar, featuring an old DC3 positioned as if about to crash-land, and with the towering tanks of Corbans' Gisborne winery a stone-throw away, the fare is 'cheese-boards, quality wine, iced beer, seafoods and music'.

Phil Parker, the owner of Smash Palace, is also a winemaker specialising in a range of high-quality bottle-fermented sparkling wines. Parker (37), who made his first experimental wines in 1987, says he prefers 'lighter styles of sparkling wines — for their drinkability'. His grapes are all hand-picked and only lightly pressed, and the wines spend a minimum of two — and up to three — years maturing on their yeast lees.

Parker's annual output of sparkling wines is low — about 1000 cases — which allows him to offer customers what he terms 'designer wines'. 'If you would like me to add more or less sugar, more or less cognac (I add about one ml to each bottle) please state it with your order and I will oblige. This is a service only I can deliver, possibly because of the small production volumes and because I'm the only one to have thought of it.'

Parker has a flair for promotion: at a function to launch his first releases in 1989, a toy yacht was smashed across a six-litre bottle of Parker Classical Brut 1987. The Classical Brut is his flagship. Its style has vascillated between a Chardonnay and Pinot Noir-predominant blend (the 1990 and 1991 vintages are fifty to sixty percent Pinot Noir) but this is consistently a classy wine — light, with plenty of yeast-derived complexity and impressive flavour delicacy and length.

Parker's non-vintage Dry Flint is a herbal, strong-flavoured, tangy sparkling, based on an unusual blend of Sémillon and Chenin Blanc. The Rosé Brut, blended from Pinotage, Pinot Noir and Chardonnay, is pink, robust and bone-dry.

First Light Red (aptly named, Gisborne being close to the international dateline) is Parker's 'cash-flow' wine. Released each year in late May — only three months after the harvest — this Pinotage-based red is produced by the 'carbonic maceration' or 'whole-bunch fermentation' technique perfected in Beaujolais. This is a buoyant light red, its accent on fresh, strong raspberryish flavours.

Some of Parker's wines are largely produced for con-

sumption at the Smash Palace wine bar. Juddy's Ruby Port, a 'light, user-friendly port', with some whole-bunch fermentation to 'jump up the fruit flavours', is drier than usual in a light, fruity, undemanding style.

Phil Parker is a specialist in characterful, full-flavoured bottle-fermented sparklings.

The Millton Vineyard
Papatu Road, Manutuke

Owners: James and Annie Millton

Key Wines: Riesling Opou Vineyard, Te Arai River Sauvignon Blanc, Chenin Blanc Barrel Fermented, Chardonnay Barrel Fermented, Clos de Ste Anne Chardonnay, Riesling Late Harvest Individual Bunch Selection, Tête de Cuvée, Cabernet Rosé, Te Arai River Cabernet Sauvignon/Cabernet Franc

Rich, honeyish Chenin Blancs and exquisitely floral, vibrant Rieslings are the leading lights of The Millton Vineyard's range. James Millton (37) runs, together with his wife Annie (36), Gisborne's most acclaimed winery on the banks of the Te Arai River at Manutuke, sixteen kilometres south of the city.

Millton initially spent two years with Montana before leaving to pursue his wine career overseas. After working on a small estate in the Rheinhessen, he returned to a vintage with Corbans and then, after opting out of a Roseworthy College winemaking course, went to work on Annie's father's vineyard in Gisborne.

Now the Milltons own four vineyards have spread out over twenty hectares in the Manutuke and Matawhero dis-

tricts. In the original, loam-clay Opou Vineyard at Manutuke, ten hectares of Chenin Blanc, Sauvignon Blanc, Sémillon, Riesling, Cabernet Franc and Cabernet Sauvignon are cultivated. The three-hectare Winery Vineyard is predominantly planted in Chardonnay, Chenin Blanc, Malbec and Merlot. In the silt loams of the six-hectare Riverpoint Vineyard — along the road from Denis Irwin in Riverpoint Road, Matawhero — the Milltons have planted Chardonnay vines. Naboths Vineyard, a 1.5-hectares planting of Chardonnay and Pinot Noir on a steep north-east facing hillside, came into production in 1992. About half of the fruit from these vineyards was formerly sold to other wine companies, but today the grapes are all reserved for the Millton range.

In the conviction that 'we are what we eat', the Milltons have set themselves the difficult task of making organically grown wines in commercial volumes. The 'Bio-Gro' status given to their vineyards by the New Zealand Biological Producers' Council affirms that the Milltons do not use herbicides, insecticides, systemic chemicals or artificial fertilisers.

James Millton readily agrees that Gisborne's warm, moist climate is not well suited to organic viticulture: 'Canterbury's vintners are sitting on a potential gold-mine. But we can do it.' For fungus control, a limited amount of copper sulphate and sulphur is sprayed on the vines, supplemented by waterglass, seaweed extract and vegetable

The Millton Vineyard

James and Annie Millton produce Rieslings and Chenin Blancs equal to any in the land. As organically grown wines they attract much publicity, but their quality is also top-flight.

oils. 'Most importantly, we use bio-dynamic herbal preparations . . . which are applied to the soil, compost and liquid manures.' Parasites and predators are also used to biologically control weeds and insect pests.

'Far from being a convenient way of marketing our wine,' say the Milltons, 'it is the protection of our own health and the environment in which we work that motivates us to pursue this direction.'

Alongside their administration building — an old shearers' quarters — the Milltons erected their coolstore winery in the summer of 1983–84. Today, with its annual output of about 8000 cases of wine, The Millton Vineyard is still a fairly small producer. James Millton oversees the vineyard and the winemaking; Annie 'keeps in contact with labelling, despatch and marketing'.

Chenin Blanc and Riesling are 'the rather unfashionable varieties I want to concentrate on,' says James Millton. In view of Matawhero's eye-catching success with the variety, why not Gewürztraminer too? 'I have a bit planted,' says Millton, 'but not enough to make it commercially. We've enough lines in the range. I didn't start off with Gewürztraminer because Matawhero was already excelling with it.'

The Millton Vineyard's Chenin Blanc Barrel Fermented vies with Collards' as the finest in the land. 'I want honey and acidity and almond flavours,' says Millton — and he gets them. Fermented in seasoned French oak barrels, this is a yellow-hued, delectable dry wine, robust and brimming with rich, pineappley, slightly honeyish flavours.

With its ravishing perfume, and intense citrus-fruit and honeyish flavours underpinned by lively acidity, the medium-sweet Opou Vineyard Riesling is equally outstanding. It is partnered in some vintages by a botrytised, gorgeously scented Riesling Late Harvest Individual Bunch Selection with a thrillingly nectareous flavour. 'Sauternes-style', full-bodied, oak matured sweet whites of high quality are also produced from such varieties as Sauvignon Blanc, Sémillon and Chardonnay.

Millton prefers 'tropical, guava' flavours in Sauvignon Blanc-based whites. His Te Arai River Sauvignon Blanc, both tank and barrel-fermented, displays appealing melon-like flavours with a herbal undercurrent.

The Chardonnay Barrel Fermented is peachy-ripe and soft — ideal for early consumption. It is overshadowed, however, by the subtle Clos de Ste Anne Chardonnay, a single-vineyard wine fermented and lees-aged in new oak casks. This is a stylish rather than powerful wine: creamy soft, with hints of mealy complexity.

In view of Gisborne's modest red wine reputation, The Millton Vineyard's reds are unexpectedly good. 'James is determined to prove Gisborne has red wine potential,' says Annie Millton. His Te Arai River Cabernet Sauvignon/Cabernet Franc is an attractive, full-flavoured amalgam of spicy and berryish fruit flavours, with soft tannins and plenty of depth.

Following a lull in the mid-1980s, when few new labels appeared, recently a host of new wineries has sprouted in Hawke's Bay. One of New Zealand's pioneer winemaking regions, Hawke's Bay has retained its traditional importance: with such prized varieties as Chardonnay, Cabernet Sauvignon and Merlot, its regional reputation is the strongest in the country.

Here is located the oldest winemaking concern in New Zealand still under the same management — Mission Vineyards, established by the Catholic Society of Mary in 1851. The oldest winery still operating, erected in stages from the 1870s, can be found at the Te Mata Estate. In 1992, with 1614 hectares (or twenty-six percent of the national total), Hawke's Bay ranked second to Marlborough in the extent of its vineyards.

The terrain of Hawke's Bay varies from the rugged inland ranges, the Ruahine and Kaweka, climbing to over 1600 metres, to the coastal Heretaunga Plains. In this sheltered environment, protected by the high country from the prevailing westerly winds, agriculture thrives: pastoralism, process cropping, orcharding and market gardening. And on the margins of the plains, at Taradale, Te Mata, Fernhill,

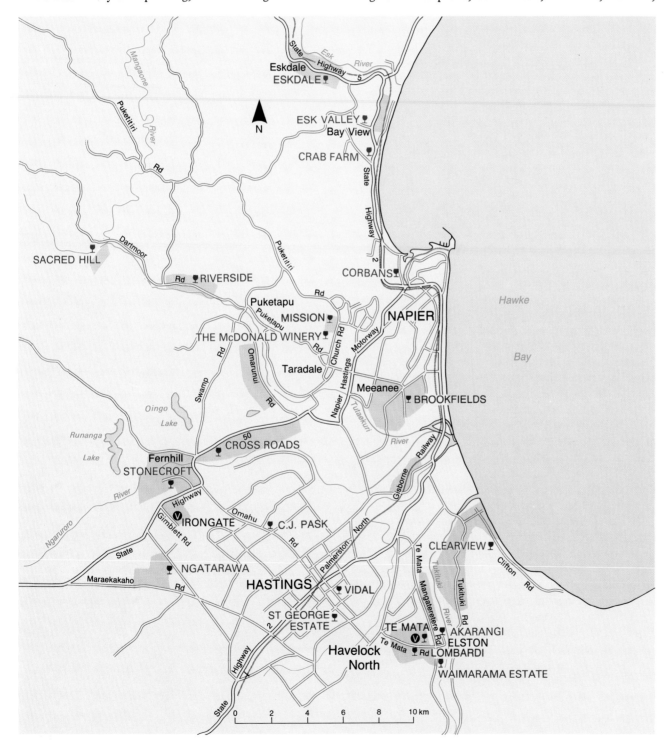

Ngatarawa, Haumoana and in the Esk Valley, the favourably dry and sunny climate supports an easy growth of the vine.

Hawke's Bay is one of the sunniest areas of the country; the city of Napier, for instance, enjoys similar sunshine hours and temperatures to Bordeaux. In summer, anti-cyclonic conditions sometimes lead to droughts; such weather can produce grapes with high sugar contents and forms a key advantage for Hawke's Bay viticulture.

One drawback is that the easterly facing aspect renders Hawke's Bay vulnerable to easterly cyclonic depressions. Some of the heaviest rains ever recorded in New Zealand have descended on the region. In bad years such as 1979 and 1988, the vineyards of Hawke's Bay can be deluged with autumn rains. Nevertheless, in most years the autumn rainfall is markedly less than at Gisborne.

One of Hawke's Bay's prime viticultural assets is its wide range of soil types: the Heretaunga Plains consist mainly of fertile alluvial soils over gravelly subsoils deposited by the rivers and creeks draining the surrounding uplands. 'Seventy percent of Hawke's Bay's vineyards are on the wrong sites,' says Steve Smith, group viticulturist for Villa Maria. 'Between Hastings and the coast — the most fertile area — grapes are too often grown just as cash crops.'

A comprehensive regional study published in 1985 by the Hawke's Bay Vintners stated frankly that 'many soils on the Heretaunga Plains are quite wet and vines grow too vigorously, giving large yields of grapes with poor balance and insufficient ripeness [notably the areas of fertile silty loams having a high water table] . . . Other more freely draining shingle soils . . . may be too dry in the growing season which would limit proper canopy development for ideal fruit maturation. This is overcome with trickle irrigation.'

Districts warmly recommended for viticulture by the regional study included the Taradale hills, river terraces along the Tukituki and Ngaruroro rivers, Havelock North and Ngatarawa (warm, dry and promising to produce grapes 'of the highest quality').

The shingly Gimblett Road area west of Hastings — site of the Irongate, C.J. Pask and other vineyards — has recently excited much interest. Steve Smith — most impressed with the inland parts of the Heretaunga Plains — has pointed out that during vintage, rainfall figures drop steeply from the coast to the Ngatarawa/Gimblett Road area.

Dr Alan Limmer, whose Stonecroft winery lies in Mere Road, just north of Gimblett Road, likens the area 'more to Australia than New Zealand. The climate is much hotter than the rest of Hawke's Bay, because the only cooling factor, the sea breeze, doesn't have the same influence here as it does on the coast. And the temperature warms up much more quickly — we can get temperatures up to 40° C.'

Steve Smith points out that almost no Hawke's Bay vineyards are grown on the hills. 'One of the problems has been the old New Zealand agriculture attitude "you have to have a big tractor", which means that people have not planted these hill sites. Within a couple of years I would certainly like to see our group owning vineyards on some of these hill sites. It would make a huge difference to the wines.'

Only in the past twenty-five years has viticulture reached significant proportions in Hawke's Bay: in the late 1930s, for instance, only twenty-five hectares of vines were grown in the province. Then in 1967 contract growing extended to Hawke's Bay. For decades the dominant force in the Bay was McWilliam's, with its dry white Cresta Doré, sparkling Marque Vue and red Bakano virtually household names. Founded in 1944, the New Zealand company was wholly Australian-owned until 1962, and grew rapidly until 1961 when it merged with McDonald's Wines to become the largest winery in the country. Following a series of mergers and takeovers, the former McWilliam's crushing and fermenting complex at Pandora has become Corbans' major facility in Hawke's Bay.

Yet, although McWilliam's, with a famous series of Cabernet Sauvignons dating back to 1965 and several extraordinarily fine Chardonnays, early proved the province's ability to produce some superb table wines, vine plantings to 1980 almost eschewed these grapes. However, in the last decade much more widespread plantings have occurred of the aristocratic grape varieties most suited to Hawke's Bay's growing environment. The 1992 vineyard survey revealed that the six most important varieties planted here are, in order: Chardonnay, Müller-Thurgau, Cabernet Sauvignon, Sauvignon Blanc, Merlot and Chenin Blanc.

A flurry of new labels has recently enlivened the wine scene in Hawke's Bay. The majority of the newcomers are grapegrowers determined to 'add value' to their crop.

Albany Lane, founded in 1988, foundered in 1992. Owned by the O'Kane family, grapegrowers at Havelock North, it marketed solid wines — mostly produced at other wineries — but never achieved a strong profile.

Hawkhurst Estate, owned by father and son team John and Jon Smith, is a Cabernet Sauvignon and Merlot vineyard in Napier Road, Havelock North. The debut 1991 Hawkhurst Cabernet/Merlot, made at the Mission, is attractively ripe, flavoursome and soft.

Dr John Loughlin, an eye specialist, is the owner of Waimarama Estate, a vineyard fronting onto River Road, Havelock North. The gold medal awarded to Waimarama Cabernet/Merlot 1991 at the Air New Zealand Wine Awards gave this fledgling producer instant acclaim. The Waimarama vineyard was planted in 1988. Today, four hectares of Cabernet Sauvignon, Merlot and Cabernet Franc have been established, and in 1994 a second, larger plot of red- and white-wine varieties will be developed at Haumoana. Loughlin freely acknowledges his debt to consultant winemaker Nick Sage: 'Without him I'd be lost.' Waimarama Cabernet/Merlot is a bold red with flashing purple-red hues, an intensely spicy, minty, plummy flavour and ripe, soft tannins. Its Cabernet Sauvignon stablemate is an equally concentrated red, packed with herbaceous, blackcurrant-like flavours. Waimarama is a major new label on the rise.

On the coast of Te Awanga, the Clearview Estate winery rose in 1992 on shingly land first planted in vines by the Vidal brothers in 1916. Tim Turvey (39) runs Clearview with his partner Helma van der Berg. The highlights of the Clearview range are the strongly herbal, complex, barrel-fermented Sauvignon Blanc; the soft, mouth-filling Chardonnay; and the rich, nutty Cabernet Sauvignon.

Linden Estate at Eskdale is a partnership between John van der Linden and his father and brother. The van der Lindens, grape-growers since 1971, produced a Franc/Merlot/Cabernet and Sauvignon Blanc from their twelve-hectare vineyard in 1991, joined by a highly promising, peachy-ripe Chardonnay in 1992.

Ian and Rachel Cadwallader's seventeen-hectare Riverside vineyard at Dartmoor is planted in Cabernet Sauvignon, Merlot, Malbec and Chardonnay. Riverside's 1992 Chardonnay is savoury and soft — a solid debut. The 1989 Cabernet/Merlot is a mellow, leafy-green red; the 1991 tastes markedly riper. Riverside will always be a small winery; the Cadwalladers plan an annual output of only 1500 cases, mostly sold at the gate and by mail order.

Moteo is a label marketed by Peter Gough, a Roseworthy graduate who grows grapes at Puketapu, near Riverside. His first wine was the robust, moderately herbal Moteo Sauvignon Blanc 1991.

Brownlie Brothers, at Bay View, is a partnership between brothers Roger, Chris and Stephen Brownlie. The first wine from their sixteen-hectare vineyard — planted in Sauvignon Blanc, Chardonnay, Gewürztraminer and Pinot Noir — and winery was produced in 1990. Devon and Estelle Lee, owners of Huthlee Estate, a six-hectare Merlot and Cabernet Franc vineyard in Montana Road, Bridge Pa, have also released their first wines — a buoyant 1991 Cabernet Franc, a fragrant, firm 1991 Merlot/Cabernet Sauvignon and 1992 rosé. Other winemaking ventures on the horizon — notably the investment by American finance company Fincen Holdings in a major new export-oriented vineyard and winery at Mangatahi — will add even greater colour to the buoyant Hawke's Bay wine scene.

Irongate

Shingly, arid, inhospitable — that's the first impression you get of the Irongate vineyard. This acclaimed site, once the bed of the Ngaruroro River, lies inland from the city of Hastings in Gimblett Road. It is named after the Irongate aquifer, which flows deep beneath the surface.

The first wine produced from this vineyard, Babich Irongate Chardonnay 1985, won a gold medal and the Vintners Trophy as the champion current vintage dry white wine at the 1985 National Wine Competition; a feat repeated in 1987. Both the 1989 and 1990 Irongate Cabernet/Merlots have also scored gold medals.

The proprietors of the vineyard — David Irving, general manager of Wattie's Foods (Hastings) and clothing manufacturer Gavin Yortt — planted their first vines in the early 1980s. 'From the start they were enthusiastic about wine quality,' points out Peter Babich, who each year buys the Irongate crop. Today Irongate is a ten-hectare vineyard planted in Chardonnay, Sauvignon Blanc, Cabernet Sauvignon, Merlot and Cabernet Franc.

Irongate is not uniformly shingly; about thirty percent of the vineyard's surface is silt, overlying shingle twenty centimetres down. 'You can see where the river once

flowed,' says Peter Babich. 'Where the silt is heaped up is the old river bend.' Irongate's soil fertility is low; before planting the vineyard was fertilised with trace elements.

In spring the deep layers of shingle warm up quickly; the vines burst their buds and flower early, giving the bunches the advantage of a long ripening season in which to develop their flavour constituents. Another asset of the Irongate gravels is their low water retention capacity. 'These gravels hold water for about two or three days,' says Babich. 'Clay holds water for six or eight weeks. At Irongate there is a danger that the vines' leaves will drop off from dehydration.'

An irrigation system installed at Irongate provides supplementary water, feeding more to the vines planted in the stonier section than those in the surface silts. 'Irongate's so dry the casuarina trees [planted as shelter belts] can die from drought,' observes Babich.

In these arid soils the vines develop small, open canopies of foliage, giving the ripening berries maximum exposure to the sun. 'Half the secret is that the soil holds so little moisture,' says Babich. 'The plants don't go crazy, the leaves are very small and the canopy is very open — you don't need to leaf-strip.' Yields are moderate — Chardonnay averages ten tonnes per hectare. The typical smallness of the berries — giving a high skin to juice ratio in the must — is a key factor behind the flavour concentration in the Irongate wines.

According to winemaker Joe Babich, the grapes from Irongate have ideal sugar, acid and pH balances and 'outstanding accumulation of flavour'. Irongate Chardonnay is stylish and taut, with steely, sustained, slowly evolving flavours. The Cabernet/Merlot is dark hued, powerful and firm-structured. Whether Irongate in the end will be renowned as a white or red-wine vineyard — or both — will be intriguing to see.

Akarangi

Only a rivulet of wine flows from Morton and Vivien Osborne's Akarangi winery on the banks of the Tukituki River. Visitors taste the Akarangi range in a century-old Presbyterian church shifted from Clive.

Akarangi Wines
River Road, Havelock North

Owners: Morton and Vivien Osborne

Key Wines: Chenin Blanc, Sauvignon Blanc, Chardonnay, Cabernet Sauvignon

The little Akarangi ('heavenly vines') winery lies in River Road, on the banks of the Tukituki River. Morton Osborne produces only a few hundred cases of wine each year: 'I don't want to get stomach ulcers,' he declares.

Osborne, raised in Ngaruawahia, graduated with an MA and Diploma in Clinical Psychology from Waikato University. He is a part-time clinical psychologist, dividing his time between private consultancy and hospital work, but also contemplating a fulltime career in wine. For several years he and his wife, Vivien, sold the crop from their three-hectare vineyard to local wineries. From 1973 they made small amounts of fruit wines, then in 1987 Osborne produced the first wine under the Akarangi label.

The vineyard is divided into two blocks. The smallest, adjacent to the Osbornes' house, consists of Sauvignon Blanc planted in heavy loams. The riverside vineyard, which is stonier and less fertile, is planted in a broad array of grape varieties. When the vines are all full-cropping, Osborne expects to produce about 2500 cases each vintage.

Akarangi's modest corrugated-iron winery started life as a boat-building shed. The wine is stored and sold in a century-old Presbyterian church, which the Osbornes bought for $600 and shifted from Clive to River Road.

Akarangi's bone-dry, partly barrel-fermented Chenin Blanc is not strongly varietal but a sound dry white; Osborne finds it 'honies up at about three years'. The Cabernet Sauvignon shows restrained French and American oak flavours: 'I think many Hawke's Bay Cabernets are oak-dominated, forcing you to wait too long for a pleasant drink,' says Osborne. This is a soft, slightly leafy-green red with light berry-fruit flavours.

Brookfields

Brookfields Vineyards
Brookfields Road, Meeanee

Owner: Peter Robertson

Key Wines: Reserve Chardonnay, Estate Chardonnay, Sauvignon Blanc, Pinot Gris, Gewürztraminer, Cabernet/Merlot, Reserve Cabernet Sauvignon, Estate Cabernet

Peter Robertson, owner of Brookfields Vineyards, only a kilometre from the sea at Meeanee, personally enjoys 'weighty reds with a bit of grip'. His muscular, richly scented, splendidly ripe-tasting Cabernet/Merlot ranks among the finest reds in the Bay; 'it sells so quickly it's frightening,' says Robertson.

Traditionally a sherry specialist, Brookfields was founded in 1937 by Hawke's Bay-born Richard Ellis. The Ellis family retained ownership for forty years. Robertson (41) who grew up in Otago, is a BSc graduate in biochemistry, whose interest in winemaking was aroused when employed as a student at Barker's fruit winery at Geraldine in the South Island. After spending a couple of years at McWilliam's in Hawke's Bay — working his way up from labourer to laboratory chemist — Robertson took over the old Ellis winery and three-hectare vineyard — then on the verge of closing down — from the founder's son Jack in 1977.

The three-hectare silty-loam home vineyard stretching from the winery to the gate is predominantly planted in Chardonnay and Sauvignon Blanc. Fruit is also drawn from growers at Tukituki, Fernhill, Bridge Pa, and a new shingle block in Ohiti Road, behind Roy's Hill.

Robertson each year produces around 7500 cases of table wine in his compact, half-century-old winery, constructed of handmade concrete blocks. In 1989 Robertson added a brick entrance-way, a new gable and colonial windows to 'put atmosphere into the place'. Visitors linger for tastings and tours: 'We dispel myths (like the "need" to rotate red wines maturing in cellars) and get them feeling better about Brookfields.'

The duo of top reds are the Reserve Cabernet Sauvignon and 'gold label' Cabernet/Merlot. Brookfields 1983 Cabernet Sauvignon, one of the best Hawke's Bay Cabernets of that year, announced this winery's arrival as a serious red wine producer, and that enticingly ripe and fragrant red has been followed by a string of equally successful unblended Cabernet Sauvignons. Working with 'punchy ripe Hawke's Bay fruit', matured in American oak casks, Robertson fashions deep-hued, weighty, flavour-packed reds, responding well to bottle age. (There is also an Estate Cabernet which is a lighter, fresh style made deliberately for early consumption.)

The Cabernet/Merlot is 'the one I really go for in the winery,' says Robertson. A blend of approximately seventy-five percent Cabernet Sauvignon, twenty percent Merlot and five percent Cabernet Franc, in top vintages this is a glorious red, dark and very substantial, with concentrated spicy, plummy, tannic flavours. Why is it so good?

The grapes are cultivated on a warm, north-facing slope across the Tukituki River from the Akarangi winery. This low-fertility vineyard, owned by David Werry, is well protected from cold southerly winds. The vines are trained on the Geneva Double Curtain trellising system (which exposes the fruit, positioned at the top of the trellis, to maximum sunlight) and the fruit is all hand-harvested. The Cabernet Sauvignon is grown on silty gravels; the Merlot, grown on a hard clay pan, ripens two to three weeks earlier.

'We get very high tannins from our Cabernet Sauvignon,' says Robertson, who leaves his wine on its skins for a lengthy period after the fermentation has subsided. Merlot is given briefer skin contact, to retain its fresh berry characters. The Cabernet/Merlot is then matured for over a year in almost entirely new French oak barriques. According to Robertson, the best vintages take ten years to peak.

Robust, savoury, peachy-ripe and complex, Brookfields Reserve Chardonnay is fermented in new French oak barrels; about twenty-five percent of the wine undergoes a softening malolactic fermentation. It is partnered by a junior label, the American oak-aged Estate Chardonnay, a ripe, soft style: 'This is the wine I drink at night,' says Robertson.

The Brookfields range also features two other absorbing white wines. With his Sauvignon Blanc, Robertson is aiming for 'a riper style with wood maturation to muscle-up the fruit'. This is a rich, tropical fruit-flavoured wine with plenty of weight and length. The Pinot Gris, its off-

dry style a concession to 'Müller-Thurgau drinkers', is spicy, soft, and packed with appealing, savoury, slightly honeyish flavours.

'Looking back,' says Robertson, 'it's been very hard. People have no idea of the stress — the heartbreak — involved when you come in with little money and don't have the right gear. But now we're on target. We're not going to grow and grow. There's a place for us at this size. If we get bigger, I'll get desk-bound and have to stop making the wine — that's not what I'm here for.'

Peter Robertson's Cabernet-based reds rank among the darkest, most powerful and rich-flavoured in Hawke's Bay. The robust, savoury Reserve Chardonnay can also reach great heights.

C.J. Pask Winery
Omahu Road, Hastings

Owner: Chris Pask

Key Wines: Cabernet Sauvignon, Cabernet/Merlot, Pinot Noir, Reserve Merlot, Roy's Hill Red, Roy's Hill White, Chardonnay, Reserve Chardonnay, Sauvignon Blanc

The taste delights of very ripe fruit shine in C.J. Pask's Cabernet-based reds, which ooze lush, sweet fruit aromas and flavours and bold alcohol.

C.J. Pask Winery is operated by Chris Pask (53). Pask, a burly former top-dressing pilot, has for twenty years been a contract grapegrower, supplying fruit to several companies. Having each year turned out a couple of barrels of wine for his friends and relatives and, he says, being 'interested in adding value', in 1985 he elected to plunge into commercial wine production under his own label.

The original C.J. Pask winery — a converted tractor shed — and vineyard in Korokipo Road, Fernhill, were in 1989 sold to Montana. A striking pastel blue and green, concrete Mediterranean-style winery, with stained glass and promi-

nent columns, has since risen on the north side of Hastings.

The 1985–90 vintages of C.J. Pask wines were made by Chris Pask himself, but his white wines lacked the consistently high quality of his reds. 'The company had grown enormously and needed a professional winemaker,' says Kate Radburnd (formerly Marris), who joined C.J. Pask in late 1990. Adelaide-born Radburnd (31) is a Roseworthy College graduate who worked as an assistant winemaker at Vidal and Villa Maria during the mid-1980s and then built a glowing reputation during her 1987–90 spell as winemaker at Vidal.

'It was time for a change,' says Radburnd. 'I wanted to see how a smaller winery works, especially one where you grow your own grapes. I'd already worked with and knew the quality of Chris's Gimblett Road fruit. From his point of view, the reputation I'd built at Vidal was a tremendous bonus.'

Chris Pask is the general manager of the company; vineyard manager Max Bixley heads the grape-growing and Kate Radburnd the winemaking.

The key to the success of C.J. Pask's reds lies in the soil. Chris Pask has a stake (with Allan Chittick) in forty-three hectares of vineyards in Gimblett Road, planted principally in Chardonnay, Sauvignon Blanc, Cabernet Sauvignon,

C.J. Pask

Chris (C.J.) Pask set up his winery in 1985 with the aid of his daughter, Tessa (right). Kate Radburnd (formerly Marris) joined C.J. Pask in 1990.

The Cabernet Sauvignon and Cabernet/Merlot (a 50/50 blend) have consistently been the highlights of C.J. Pask's annual output of about 12,000 cases. The winery's trump card is the optimal ripeness of its Gimblett Road grapes, giving its reds impressive concentration. 'Richness and a ripe, sweet fruit flavour are the goals,' says Radburnd, 'together with elegance and very fine French oak handling.' Both reds are delicious drinking: muscular and supple and bursting with sweet-tasting fruit.

Roy's Hill Red — a blend of forty percent Cabernet Sauvignon, forty percent Cabernet Franc and twenty percent Merlot — is a soft and fleshy red. Briefly oak-aged, it is designed for 'early consumption'. The Reserve Merlot, launched from the 1991 vintage, is a voluptuous, perfumed red with a delectable surge of complex, plummy, spicy flavours.

Roy's Hill White is a low-priced, non-wooded dry Chenin Blanc for no-fuss drinking. Fresh, vigorous and lush, with a trace of sweetness, the Sauvignon Blanc is very appealing. C.J. Pask Chardonnay is a blend of tank and barrel-fermented wine with some lees-aging but its accent is on Hawke's Bay's citrus-fruit flavours; 'the fruit should shine through,' says Radburnd. From 1992 flowed the first fully barrel-fermented Reserve Chardonnay.

Cabernet Franc and Merlot, with small plots of Pinot Noir and Malbec. First planted in 1982, these dry, stony vineyards excite Radburnd: 'They're very free-draining and shingly, but with enough fertile soil on top to establish young vines without irrigation. In years of extreme stress we hire an irrigator which spans about twenty rows. The red bunches are beautifully exposed to the sun.'

Kate Radburnd is clear about C.J. Pask's future. 'Our aim is to utilise our vineyards as well as we can, with a strong emphasis on red wines because those sites grow reds best. We also want to build strong export markets.'

Crab Farm

Crab Farm Winery
125 Main Road North, Bay View, Napier

Owner: James Jardine

Key Wines: Jardine Cabernet, Cabernet/Merlot, Petane Red, Pinot Noir, Chardonnay, Sauvignon Blanc, Gewürztraminer

Hamish Jardine (in the cab) has in recent vintages relied increasingly on estate-grown fruit, with a resulting upswing in his wines' quality. The Reserve Chardonnay, which Jardine describes as 'a fat, boisterous wine, jumping with hot Hawke's Bay fruit', is his finest achievement yet.

When winemaker Hamish Jardine's great-grandfather first acquired land at Petane, Bay View, it was a mud-flat covered with rushes and crawling with crabs. The Hawke's Bay earthquake of 1931 later lifted 'Crab Farm' above sea level, creating the site for the present vineyard and winery.

James Jardine, Hamish's father, who planted his first vines on the main highway just south of Esk Valley in 1980, for several years sold the grapes to local wineries. In 1987 an implements shed was converted into a small, vertical-timbered winery. Hamish Jardine (32), who gained his early winemaking experience at Matawhero in Gisborne and Chateau Reynella in Australia, was appointed winemaker.

Only 3000 cases of wine flow each year under the Crab Farm label. The twelve-hectare vineyard adjoining the winery, which contains pockets of shingly and silty soils, is being replanted in Chardonnay, Sauvignon Blanc, Cabernet Sauvignon, Merlot, Cabernet Franc and Pinot Noir. 'Sea breezes cool the vineyard, making this a relatively cool site for Hawke's Bay, especially at night,' says Jardine.

Crab Farm's initial releases were unmemorable, but since 1991 a move towards increased reliance on estate-grown rather than bought-in fruit has paid dividends. The Cabernet Sauvignon 1991 is a gutsy, firm red with strong, chocolatey, herbal flavours. The Reserve Chardonnay 1992 is fat, ripe-flavoured and rich — Crab Farm's finest achievement to date.

Cross Roads Wine Company
Korokipo Road, Fernhill

Owners: Lester O'Brien and Malcolm Reeves

Key Wines: Chardonnay, Reserve Chardonnay, Riesling, Riesling Dry, Gewürztraminer, Sauvignon, Cabernet Sauvignon/Merlot

Malcolm Reeves, a Massey University lecturer in food technology and wine columnist, has long been an indefatigable researcher into various aspects of winemaking, and acted as a technical consultant to several wineries. When he launched the first 1990 wines under his own Cross Roads label, it was therefore no surprise that their quality instantly won respect.

Why was the winery named Cross Roads? 'If the wine venture succeeds,' says Reeves, 'my academic career will be at the cross roads.'

Reeves (47) has set up his Fernhill winery in partnership with Lester O'Brien (52), a Victoria University chemistry graduate he met in the early 1970s, when both were lecturing at Massey University. O'Brien, with another New Zealander, later established a computing company in Belgium to service the European textile industry. Having sold his computing interests, O'Brien is now living in Napier as Cross Roads' general manager.

Cross Roads' 1990–93 vintages have been based on grapes purchased from several Hawke's Bay growers, and the wines have been produced at a variety of premises. A 'Spanish mission style' winery rose on State Highway 50, one kilometre on the Taradale side of Fernhill, in time for the 1993 vintage. In the company's adjacent five-hectare vineyard, which has blocks of shingly loams and sandy loams, the existing Müller-Thurgau vines are being uprooted and the vineyard gradually replanted, at first with red varieties.

The white wines are uniformly impressive with their clear, deep varietal flavours. Cross Roads Chardonnay is a robust, fleshy style, its barrel fermentation, partial malolactic fermentation and lees-aging imparting an appealing buttery-soft complexity. The medium Riesling and

Riesling Dry are awash with this variety's classic lemon and grapefruit-like flavours and hints of botrytis-derived honeysuckle. Cross Roads Sauvignon is a moderately herbal style, its touch of complexity reflecting its partial fermentation in oak casks. The Gewürztraminer, packed with citric/spicy flavours, balances its whisker of sweetness with lively acidity.

Reeves, having conducted research into colour extraction and malolactic fermentation, is also naturally eager to pursue red wines. He produced an enjoyably fruity, soft and undemanding Cabernet Sauvignon/Merlot in 1991 and an unblended Cabernet Sauvignon in 1992. When the fruit quality warrants it, a Reserve label of any of the wines will also be marketed.

Cross Roads' output is climbing steadily: from 1500 cases in 1990 to over 5000 cases in 1993. If production reaches its planned peak of 10,000 cases, the Cross Roads label will be widely seen.

Cross Roads

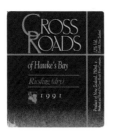

Malcolm Reeves (left) and Lester O'Brien met twenty years ago while lecturing at Massey University. Cross Roads has swiftly achieved recognition for the standard of its white wines, especially its full, lusciously fruity Rieslings.

Esk Valley Estate
Main Road, Bay View, Napier

Owner: Villa Maria Estate

Key Wines: Reserve Chardonnay, Merlot/Cabernet/Franc, Merlot; Private Bin Chardonnay, Chenin Blanc, Sauvignon Blanc, Merlot/Cabernet Sauvignon, Merlot/Cabernet Franc Dry Rosé

A Mediterranean-style winery has recently sprouted in Hawke's Bay. Under its latest name, Esk Valley, the ram-

shackle old Glenvale winery on the coast just north of Napier has been transformed into a tourist drawcard. The Mediterranean theme is reflected in Esk Valley's fresh-painted concrete building, its vine-draped terraces, olive grove and outdoor café. The wines are among the best in the Bay.

The Glenvale winery's rebirth as Esk Valley is the latest idea of one of the industry's greatest entrepreneurs, George Fistonich. Fistonich, co-owner of Villa Maria, bought Glenvale in 1987 from Robbie and Don Bird, grandsons of Glenvale's founder, Englishman Robert Bird.

In 1933 Bird bought five hectares of land at Bay View, planning to establish a market garden and orchard. But

Esk Valley

during the Depression the return for grapes of under two-pence per pound soon encouraged the fifty-one-year-old Bird to enter the wine industry. During his childhood in Dorset, he had watched his mother make damson and blackberry wines, and Government Viticulturist Charles Woodfin offered his assistance. Bird's first crusher was a mangle. In the original cellar, a tunnel scooped out of the hillside, early Glenvale wines were vinted using the humble Albany Surprise variety.

By the time of his death in 1961, Robert Bird owned a twenty-eight-hectare vineyard and a large modern winery. His son — the second Robert — who worked at the Te Kauwhata Viticultural Research Station before joining Glenvale in 1947, extended the vineyards to over one hundred hectares of principally hybrid varieties. He retired in 1979, opening the way to the top for Robbie and Don while they were still in their mid-twenties.

Under three generations of Bird family management, Glenvale grew steadily into one of New Zealand's largest family-run wineries. Its production traditionally emphasised fortified wines; and in particular its Extra Special Sherry had many fervent supporters. But in 1976 the release of two varietal table wines — Müller-Thurgau and Sonnen-gold (Chasselas) — marked the company's serious move into the table-wine market. Earlier Glenvale table wines had had a deservedly low image.

The far superior releases from the 1983 vintage were led by three silver-medal wines — Chenin Blanc, Chardonnay and Claret — followed in 1984 by a marvellous Sauvignon Blanc. Other premium bottled wines were reliable and competitively priced. The Bird brothers' undoing, however, was to over-expand in the highly price-sensitive cask-wine market. The ferocious wine price war of 1985–86 brought the company to its knees, leading to Villa Maria's takeover a year later.

Fistonich immediately set about repositioning Glenvale — swiftly renamed Esk Valley — as a top-end-of-the-market, 'boutique' producer. Overseeing the winery's about-turn from 1989 until 1993 was Grant Edmonds, who produced a flow of excellent white and red wines until his recent appointment as chief winemaker of the Villa Maria/Vidal/Esk Valley empire.

On terraces carved thirty-five years ago out of the north-facing slope alongside the winery ('with the aid of a bull-dozer and a very brave and talented driver', according to Esk Valley), Merlot, Malbec, Cabernet Sauvignon and Cabernet Franc vines have been planted for a premium estate red. The founder, Robert Bird, originally planted these terraces with the Albany Surprise variety; the vines flourished and produced very early-ripening grapes, but owing to high labour costs, in the late 1950s they were replaced by pines.

The new hillside vineyard is densely planted, and irrigated to ensure adequate vine growth on this very drought-prone site. Esk Valley believes its terraced vineyard will 'set the standard for New Zealand red wines in the twenty-first century'. In 1991 the steep dry slope yielded its first grape crop for thirty years, and Edmonds was openly excited by its quality. 'It's superior to anything else in the winery,' he reported.

Terraced vineyards, a frequent but magical sight in Europe, are rare in New Zealand. Esk Valley predicts its terraced vineyard, planted in 1989 in Merlot, Malbec, Cabernet Sauvignon and Cabernet Franc, will 'set the standard for New Zealand red wines in the twenty-first century'.

The vast majority of Esk Valley's grapes, however, are drawn from contract growers scattered throughout Hawke's Bay. Only about one-fifth of the annual grape crush is reserved for the Esk Valley label; much of the rest is earmarked for release under Villa Maria's bulk-wine brand, Maison Vin.

The unsightly clutter of concrete and iron buildings that formed the Glenvale winery — once described by the *Auckland Star* as 'looking like a bomb site' — sat on one of the most glorious winery sites in the country, with magnificent views of the Hawke's Bay coastline and the Pacific Ocean. Fistonich, who from the start saw the potential, has greatly enhanced the old winery's efficiency and visual appeal.

The heart of the Glenvale winery, a strong concrete structure about fifty years old, has been preserved. But the changes have been radical: the former bottling-line was scrapped; outside tanks were enclosed; a new press was installed; refrigeration capacity was increased five-fold; a new, half-underground, temperature-controlled barrel fermentation room was added; more than 500 casks were purchased; and a side was carved out of old concrete storage tanks to transform them into storage bins for bottled wines. The entire complex has been modernised.

Esk Valley is 'Merlot country', according to the winery's newsletter, and at the Air New Zealand Wine Awards in 1991, Esk Valley's voluptuous, silky Reserve Merlot/Cabernet/Franc 1990 was crowned the champion wine of the show. In their pursuit of a 'mouthwatering, not mouthpuckering' style of Merlot, Esk Valley's winemakers ferment the wine in open vats and hand-plunge the grapes' skins into the juice. This traditional approach, says Edmonds, 'extracts flavour and colour without excessive tannins.' Esk Valley's other top red-wine label, the unblended Reserve Merlot, also ranks among Hawke's Bay's plumpest and most stylish reds.

The Private Bin Merlot/Cabernet Sauvignon has a fragrant blackcurrant-like, plummy, cigar-box bouquet, with strong, spicy, red berry-fruit flavours and firm tannin. An elegant middle-weight red, it responds very well to two-three years' bottle-aging.

Esk Valley's Private Bin Merlot/Cabernet Franc Dry Rosé is a 'serious', but fragrant and delicious wine, with fresh, strong berryish flavours and a whiff of oak. The non-wooded Private Bin Sauvignon Blanc is mouthfilling and ripely herbal in warmer vintages; lighter and brisker in cooler years.

The Private Bin Chenin Blanc, which is matured for four months in seasoned French and American oak casks, typically lacks charm in its youth, but after a couple of years' bottle-aging, rich tropical-fruit and honeyish flavours unfold.

Esk Valley Private Bin Chardonnay is predominantly barrel-fermented and then matured for four months in French and American oak casks. Crisp and lemony in lesser years, in warm vintages it is markedly fatter and richer. The Reserve Chardonnay, a very stylish wine, is made from hand-picked fruit which is whole bunch-pressed, barrel-fermented and lees-matured. This is an intense, citric-flavoured wine, subtle and sustained.

Eskdale Winegrowers
Main Road, Eskdale

Owners: Kim and Trish Salonius

Key Wines: Chardonnay, Late Harvest Gewürztraminer, Cabernet Sauvignon

'I have to make wine a little differently,' says Kim Salonius. Eskdale Winegrowers, lying unobtrusively alongside the Napier-Taupo road through the Esk Valley — its presence marked only by a carved wooden sign at the gate — is one of Hawke's Bay's smallest and most individual vineyards.

Kim Salonius (49) came to Auckland from his native Canada in 1964 to read for a degree in history. As a child he watched while his father made wine for home consumption from grapes brought north from California. Later, while advancing his medieval studies in Germany, his interest in winemaking was rekindled. By 1973 his first vines were in the ground. From 1973 to 1977, when he made his first commercial release of wine, Salonius and his wife, Trish, survived on her teaching salary and lived for part of that period in the local schoolhouse. His first wines were produced out in the open on a concrete pad. The award of silver medals to two early vintages, Cabernet Sauvignon 1977 and Chardonnay 1978, was an auspicious debut.

Salonius adopts a low profile, does not advertise his wine or regularly participate in industry affairs, enjoys deep, wide-ranging conversations with his customers and, he states, tries not to work more than four hours per day — which he says gives him the time to go hang-gliding and devour several books weekly.

Eskdale's sole source of fruit is the silty, three-hectare home vineyard between the winery and the road. Established in Cabernet Sauvignon, Chardonnay and Gewürztraminer, viruses have forced its gradual replanting. Pinot Noir was uprooted after only one vintage, 1981, met Salonius's approval. 'I pulled my Pinot Noir out because its wines were very light,' he recalls. 'Now light Pinot Noirs are winning acclaim.'

His tiny winery, with wooden trusses, white plastered walls and stained-glass windows, looks every inch like a shrine to Bacchus. Salonius built it himself, using bricks and Douglas fir, placing his storage tanks underground to preserve the winery's beauty. A second building, for the storage of casks and bottled wines, has also recently risen. Everything is planned to last: 'This is home, I'll die here,' he says contentedly.

Salonius favours robust, strong-flavoured wine: 'As

Eskdale

ESKDALE
WINEGROWERS

GEWÜRZTRAMINER
LATE HARVEST

Kim and Trish Salonius own one of the lowest-profile quality-oriented wineries in New Zealand. Eskdale is a tiny winery — and the Saloniuses plan to keep it that way.

big as I can get it.' His grapes are held on the vines to ripen to very advanced levels before the hand-pickers move in. The juice is held on its skins as long as possible after crushing to get maximum flavour. The wines are barrel-matured longer than at any other winery in the country — two years for Chardonnay, three for Cabernet Sauvignon.

The arched brick storage bins near the winery door hold an absorbing trio of wines for sale. Eskale wines are scarce; Salonius' annual output is only about 1250 cases and he has no interest in making more: 'I'm slowing down — the work is taking longer.'

The Cabernet Sauvignon is a bold, chunky, slightly austere red with powerful cassis-like and spicy flavours and a touch of leafy-green herbaceousness. Salonius' personal fondness for heavy red wines — he has enthused over a Rumanian Pinot Noir to me — affords an insight into his own red style.

The Chardonnay is mouth-filling and deep-flavoured, with strong oak and plenty of barrel-ferment complexity. The Late Harvest Gewürztraminer, stop-fermented with a trace of residual sugar, is robust, pungently spiced and lingering.

These wines display an individuality that can be quite absorbing.

Lombardi

Lombardi Wines
Te Mata Road, Havelock North

Owners: The Green family

Key Wines: Marsala, Dry Sherry, Riesling-Sylvaner, Dry Red Pinotage

'More body, more flavour, more alcohol per dollar than all other drinks', recently boasted the sign in this little winery's shop. Lombardi, the oddity of the Hawke's Bay wine trail, still preserves the fortified-wine traditions of an almost-vanished era.

Italian-born Concetta 'Tina' Green and her son, Tony, run the only well-known winery south of the Waikato to retain a heavy emphasis on fortified wines.

In 1948 English-born W.H. Green and his wife Tina planted a 1.2-hectare vineyard on a soldier's rehabilitation block in Te Mata Road, near Havelock North. Green — who until his death in 1992 still carried out all the spraying and pruning — learned about wine when he joined the Vidal brothers' staff in 1937. After Tina Green, born in the Bay of Naples (and still maintaining a strong Italian accent) turned to her grandparents in Italy for additional winemaking advice, 1959 brought the first Lombardi vintage.

Today the founders' son, Tony (41), a Massey University chemistry graduate, runs the winery: 'He's the mainstay,' says Tina Green. Following the recent uprooting of forty-year-old vines — their productive capacity had declined — the estate vineyard has shrunk to only a half-hectare of Muscat Italia. Fruit is also purchased from Hawke's Bay growers.

Production is still centred on fortified wines; dry sherry is a strong seller. But the vineyard specialties are Italian-style liqueurs and vermouths. The vermouths range from Dry (white) to Vermouth Di Torino (medium red) and Vermouth Bianco (sweet white). The range of liqueurs — packing only a soft punch at just below twenty-three percent alcohol, the legal maximum — features Anisette (aniseed), Astrega ('Drambuie type'), Caffé Sport (coffee) and Triple Sec (orange).

Marsala, a deep amber-green, full-flavoured, mellow drink flavoured with herb essence imported from Italy, is one of Lombardi's most popular labels. Demand for Lombardi's fortified wines is falling, however, due to changing consumer preferences and government policy. 'Excise is a major problem,' says Tina Green. 'The Government's ruining the wine industry. You work for nothing. When they've closed all the businesses down, where are they going to get the tax?'

Table wines sold at the winery include Hock (dry white), the light, slightly sweet Riesling-Sylvaner, Sauternes (medium sweet), Pinotage (a medium-dry rosé) and the gutsy, non-wooded Dry Red Pinotage — these are clean quaffing wines. Most of the vineyard's output is sold at the gate and the Lombardi range is rarely seen beyond Hawke's Bay.

What of the future? 'We can't predict,' says Mrs Green. 'We're playing it year by year. I have to retire some time. If we get a good price, we might sell. There's no need for hasty decisions.'

Mission Vineyards
Church Road, Taradale

Owner: Greenmeadows Mission Trust Board

Key Wines: St Mary Riesling-Sylvaner, Sauvignon Blanc, Fumé Blanc, Sémillon/Sauvignon Blanc, White Meritage, Müller-Thurgau, Rhine Riesling, Pinot Gris, Gewürztraminer, Chardonnay, Chanel Chardonnay, Estella Sauternes, Sugar Loaf Sémillon, Claret, Cabernet Sauvignon, Cabernet/Merlot; Reserve Chanel Chardonnay, Gewürztraminer, Sémillon, Cabernet Sauvignon; Marinella Sherry, Fine Old Port

Twin columns of English plane trees create a memorable entrance to the Mission winery, nestled at the foot of the Taradale hills. On the left lies the Chanel vineyard, the source of Mission's top-end Reserve Chanel Chardonnay. Above rises the graceful cream and green wooden building that for many decades served as the Society of Mary's seminary. Following a steep decline in the number of young men attracted to the priesthood, the old seminary now houses a restaurant, language school and student hostel. The few remaining Marist brothers live in an adjoining, modern building.

The Society of Mary is New Zealand's only nineteenth-century producer of wines still under the same management. The present site is the last of several occupied by the Marist mission during its long history in Hawke's Bay. Father Lampila and two lay brothers, Florentin and Basil, after mistaking the Poverty Bay coast for their real destination, Hawke's Bay, planted vines near Gisborne in 1850. A year later they moved south and planted more vines at Pakowhai, near Napier. The story goes that in 1852, on a return visit to Poverty Bay, Father Lampila found the abandoned vineyard bearing a small crop of grapes, made a barrel of sacramental wine and shipped it to Napier. But the seamen broached the cargo, drank the wine — and the cask completed its journey full of sea water.

A Maori chief, Puhara, took the French missionaries under his protection at Pakowhai. The brothers taught and nursed the local Maori, and gardens and vineyards were laid out. After Puhara was killed, however, in an inter-tribal clash in 1857, the brothers were forced to move again, this time to Meeanee.

For several decades wine production at Meeanee was very limited, sufficient only to supply the brothers' needs for sacramental and table wines. A son of a French peasant winemaker, Brother Cyprian, arrived in 1871 to take charge of winemaking — but not until around 1895 were the first recorded sales made, mainly of red wine.

Two years later, local rivers burst their banks, flooding the Meeanee plains and inundating the Mission cellars. After deciding to shift to higher ground, the Society of Mary bought 240 hectares of Henry Tiffen's estate at Greenmeadows — including 0.2 hectares of Pinot Noir — and established a four-hectare vineyard there. But not until 1910, after further disastrous floods, was the seminary itself moved to Greenmeadows; the wooden building was cut into sections and hauled there by steam engine. Fire almost destroyed the wine vaults in 1929, and thousands of gallons of wine were lost in the Napier earthquake of 1931, but of late nature appears to have made its peace with the Mission.

Under the guidance of Brother John, the winemaker from the 1960s until 1982, the Mission acquired a reputation for sound wines. John Buck, writing in 1969, praised the red Pinot Reserve for its 'beautiful varietal nose', though it tended to be 'a little thin and sharp', and the Sauternes as 'a sweet wine, clean, light and well balanced'. Fontanella, Mission's bottle-fermented sparkling wine — now phased out of the range — also enjoyed a reputation as this country's top, although rarest, sparkling.

By the end of the 1970s, however, Mission wines were often mediocre and lacked depth, reflecting a lack both of finance and of advanced winemaking equipment. To many consumers, a halo appeared to encircle bottles of Mission wine, yet medal successes were few. The improved industry-wide standards left the church authorities with a basic decision: to be left further behind, or to compete.

The result was a programme of major expansion, designed to lift both the production level and the standards of Mission table wines. $250,000 was allocated for winery equipment and vineyard improvements.

Today the annual output of the Mission winery is around 50,000 cases, making it a medium-sized winery by New Zealand standards. The affable Brother Martin hosts the Mission's never-ending flow of visitors. His conversation flits from Roman viticulture to the widow Clicquot to the Napier earthquake as he steers tourists past dark oval German oak casks dating back to last century.

For the past eleven years, thirty-eight-year-old Paul Mooney has held the winemaking reins at the Mission. A science graduate from Waikato University, Mooney served

Brother Osika, a Tongan who has spent more than twenty years as a lay brother, heads the Mission's viticultural activity. The Mission's finest fruit comes from the warm estate vineyard nestled against the flanks of the Taradale hills.

as a DSIR technician on Campbell Island until he met Brother John. 'John needed an assistant,' recalls Mooney. 'He'd trained in Bordeaux and in the sixties was regarded as one of the best winemakers in the country.'

At $8.50 for Sauvignon Blanc, and $8.95 for a Pinot Gris, Sémillon or Gewürztraminer, the Mission's wines are some of the lowest-priced around. 'We see our role as that of a value-orientated producer of well-made, affordably priced wines,' says Warwick Orchiston, Mission's winery manager. 'Our prices are modest because of our large production, because we don't have to pay heavy dividends to shareholders, and because our image is still a bit lean at the top end of the market.'

When Orchiston (51) arrived at the Mission in 1986 — after more than a decade as the winemaker at Vidal — serious doubts hung over the winery's future. 'There were surplus stocks of several lines and profitability was marginal. I knew a lot about vineyards and winemaking, but very little about wine marketing,' he says. After observing Montana's and Babich's strong, profitable sales, Orchiston endeavoured to set the Mission on a similar course of 'well-made, affordably priced wines'.

'The first thing was to tidy up the vineyards,' he recalls. Old, heavily virused Cabernet Sauvignon and Müller-Thurgau vines, cropping at a miserly two tonnes per hectare, were uprooted. The Mission now grows fifty percent of its fruit intake, either at its fertile, twenty-hectare Meeanee site, which Orchiston views as a 'bulk' vineyard, or in the 'cellars block' and premium Chanel vineyard

flanking the winery. Brother Osika, a Tongan who has spent over twenty years as a lay brother, oversees the Mission's viticultural activity.

Mooney's top wines have been marketed as a Reserve range: oak-aged Sémillon, dry Gewürztraminer, Botrytised Gewürztraminer, Chanel Chardonnay (there is also a non-reserve version of this label), and Cabernet Sauvignon. My favourites are the Reserve Gewürztraminers, which are both packed with varietal spice; the ripe, deep-flavoured Reserve Sémillon; and the rare, but classy, Reserve Chanel Chardonnay. The Reserve Cabernet Sauvignon is a spicy, ripe, plummy red with excellent flavour depth and touches of complexity.

Mission's low-priced 'commercial' range features a broad array of sound wines. The Pinot Gris is typically an attractive, full-bodied wine with ripe stone-fruit and slightly earthy flavours and a medium-dry finish. The Sémillon/Sauvignon Blanc, one of the first blends produced in this country from these traditional partners in white Bordeaux, is usually — but not always — austerely crisp and piquant. Other tangy, herbaceous whites include a quaffing, non-wooded, slightly sweet Sauvignon Blanc and a nutty, full-flavoured, partly barrel-fermented Fumé Blanc. Sugar Loaf Sémillon, a medium style with notably restrained varietal grassiness, is promoted as 'a darned nice alternative to Müller-Thurgau'.

Mission's bottom-tier Chardonnay displays moderate depth of citrus-fruit flavours and very restrained oak influence. Chanel Chardonnay (the non-reserve, widely available version) is a barrel-fermented wine with increasing flavour intensity. It is no match, however, for its big brother, the Reserve Chanel Chardonnay, based on the finest hand-picked fruit fermented in 100 percent new oak barrels. Succulent, oaky and peachy-ripe, this is clearly Mission's top white wine.

Mission White Meritage — the new name for the old White Burgundy — is a fresh and crisp, slightly herbal, dryish rather than bone-dry wine for everyday drinking. St Mary Riesling-Sylvaner is one of this vineyard's biggest selling lines. Pleasantly fruity, it is a medium style, sweetened with Muscat Dr Hogg juice; a Müller-Thurgau in everything but name. It is partnered by a similar wine labelled as Müller-Thurgau, but sweetened by back-blending with Sauvignon Blanc. Sweeter again is the long-popular Estella Sauternes, light and mild.

Mission's Rhine Riesling debut, from the 1992 vintage, was of eye-catching quality — packed with marmalade-like, crisp, lingering flavours. With its delicious concentration and botrytis-derived honeyishness, this will be a hard act to follow.

Three widely available Cabernet-based reds are marketed. The Claret is a fresh, simple, raspberryish quaffer. The Cabernet Sauvignon and Cabernet/Merlot are both good, honest reds, matured in seasoned oak casks. They lack flavour generosity in cooler years, but in warmer vintages like 1989 and 1991 display medium–full body, strong, attractive spicy and blackcurrant-like flavours and soft, easy tannins.

Mission still enjoys a following for its fortified wines, including Marinella Sherry, Fine Old Port — sweet and mellow but not complex — and Altar Wine — a medium-sweet style with simple, orange-like flavours 'vintaged in accordance with Church requirements'.

Ngatarawa Wines
Ngatarawa Road, Bridge Pa, Hastings

Owners: Alwyn Corban and Garry Glazebrook

Key Wines: Stables Late Harvest Riesling, Stables Chardonnay, Stables Sauvignon Blanc, Stables Red; Alwyn Chardonnay, Glazebrook Cabernet/Merlot, Penny Noble Harvest Riesling Selection; Old Saddlers Sherry, Old Saddlers Port

Alwyn Corban in the early–mid 1980s earned Ngatarawa wide respect with his robust, dry wine styles. Recently, reflecting the Riesling variety's botrytis-proneness in the Ngatarawa estate vineyard, gorgeous honey-sweet Rieslings have also moved centre-stage.

Ngatarawa (meaning 'between the ridges') winery, ten minutes' drive west of Hastings near Bridge Pa, is a partnership formed in 1981 between Corban and the Glazebrook family, of the 2400-hectare Washpool sheep station, who have owned the site of the present vineyard for over half a century. Ngatarawa's initial vintage, 1982, was based on Te Kauwhata fruit; the first Hawke's Bay wines flowed in 1983.

Alwyn Corban (41), a reserved and gentle personality, is the son of Alex Corban, the Wine Institute's first chairman. After capping his impressive academic record with a master's degree in oenology at the University of California, Davis, Corban spent a year at the Stanley Wine Company in South Australia, followed by four years at McWilliam's in Napier, before founding Ngatarawa with Garry Glazebrook. Glazebrook, now in his late sixties, is the chairman of Ngatarawa's board, but the day-to-day running of the winery is in Corban's hands.

The Ngatarawa winery is based on a converted stables built of heart rimu and totara in the 1890s. While the building's soft exterior lines have been preserved — creating 'a winery that doesn't look like a winery,' says Corban

contentedly — the internal walls have been gutted to free up space for wine storage and a new concrete floor laid. Barely visible at the rear of the old stables winery, a newer building has arisen to accommodate most of the wine-making equipment.

The Ngatarawa vineyards are in the 'Hastings dry belt', a recognised low-rainfall district. The seven-hectare irrigated estate vineyard, on sandy loam soils overlying alluvial gravels — which dry out swiftly in summer — is where Riesling, Sauvignon Blanc and young Cabernet Sauvignon vines are planted. A second, eight-hectare vineyard to the west is established in Chardonnay, Cabernet Sauvignon, Merlot, Malbec and Cabernet Franc. Only a small proportion of the crop is purchased from other growers.

Alwyn Corban's growing devotion to Riesling has few parallels in the region. 'Riesling isn't an ideal contract growers' grape in Hawke's Bay,' he says. 'The large companies pay for Riesling as if it's a bulk variety, assuming the best stuff comes from Marlborough. But we don't see Riesling as a bulk variety. We're successful because we have a close vineyard/winery integration and an ability to handle botrytis.'

To encourage the onslaught of 'noble rot' on his Riesling grapes, Corban doesn't use anti-botrytis sprays — thus risking ignoble rot — and hangs the ripening bunches late on the vines. Sweet styles are 'not difficult to make', he insists. Ngatarawa's Riesling Dry was recently — and reluctantly — dropped from the range ('We have two other dry whites and there's a niche for dessert styles') and all the Riesling fruit is now used for two sweeter styles.

Stables Late Harvest Riesling displays a pure, delicate, intensely floral bouquet and a medium-sweet, appealingly concentrated lemony/honeyish palate. Corban sees it as his 'fruit style' — meaning it doesn't have the qualities of a heavily botrytised wine.

The delectably sweet, gorgeously botrytised Penny Noble Harvest — named after Penny Reynolds, Garry

Ngatarawa

Alwyn Corban is a quiet personality — but his wines can speak for themselves. The stylish, steely Alwyn Chardonnay and rampantly botrytised Penny Noble Harvest rank among Hawke's Bay's finest wines.

Glazebrook's younger daughter, a shareholder and director of the winery — is made from a high percentage of shrivelled, nobly-rotten fruit, for which in 1991 the hand-pickers made four separate sweeps through the vineyard. This gloriously perfumed and nectareous beauty is one of New Zealand's most extraordinary 'stickies'. As Oz Clarke, the British wine writer, recently put it, this 'is a wonderful, golden orange wine, so thick and syrupy it plops out of the bottle like oil and swims lazily from side to side as you swirl your glass.'

Ngatarawa's output is low: only 10,000 cases annually. The three top-end-of-the-range wines — Penny Noble Harvest, Alwyn Chardonnay and Glazebrook Cabernet/Merlot — are marketed under a yellow label; the 'commercial' selection of four 'Stables' wines carry a grey label; and two black-labelled fortified wines complete the lineup.

Stables Red has long enjoyed a strong following. With its typical excellent fruit ripeness and soft tannins, this is an ideal red for drinking in its youth. The style originated in 1982 — when Ngatarawa simply could not afford to mature it in wood — and proved instantly popular. The label, based on Cabernet Sauvignon, Merlot and Cabernet Franc, is still 'balanced more towards fruit than wood,' says Corban.

Stables Sauvignon Blanc is a limey, non-wooded, bone-

dry style with ripe, persistent flavours. Its Chardonnay stablemate (spot the pun?) is a robust barrel-fermented wine with strong lemony fruit and oak flavours and touches of lees-aging complexity.

Alwyn Chardonnay is a very stylish, creamy, mealy wine, fermented in new French oak barriques and matured on its yeast lees for a full year. It displays deep, lush fruit flavours interwoven with steely acidity and a subtle, sustained finish.

Glazebrook Cabernet/Merlot is one of the winery's flag-ships. A deep-hued, tautly structured red with Cabernet Franc in the blend, it is matured in new French oak barriques. With its concentrated, spicy/leafy flavours and powerful tannins, this is a classic claret-style red which cries out for lengthy bottle-aging. It is only produced in the best vintages.

The pick of Ngatarawa's two fortified wines is the Old Saddlers Sherry, based on a solera started in 1982. This is a mature, amber hued, nutty medium sherry with excellent 'rancio' complexity.

Corban is facing the future with confidence. 'You haven't seen the best of Ngatarawa yet,' he says. 'It's taken us a decade to get to know our vineyard site and settle on our wine styles and market niches. But our profile needs to be higher. In the future I'll be getting out in the marketplace more.'

Sacred Hill Winery
Dartmoor Road, Puketapu

Owners: The Mason family

Key Wines: Sacred Hill Fumé Blanc, Chardonnay;
Whitecliff Sauvignon Blanc, Chardonnay,
Gewürztraminer; Dartmoor Cabernet Sauvignon,
Pinot Noir, Merlot

A lush, tropical fruit-flavoured, oaky Fumé Blanc is the most eye-catching wine flowing from this tiny winery in the back-country hills of the Dartmoor Valley.

David and Mark Mason produced Sacred Hill's first wines in 1986. David (35) oversees the company's administration and Mark (32) who holds a wine marketing diploma from Roseworthy College, is immersed in viticulture and winemaking. Their father planted the family's first vines near Puketapu ('sacred hill') in 1982.

The brothers enjoy first choice of the grapes cultivated in the family's vineyards. The fertile, silty twelve-hectare Dartmoor vineyard, just down the road from the winery, is established in Sauvignon Blanc, Chardonnay, Gewürztraminer and Pinot Noir; Cabernet Sauvignon, Cabernet Franc and Syrah have recently been added for a 'Stony Broke' red, to be launched from the 1994 vintage.

A couple of kilometres further up the valley, on a spectacular site overlooking white limestone cliffs carved by the Tutaekuri River, the six-hectare Whitecliff vineyard is close-planted in a wide array of varieties. Here the soils — red volcanic ash overlying limestone — are of very low fertility, and the lighter crops better suited to premium wines. 'We're so sheltered here,' says Mark Mason. 'We don't get the coastal breezes; our soils are warmer than on the plains and the fruit ripens earlier.' Is the rainfall heavier? 'I don't know. We have to irrigate at Whitecliff.'

Surrounded by trees alongside the main road through the Dartmoor Valley, the modest Sacred Hill winery was originally a farm building, extended in 1991 by excavating into the hillside and adding a gravity-fed pressing area and refrigerated cellar for barrel maturation. At about 5000 cases per year, Sacred Hill's output is still low.

Vinifera Ventures, the Masons' first company that produced the Sacred Hill wines, slid into receivership in 1989, but under a new trading name, Dartmoor Vineyards, the Mason brothers have bounced back. Their top wines are still marketed under the Sacred Hill label; lower-tier white wines are labelled Whitecliff and lower-tier reds Dartmoor.

Sacred Hill Fumé Blanc is the brothers' favourite wine. 'We're doing something a bit different by looking for a Graves style: full, complex and long-lived,' says Mark Mason. Half barrique and half stainless steel-fermented, this is a big wine with powerful oak and well-ripened peach and nectarine-like flavours. Sacred Hill Chardonnay, a barrel-fermented and lees-matured style with a strong malolactic influence, is robust, peachy-ripe, toasty and soft.

Whitecliff Chardonnay and Sauvignon Blanc are less obviously wood-influenced styles, placing their accent on fresh fruit flavours. Dartmoor Cabernet Sauvignon and Pinot Noir are both light, savoury and mellow.

'The past five years have been dizzy,' says Mark Mason. 'We haven't had time to sit down. But it's slowly coming together. In the future we want to reduce cropping levels. Reds will be more important, too, with mouthfeel and extract.' An outdoor café, serving a platter lunch of 'smelly cheeses, fresh breads and smoked meats', opens from late October until Easter.

Sacred Hill

David (left) and Mark Mason run the small Sacred Hill winery in the tranquil upper reaches of the Dartmoor Valley, inland from Taradale. Their most distinctive wine is Sacred Hill's barrel-fermented, moderately herbal, ripe-tasting Fumé Blanc.

St George Estate
St George's Road South, Hastings

Owners: Michael Bennett and Martin Elliott

Key Wines: Cabernet/Merlot, Petite Syrah, Rosé,
Chardonnay, Sauvignon Blanc, Fumé Blanc,
Rhine Riesling, Gewürztraminer, Cheval Blanc,
Vintage Port

St George Estate is a genuinely boutique-scale operation on the Havelock North-Hastings highway. Diners in the humming vineyard restaurant consume much of the output, which is spearheaded by a gutsy, full-flavoured and soft Cabernet/Merlot.

The company is a partnership founded in 1985 between winemaker Michael Bennett (53) and Martin Elliott (61), formerly proprietor of the local wineshop, who has recently retired. 'It was the worst possible time to start a winery,' says Bennett. 'Two years later all the yuppies disappeared.'

Born in England, Bennett made 'vinho verde' in Portugal

St George

Michael Bennett principally produces 'soft, early-drinking wines' under the St George label. 'Very few people ask for a wine they can cellar.'

selves. The annual output is still very low at 2500 cases. In the cosy vineyard restaurant, which opens every day of the week for lunches, the 'seasonal platter' — in spring featuring smoked salmon, avocados and asparagus — is popular. Bennett interrupts his winemaking to work in the restaurant two days per week: 'Ringing up the till is the fun part,' he grins.

Bennett favours soft, easy-drinking wines. 'My whole aim with our wines — steered by the response I get across the winery counter — is to make them drinkable early. Very few people ask for a wine they can cellar.' St George's three most popular wines are its Cabernet/Merlot, Chardonnay and Sauvignon Blanc.

The Cabernet/Merlot, blended from seventy-five percent Cabernet Sauvignon and twenty-five percent Merlot, used to be a dark, strapping red with the sweet, lifted perfume of American oak. More recently the wine has been matured in French oak casks. 'This is the wine I enjoy making the most,' says Bennett.

Robust and vigorous, the Sauvignon Blanc is packed with ripe, gooseberry-like flavours, and has recently been partnered by a subtly wooded Fumé Blanc. The Rhine Riesling is tangy, full bodied, lemony and bone-dry. The equally dry Gewürztraminer is weighty and full-spiced.

The barrel-fermented Chardonnay is a very upfront, buttery, toasty style with a creamy-soft finish — a delicious mouthful in its youth. Cheval Blanc — the name refers to the white horse on the label — is a musky, fruity medium style blended from Sauvignon Blanc, Riesling and Muscat; this is 'our Müller-Thurgau,' says Bennett. Another St George specialty is the Rosé, based on an unusual blend of Sauvignon Blanc and Malbec — dry, light but delivering plenty of flavour.

In the future, Bennett aims to 'keep producing the same wines, but I'll be working on their quality all the time. I won't get bigger; this is the ideal size for a one-man band.'

while trying to eke out a living as a freelance photographer before joining Villa Maria/Vidal for three years and then Te Mata between 1979 and 1984. Assured and amiable, he observes that each time he has shifted employment within the wine industry he has gone to a smaller-scale operation until eventually, he says, he 'will disappear altogether'.

On flat, silty soils in front of the winery, the three-hectare home vineyard owned by Martin Elliott and leased to the company is planted in Gewürztraminer, Muscat, Sauvignon Blanc, Riesling, Merlot and Cabernet Franc. Premium Chardonnay, Sauvignon Blanc and Cabernet Sauvignon grapes are also sourced from Bob Taylor's vineyard at Haumoana, planted in light silts overlying river gravels.

The elegant Cape Dutch-style winery was built out of concrete blocks and timber by Bennett and Elliott them-

Stonecroft

Stonecroft Wines
Mere Road, Hastings

Owner: Dr Alan Limmer

Key Wines: Chardonnay, Gewürztraminer, Sauvignon Blanc, Cabernet/Merlot, Syrah

'No sherries, no sparklings, no bus loads', advises the sign outside the tiny Stonecroft winery. Dr Alan Limmer's ambitions lie elsewhere: 'We've got one of the best vineyard sites in the country and therefore the potential to make some of New Zealand's best wines.'

Stonecroft (the name means 'stony small farm') lies in Mere Road, on part of the same shingle block as Gimblett Road in the Roy's Hill area west of Hastings. Limmer (41) studied earth sciences and chemistry at Waikato University, where he earned his doctorate, and then in 1981 moved to Hawke's Bay to manage a private analytical

laboratory servicing agricultural needs. He planted the first vines in his stony, extremely free-draining gravels in 1983 and made his first wines — a Chardonnay and Cabernet Sauvignon — in 1987.

Why did he plunge into the wine industry? 'I'd always enjoyed wine,' recalls Limmer. 'As a child I made fruit wines — jars of stuff were always bubbling away under my bed. I enjoy the land and Hawke's Bay and was keen to be involved in horticulture rather than pastoral farming because with horticulture you're closer to the crop. Wine was the perfect marriage of all those aspirations.'

Stonecroft's four-hectare vineyard is planted in Chardonnay, Gewürztraminer, Sauvignon Blanc, Cabernet Sauvignon and Syrah. These silty, sandy gravels running ten metres deep can change from a state of water saturation to vine-stressing aridity in only four or five days. 'This gives consistent vintages,' says Limmer. 'The weather changes from one year to another but my fruit is much the same. With trickle irrigation [installed to avoid excessive water stress] I've got amazing control — I can almost dial up the vintage I want.'

Syrah was planted as 'a punt'. In Stonecroft's warm soils the vintage is usually finished by early April; on occasions the harvest, including the picking of the late-ripening Cabernet Sauvignon variety, is over by the end of March. 'I am trying to work out what we can do with the three or four weeks we have left of the normal vintage,' says Limmer. 'Syrah fits in well. We can certainly get it ripe.'

Limmer has to employ temporary staff to assist with the pruning and picking, but he and his wife, Glennice (a full-time teacher) and their two children perform all the other vineyard and winery tasks. 'We'll sacrifice anything to get the fruit quality,' says Limmer. The vineyard yields low crops — around six tonnes per hectare — and the fruit is left on the vines as long as possible, ripening to very high sugar levels and sometimes shrivelling.

In his rustic, vertical-timbered, stained glass-windowed winery, which Limmer built himself, each year he produces only about 1250 cases of wine, which may climb to a maximum of 1750 cases. A winemaker who understands wine chemistry, he nevertheless favours 'a very hands-off approach. I recently produced a Chardonnay without having any idea of its pH or total acidity.'

Limmer also harbours very few preconceptions about Stonecroft's wine styles. 'I am definitely not trying to make to a formula — it doesn't worry me if the wines are different from season to season. If we decide to leave fruit on the vine and it comes in too ripe to make a dry wine I will make a sweet one. I would hesitate to make a sweet Cabernet, but that's about where I'd draw the line!'

Stonecroft Syrah is a strapping red with flashing purple-black hues and concentrated peppery, chocolatey flavours, rich and lingering. Matured for twenty months in half new French oak barriques, it is a clear signpost to Syrah's potential in New Zealand's warmer and less fertile vineyard sites.

The dark, powerful, fragrant and supple 1991 vintage is the finest Syrah yet made in New Zealand. 'Syrah is a very seductive wine,' says Limmer. 'I'm seduced.' His Cabernet/Merlot is also impressively robust, dark and chewy.

Stonecroft Chardonnay is a bold, richly alcoholic, citric-flavoured wine with a steely finish. Fermented and lees-matured for over a year in French oak casks, Limmer sees it as an 'austere, slowly-evolving' style. The Gewürztraminer and Sauvignon Blanc have ranged in style from sweet to medium to dry, with the odd disappointment along the way, but these wines are usually full of character.

'My last nine years have been 100 percent focused on the winery,' says Limmer. 'I've spent twelve to fourteen hours per day, seven days per week on the job. That's the price you pay — every penny, every spare hour.'

Dr Alan Limmer produces dark, strapping, full-flavoured reds. His perfumed, peppery and supple Stonecroft Syrah 1991 set a new standard for the variety in New Zealand.

Te Mata Estate Winery
Te Mata Road, Havelock North

Owner: Te Mata Estate Winery Limited

Key Wines: Coleraine Cabernet/Merlot, Awatea Cabernet/Merlot, Cabernet/Merlot, Rosé, Elston Chardonnay, Cape Crest Sauvignon Blanc, Castle Hill Sauvignon Blanc, Oak Aged Dry White

Discussion of Te Mata until recently invariably turned to the much-vaunted Coleraine Cabernet/Merlot and its slightly less well-known stablemate, the Awatea Cabernet/Merlot. New Zealand's international reputation hinges almost exclusively on its white wines, but Te Mata was initially acclaimed for its magnificent reds. Now its formidable Elston Chardonnay and — to a lesser extent — Cape Crest and Castle Hill Sauvignon Blancs have also moved centre-stage.

Bernard Chambers, a member of the wealthy landowning family that ran Te Mata Station, as a hobby in 1892 planted a hectare of vines supplied by the Mission brothers. The vines flourished, leading Chambers to convert a stable erected in 1872 into his cellar, to employ an Australian winemaker and, in 1896, to plunge into commercial wine production.

By 1909 the Chambers vineyard was the largest in the country, annually producing 54,000 litres of wine from the fourteen hectares of Meunier, Syrah, Cabernet Sauvignon, Riesling and Verdelho vines. Commented the *New Zealand Journal of Agriculture* in May 1914: 'Mr Chambers' wines are principally hocks, claret and sweet, and are commanding a large sale.' A bottle of Te Mata Madeira 1906 was broached in Auckland in 1990; according to wine writer Ron Small, 'it smelt and tasted like very old prune juice, but it was quite drinkable.'

But from this early peak, production declined and eventually during the Depression the vineyard went into

Te Mata

receivership. A series of new owners failed to restore the vineyard's fortunes.

The revival of Te Mata's reputation began in 1974, when Michael Morris and John Buck, both then active as wine judges, acquired the run-down company. Morris, a Wellington business executive, is a non-working partner. Buck, the managing director, has enjoyed a high profile on his career path leading finally and triumphantly to the hills of Havelock North.

After a two-year career in the United Kingdom wine trade, Buck returned to New Zealand in 1966. 'I drove up to Hawke's Bay, spent some time with Tom McDonald, was given a couple of Tom's Cabernets which were served up blind at a luncheon with a Château Haut-Brion, and I also got a thing called McWilliam's Tukituki Rhine Riesling . . . I saw these wines and I thought, ye gods! There's actually some colossal potential here.' Buck was soon convinced that in Hawke's Bay, given the correct combination of site, soils and grape varieties, wines of world class could be made.

After an eight-year search for the right site, during which Buck and Morris looked at 150 properties, they purchased the old Te Mata winery; the cellars, built of brick and native timbers, were restored to their original condition and equipped with stainless steel tanks and new oak casks. For the new owners' first vintage, 1979 — an unusually wet one — the winemaker of that period, Michael Bennett, had to work with such limited fruit material as Chasselas, Baco 22A, Palomino and Müller-Thurgau. 'I really thought after that vintage, you know, if this is winemaking, what the hell am I doing here?' recalls Buck.

The year 1980 brought a rapid change of fortune. 'We were fortunate to acquire the Awatea vineyard and it had a small block of old Cabernet on it. I guess we picked three or four tonnes off it and the moment we crushed those grapes we knew that our assertion as to the right variety to grow on these hills was correct.' Te Mata 1980 Cabernet Sauvignon then carried off the trophy for the best red table wine at the 1981 National Wine Competition — a feat repeated by the 1981 vintage in 1982 — and Te Mata was on its way.

The winery until recently marketed all its top wines with a vineyard site designation. Coleraine Cabernet/Merlot, for instance, was sourced from John and Wendy Buck's own two-hectare vineyard called Coleraine, planted with a mix of varieties modelled on the Médoc: Cabernet Sauvignon seventy-five percent, Merlot twenty-two percent and Cabernet Franc three percent. Since the 1989 vintage, however, Te Mata's red wines have been produced in a tiered group, with Coleraine at the top, closely followed by Awatea, and then a third wine called Te Mata Cabernet/ Merlot. 'All three will be a blend of wines from our spectrum of vineyards,' says Buck. 'This development gives us access to a far greater range of flavours when assembling the wines, providing more flexibility . . . to craft even finer wines.'

John Buck (left) and Peter Cowley's (and indeed New Zealand's) most prestigious red wine is Coleraine, named after the Buck family property (at rear). Since 1989 Coleraine Cabernet/Merlot has been no longer a single-vineyard wine, but Te Mata's top-tier claret-style red.

John Buck

The driving force behind New Zealand's most prestigious red wine — Coleraine — and one of its most acclaimed Chardonnays — Elston — John Buck was in 1991 elected chairman of the Wine Institute. He is clearly one of the wine industry's dominant figures.

Buck loves the limelight — and performs brilliantly in it. 'I don't think I'm a person who has ever lacked self-confidence,' he admits. He is learned ('The lessons of wine are all there in the books and magazines'), sharp-witted and extroverted. He came to Hawke's Bay because he wanted to produce Cabernet-based reds and the 'richer' style of Chardonnay. 'This is the logical place to make them,' he says.

Buck has long been intrigued by Cabernet-based reds. 'I think it is the most difficult wine to make properly. I don't accept the argument that Pinot Noir is the hardest to make. The French concepts of "elevage" [raising the wine, as one would children] and "assemblage" [final selection and blending] really come into play when you are using a multitude of varieties from different soils. Cabernet-based reds therefore have a lot of human input to put the best wines together and often the better the people the better the wine. The *creating* of a wine represents a challenge; perhaps that is what this winery is all about.'

In practice this involves broadening the range of blending options, initially by handling the fruit from each block separately. 'You then introduce the components of wood — type, age and so on. After maturation you start bringing the various components together. Your success then depends on knowing beforehand what you are setting out to make, and on how ruthlessly you cull. It strikes me that if you look at the good red winemakers in Australasia the best ones cull ruthlessly.'

Tasting ability is also crucial. 'The best wine in the world is made by people who understand what technology can do but still have the palate. It's like the chef dipping his wooden spoon in the stock pot, licking it and deciding it still needs more of this or that.'

Coleraine Cabernet/Merlot sells for about $35 directly from the winery; Elston Chardonnay for $27.50. Does Te Mata strike any price resistance? 'No, we price to a level we think the market will sustain. Our prices in real terms have not gone up for several years. As we get more firmly established we are gaining efficiencies, which are then passed on to our customers. We know what rate of return we want and if we achieve that we are happy — any other advantages go to the customer. We were regarded as operating at the upper end of the price market — we are certainly not now.'

Buck refuses to use other New Zealand wineries as yardsticks. 'I think there are dangers inherent in this, using local standards. We are very tightly focused on our own business, and we evaluate ourselves in the market, especially the export market, where nobody is doing us any favours.'

Buck is still only fifteen years into his first twenty-year plan. 'We intend to be here for many years — Te Mata is not for sale to the first person with the money to buy. It's a lifetime commitment.'

Elston vineyard, the source of Te Mata's powerful Chardonnay, is established in tan-coloured gravels (or 'red metal') topped with sandy loams. Cape Crest, planted on a terrace overlooking Te Awanga, is another 'red metal' vineyard. These are not company-owned vineyards. For winemaker Peter Cowley (39), the answer to Te Mata's success lies here, in these vineyards sited up off the alluvial flats on well-drained, low-vigour soils sloping north. An Auckland University BSc graduate, Cowley gained a Roseworthy diploma and then worked with Larry McKenna at Delegat's before joining Te Mata. 'It's not a bad place to work if you have to work,' he says grinning. Cowley is now a winery shareholder: 'Our prosperity is linked,' says Buck.

Buck, Morris, Cowley and their families have also purchased a sixteen-hectare, 'red metal' block midway between Ngatarawa and Maraekakaho. The partners' first commercial crop of red-wine grapes was harvested in 1993.

Te Mata's Ian Athfield-designed, plastered-concrete headquarters, painted throughout in cool pastel shades, houses a boardroom and kitchen on the upper floor, with the offices, sales area and a tasting room at ground level. A landscaped courtyard has recently been added featuring a cloistered walkway, sitting steps and fish ponds. A concrete temperature-controlled red wine 'cuverie' (fermentation room) with pneumatic plungers was also introduced for the 1991 vintage.

With its annual output of about 16,000 cases, Te Mata is still by New Zealand standards a fairly small winery; Buck estimates production will climb to 24,000 cases by 1997. Exports are of growing importance, especially to Australia, Great Britain and Switzerland.

Buck and Cowley's determination to produce long-lived wine styles is a central strand in Te Mata's approach to winemaking. 'To gain true international recognition, an industry has to be capable of making wines that improve with age — that's the ultimate quality factor,' says Buck. People need to be able to 'put wine into their cellars with confidence and know that when they pull them out they will be a damn sight better than when they put them in.'

Buck recommends drinking Elston Chardonnay at between three and five years old. In their extreme youth, the Coleraine and Awatea Cabernet/Merlots 'show

Elston

Since its first 1984 vintage, Elston Chardonnay has stood out for its power and ability to flourish for several years in the bottle, evolving satisfyingly rich, complex, citrusy, savoury flavours.

When John Buck and Michael Morris bought the Te Mata winery, the fruit from two vineyards was available with the purchase, but other sites had to be identified and planted in accordance with the partners' quality aspirations.

'Hawke's Bay has a very confused geology,' says Buck. 'You have to get the maps and the scientists to help select the site, and then plant the varieties to match the soils. Then we got a guy, John Edwards, to buy the land and grow the grapes at Elston.'

The Elston vineyard lies on a north-east facing slope in Te Mata Road, very close to the Te Mata winery. In low-fertility, non-irrigated soils, three hectares of Chardonnay (the Mendoza clone, known for its 'hen and chicken' formation of large and small berries) have been planted. The vines' spacing (1.5 metres between plants, three metres between rows) and trellising (mostly standard upright, with some Scott-Henry) are both conventional.

A hard silica pan runs down the hills from Te Mata Peak, thinning out down the terraces at the Awatea vineyard, directly below Elston. To protect the vines from extreme water stress, the pan one metre below the surface at Elston has been ripped and shattered, allowing water to penetrate the sub-soil. Elston's soils also have a high content of lighter calcareous matter. 'We wanted this for Chardonnay,' says Buck. 'Calcareous [limestone] soils give you "nervosite" — racy, steely acids.'

The grapes are usually harvested very ripe, at 23.5 brix or above. Reflecting the vineyard's low soil fertility, yields are very low, averaging only five tonnes per hectare.

Winemaker Peter Cowley describes the fruit as 'fat and ripe, with grapefruit flavours, and I don't want to modify that too much. I want to keep that flavour, but add to it, rather than overpower it.' In the winery this involves 'not too much of anything. Nine months in wood and lees contact. Fermentation in barrel, the usual things, but always checking, tasting, being careful not to lose anything. Some proportion of malolactic fermentation helps. I am adding complexity, not killing it.'

"primary" characters — fruit, oak, tannin, acid and so on — all being separately identifiable. They then go into a quieter phase at between three and a half and five years from harvest, following which they emerge with all the separate factors integrated, that is they become claret-like. From this point on they will continue to develop, gaining in aroma and softening in palate . . .'

Coleraine, compared with Awatea, is more new oak-influenced and more slowly evolving. The 1991 vintage is a blend of fifty-nine percent Cabernet Sauvignon, twenty-nine percent Merlot and twelve percent Cabernet Franc. The 1989–91 vintages of Coleraine are glorious reds: the 1989 dark, concentrated, spicy and tight-knit; the 1990 deliciously perfumed, ripe and supple; the 1991 deep-scented, robust and superbly full flavoured. Awatea is usually voluptuously fragrant, deep-flavoured and rich, and markedly more forward than the Coleraine of the same vintage. Both are great wines, firmly in the vanguard of New Zealand's reds.

Elston Chardonnay is a very mouthfilling, citric-flavoured, mealy style, wholly barrel-fermented and matured on its yeast lees for 'ages', according to Peter Cowley. At about four years old it is a notably 'complete' wine, powerful and complex. Elston, too, is consistently one of the finest New Zealand examples of its style.

Of Te Mata's Cape Crest and Castle Hill Sauvignon Blancs, more often than not the Cape Crest has the edge in weight and flavour depth. Since the 1991 vintage, Cape Crest has been whole bunch-pressed and one-third barrel-fermented, in a bid to more clearly differentiate its style from that of Castle Hill. Both wines are classic examples of the Hawke's Bay Sauvignon Blanc style: robust, ripely herbal and sustained. Castle Hill in its youth is the more obviously 'varietal' of the two; Cape Crest is potentially the more complex.

Easily overlooked in the Te Mata range is the Oak Aged Dry White, a fifty-fifty blend of Sauvignon Blanc and Chardonnay. This is a full, fresh and unexpectedly flavour-packed wine, bargain priced. The Rosé, blended from Cabernet Sauvignon, Cabernet Franc and Sauvignon Blanc, is bright pink, dry and tangy, with Sauvignon Blanc's herbal flavours clearly evident.

'Nothing is spur of the moment,' says John Buck. 'Our next tasks are to finish the new vineyard and build a barrel hall, packaging and warehousing centre. We came here with a very clear twenty-year plan. We've since "downsized" the winery to a 24,000-case goal, but otherwise we've rigidly adhered to the plan.'

The McDonald Winery
200 Church Road, Taradale

Owner: Montana Wines

Key Wines: Church Road Chardonnay, Church Road Cabernet Sauvignon

With much fanfare, in 1988 and 1989 Montana thrust into Hawke's Bay, snapping up $6 million worth of existing vineyards and the old McDonald winery at Taradale. Its goal: to elevate the standard of the giant company's previously unspectacular red wines and — to a lesser extent — Chardonnays.

Although deeply rooted in the Gisborne and Marlborough regions, New Zealand's largest wine company had previously been conspicuously absent in Hawke's Bay. Its belated but heavy investment in the Bay — the country's foremost red-wine region — is designed not only to enhance Montana's wine quality but also its reputation.

'The key to The McDonald Winery will be image,' observes Peter Hubscher, Montana's managing director, acknowledging the fact that wineries with a huge output consistently struggle to win consumer acceptance at the top end of the market. Montana's ownership of the rejuvenated McDonald winery has given it a small-scale production facility ideal for making limited volumes of top-flight wines and a golden opportunity to market them in direct competition with the boutiques' top labels.

Montana's purchase of Penfold's (NZ) in late 1986 first triggered its involvement in Hawke's Bay, by linking it with Penfold's contract growers. By also purchasing 238 hectares of vines, Montana swiftly staked out a substantial share of the region's vineyards.

Montana's operation in Hawke's Bay is double-pronged. Most of its Hawke's Bay fruit, destined for the company's commercial labels, is trucked to Auckland and processed at the company's Glen Innes winery. Only the finest, hand-harvested grapes are crushed, fermented, and matured at The McDonald Winery.

'Red wines will be the focus here,' says winemaker Tony Prichard (34), a Massey University food technology graduate who joined Montana at its Gisborne winery ten years ago. The McDonald Winery's first red wine, 1990 Church Road Cabernet Sauvignon — a mid-range label — was highly impressive, but Prichard is adamant that the company won't be launching its top-end-of-the-range red 'until we're absolutely confident it's competitive with New Zealand's best. That could take five to ten years.'

Montana's ambition at The McDonald Winery is crystal-clear: 'to produce a small quantity of wine . . . of the highest quality ever produced in this country.' To boost its red-wine chances it has formed a close relationship with Cordier, one of Bordeaux's major wine firms with extensive cru classé holdings — notably Châteaux Talbot and Gruaud-Larose in St Julien — and several excellent petits châteaux under its wing.

Cordier's Bordeaux-based winemakers travel regularly to New Zealand; Montana's winemakers go to Bordeaux. 'The work covers viticulture, fermentation techniques, cellar maturation and blending,' says Hubscher. The fruits of this venture, based at The McDonald Winery, are awaited with extreme interest by wine lovers.

The extended roots of The McDonald Winery run back to the closing years of the nineteenth century. In 1896 Bartholomew Steinmetz, a native of Luxembourg, resigned his position as a lay brother at the Society of Mary's Marist Mission to settle in Taradale and marry. Steinmetz purchased five acres (2 ha) from the estate of pioneer winemaker Henry Tiffen, planted vine cuttings supplied by the adjoining Mission, and by 1901 the first vintage of Taradale Vineyards' wines was in the barrel.

For the next twenty-five years Steinmetz made his living selling table grapes and wine. The 22 October 1926 edition of the *Daily Telegraph* carried this advertisement:

Steinmetz's Vineyard, Taradale
Best N.Z. wines on the market. Phone 23. Started 1896 and still going strong. First grapes 1897. Port, Madeira, etc. etc. all at two pounds per dozen (Quarts). Fine invalid Port 10 years old 2 pounds 10 shillings per dozen. Draught. Fine old port 18/- per gallon. Customers to provide own jars. Delivered free any part of Napier. Cash on delivery. Samples and terms to Hotels on application.

The McDonald Winery

The manager of The McDonald Winery is Barrie Browne (left). Tony Prichard, the winemaker, aims 'to make, consistently, the best white and red wines in New Zealand'.

Also in 1926, Steinmetz leased his winery to nineteen-year-old Tom McDonald, whose later exploits with Cabernet Sauvignon were to indivisibly link the name McDonald with fine quality reds (see page 10).

The nine vineyards Montana purchased in Hawke's Bay in the late 1980s have since been grouped into four estates — Phoenix, Twin Rivers, Korokipo and Fernhill. Not all the vineyard sites are outstanding; according to Montana, some have 'good moisture retention and are suitable for high-yielding varieties . . .' Montana expects 'top quality' fruit, however, from its Phoenix Estate Vineyard alongside the Tukituki River. Named after a striking avenue of phoenix palms in its midst, this vineyard has a mix of soil types — gravelly, silty, sandy and heavy.

Vineyards composed of a 'fruit salad' of varieties have been rationalised, with the new planting emphasis on Cabernet Sauvignon, Merlot, Pinot Noir and Chardonnay. Reduced crop levels and improved fruit exposure 'meant heaps in terms of fruit quality improvement between the 1990 and 1991 vintages,' says winemaker Tony Prichard.

Refurbishing and re-equipping The McDonald Winery cost Montana $2 million. Most of the original buildings, many featuring large Oregon beams and rimu panelling, have been retained. Following the installation of sixty stainless steel tanks of varying sizes, a German Willmes press, a basket-type Coquard Champagne press (for a 'serious' bottle-fermented sparkling) and refrigeration plant, The McDonald Winery now has the capacity to crush about 1200 tonnes of fruit, making it a medium-sized winery by New Zealand standards. One thousand French and American oak casks lie in the insulated cooperage, where The McDonald Winery's early vintages are maturing.

The refurbishment of 'The Cellar Door' winery shop, the creation of a wine museum, and development of a landscaped garden, have transformed The McDonald Winery into one of the highlights of the Hawke's Bay wine trail. Renovated in the art deco style popular in Hawke's Bay,

'The Cellar Door' boasts an absorbing array of wine-related items — books, glasses, corkscrews — and old vintages of Montana wines.

Upstairs, in the country's first wine museum, strikingly lifelike models toil with pioneer winemaking equipment. In the gardens below, a collection of grape vines is being planted to enable the public to see at close quarters the varieties that form the foundation of the winery's range.

The first grapes were crushed at The McDonald Winery on 6 March 1990 and the first release under The McDonald Winery label, Church Road Chardonnay 1990, was of eye-catching quality for its modest $13–$15 price-tag. The 1991 proved even better. Fermented and then lees-matured for nine months in French and American oak casks, it is rich and complex, with a savoury, buttery, well-sustained flavour. The 1992 is equally good. This label is shaping up as one of the best-value Chardonnays in the country.

Church Road Cabernet Sauvignon 1990 was the first red marketed by The McDonald Winery. Matured for almost a year in new French and American oak barriques, this is a well-structured red, with impressive weight and layers of complex, minty, spicy flavours, stylish and sustained. In the mid-price range, it too delivered almost unrivalled value for money. The 1991 vintage is even better.

The purchase of Tom McDonald's old winery has undoubtedly been a public relations coup for Montana, enabling it to bask in the reflected glory of the legendary winemaker's prestige (although 1992 brought a legal struggle between Montana and Corbans over rights to the McDonald brand. The outcome has been that Montana continues to call the property The McDonald Winery, but markets its output solely under the Church Road brand.)

Montana declared in 1989 that a Tom McDonald commemorative wine would be produced, 'New Zealand's answer to Australia's prestigious Grange Hermitage'. The next few years will show whether The McDonald Winery can rock the world with its reds, the way Tom McDonald electrified winemakers and wine lovers several decades ago.

Vidal

Vidal of Hawke's Bay
913 St Aubyns Street East, Hastings

Owner: Villa Maria Estate

Key Wines: Reserve Chardonnay, Fumé Blanc, Gewürztraminer, Cabernet Sauvignon, Cabernet Sauvignon/Merlot; Private Bin Chardonnay, Fumé Blanc, Gewürztraminer, Müller-Thurgau, Cabernet Sauvignon/Merlot; Méthode Champenoise Brut, Blanc de Blancs

The quintet of wines marketed under Vidal's reserve label are extraordinarily good, often achieving truly exciting heights, and capable of footing it in the quality stakes with

any other wines in Hawke's Bay. Vidal is an old-established Hastings winery which — like Esk Valley at the far end of the Bay — is now an integral part of the Villa Maria conglomerate.

Anthony Vidal, the founder, came to New Zealand from Spain at the age of twenty-two in 1888. After eleven years working with his uncle, Wanganui winemaker Joseph Soler, Vidal experimented with viticulture at Palmerston North before shifting to Hawke's Bay. In 1905 he bought a half-hectare property at Hastings, converted the existing stables into a cellar and planted grapevines.

The winery flourished; a new, three-hectare vineyard was established at Te Awanga in 1916 and, a few years later, another three hectares was acquired from Chambers' Te Mata vineyard. After Anthony Vidal's death, control of the company passed to his three sons: Frank, the winemaker; Cecil, who concentrated on sales; and Leslie, who supervised the vines. For decades the winery enjoyed a solid reputation. John Buck, now of Te Mata Estate, in 1969 stated in his book *Take a Little Wine*, that Vidal's Claret and Burgundy were 'the two finest, freely available dry reds on the New Zealand market'. Using Cabernet Sauvignon, Meunier and hybrid fruit, the brothers produced a Burgundy of 'style, good colour, body and balance' and a Claret 'lighter in body and more austere to taste'.

But after 1972, when Seppelt's of Australia acquired a sixty percent share of Vidal, standard lines were dropped, labels changed and the quality of the wine began to fall away. The slide continued under another owner, Ross MacLennan, from 1974 to 1976.

The steady restoration of Vidal's reputation began in 1976, after George Fistonich of Villa Maria bought the company. In 1979 the first vineyard restaurant in the country opened at Vidal, to which a late-night café, 'Just Desserts', has been added. The company has retained its separate identity under management policy and the grapes, all contract grown, are drawn entirely from the Hawke's Bay region.

Following the 1990 departure of high-profile winemaker Kate Radburnd (formerly Marris) — who in her four vintages at Vidal won a raft of major trophies — for C.J. Pask, Elise Montgomery has taken over the reins. Auckland-born, Montgomery (30) graduated in 1985 as a Bachelor of Horticultural Science from Massey University, and in 1986 from Roseworthy College with a Graduate Diploma in Wine (Oenology). After working successive vintages in Victoria, California and France, Montgomery joined Villa Maria as an assistant winemaker in early 1990. 'The big joke at Villa Maria was that for the first time I'd have to work two vintages at the same place. It didn't happen — in late 1990 I was appointed winemaker at Vidal.'

Vidal has recently enjoyed glowing success in show judgings. Vidal Reserve Chardonnay 1990 was the champion Chardonnay at the 1991 Air New Zealand Wine Awards, and its Reserve Cabernet Sauvignon/Merlot 1990 won the trophy for champion wine at the 1993 Liquorland Royal Easter Wine Show. At the 1990 Smallmakers' (now Sydney International Wine) Competition, Vidal Reserve Cabernet Sauvignon/Merlot 1987 won the trophy for champion wine of the show, forcing Australian wine judges to sit up and take notice of a New Zealand red.

Vidal is a medium-sized winery, with an annual output of about 50,000 cases. Under its Private Bin label, Vidal markets a range of solid Hawke's Bay varietals, including a spicy/citric-flavoured, slightly sweet Gewürztraminer; a very lightly wooded Fumé Blanc which is a tasty amalgam of grassy-green and ripe tropical-fruit flavours; a full-bodied, lemony Chardonnay in the typical Hawke's Bay mould; and a robust, deep-flavoured Cabernet Sauvignon/Merlot which in top vintages can be an unexpectedly classy, irresistibly good-value red. The Méthode Champenoise Brut, blended from seventy percent Pinot Noir and thirty percent Chardonnay, matured on its yeast lees for

two years, is light, moderately complex, firm and crisp.

The jewels in the Vidal crown are its formidable lineup of Reserve wines. Bursting with powerful, ripe, intense Hawke's Bay fruit flavours, they are also stylish, extremely well-balanced wines.

Vidal Reserve Chardonnay, barrel-fermented and lees-matured in new French oak barriques, is mouthfilling with an explosion of ripe stone-fruit, oak and yeasty flavours, all in harmony. The Reserve Gewürztraminer is a big, lush style with pungent spiciness. The Reserve Fumé Blanc, fermented and briefly matured in new French oak barriques, bursts with seductively ripe melon-like fruit flavours and a deft touch of wood.

Vidal's superb Reserve Cabernet Sauvignon serves as a reminder that New Zealand's top Cabernet-based reds don't always have to be blended with Merlot to achieve distinction. Based (since 1990) on fruit grown at Clive, matured up to eighteen months in new and one-year-old French oak barriques, this is a power-packed, spicy, cedary, tannic red of great richness and length. Its stablemate, the marvellously fragrant Reserve Cabernet Sauvignon/Merlot, from Gimblett Road, is a more voluptuous style, its delicious surge of plummy, minty fruit and strong oak flavours underpinned by firm tannin. These are great reds.

Elise Montgomery, who has made wine in California and France, was appointed winemaker at Vidal in 1990. The Cabernet-based reds and Chardonnay under the Vidal Reserve label are consistently stunning.

WAIRARAPA

The Wairarapa — in particular its most famous wine district, Martinborough — has in less than a decade emerged as one of New Zealand's most prestigious winegrowing regions. The first modern-era Wairarapa wines were ensconced in bottles in 1984; since then the flow of Chardonnays, Rieslings, Pinot Noirs and other varieties has been of exciting quality. New arrivals are mushrooming: the Wairarapa wine trail now boasts over a dozen wineries.

The southern part of the North Island has inherited its own winemaking legacy. In 1883 a wealthy Wairarapa landowner, William Beetham, planted the first vines at his tiny Masterton vineyard. Romeo Bragato — who was later appointed New Zealand's first Government Viticulturist — visited Beetham during his 1895 national vineyard tour, and reported tasting a Hermitage wine of 'prime quality'. Following the prohibitionists' no-licence victory in Masterton in 1905, Beetham's vineyard was uprooted.

One of the first vineyards in Martinborough (recently described in the *Evening Post* as 'a picturesque and drowsy country town on the road to nowhere') was established by the then prominent publisher and wine lover Alister Taylor, on flats near the Huangarua River in 1978. 'It reminded me of Blenheim,' recalls Taylor. 'I thought: "Grapes will grow here".' He planted Chardonnay, Chenin Blanc and Gewürztraminer, but eventually the vineyard fell prey to the depredations of rabbits, possums and mortgagors. The land was taken over by Tom and Robin

Draper, who later founded the Te Kairanga winery.

The major impetus for the recent surge of interest in Martinborough winemaking, however, came from Dr Derek Milne's 1979 report, pinpointing similarities between Martinborough's climate and soils and those of premium French wine regions. 'I had joined a serious [Wellington] wine tasting group,' Milne recalls. Following a 1977 visit to Germany and Alsace, which gave him 'an insight into actual site selection', Milne looked closely at New Zealand's climate regions. 'Climatic comparisons with Europe indicated that there were many areas well suited to *vinifera* grape production as yet unexploited, most being in areas which traditionally have been regarded as too cold to ripen grapes. Martinborough was one of those areas and it was developed because it was closest to Wellington, where the original pioneers [of Martinborough wine] were based.'

A 'gang of four' pioneered the planting of commercial vineyards in Martinborough: Dr Neil McCallum of Dry River in 1979, followed in 1980 by Clive Paton of Ata Rangi, Stan Chifney and Derek Milne himself — who put his money where his mouth was as a founding partner in Martinborough Vineyard. As Chifney was the only one who could initially afford a winery, the earliest Martinborough wines were all crushed and fermented at his tiny winery.

'Looking back,' says Chifney, 'I now realise how green everyone was. Neil [McCallum] knew about the theoretical

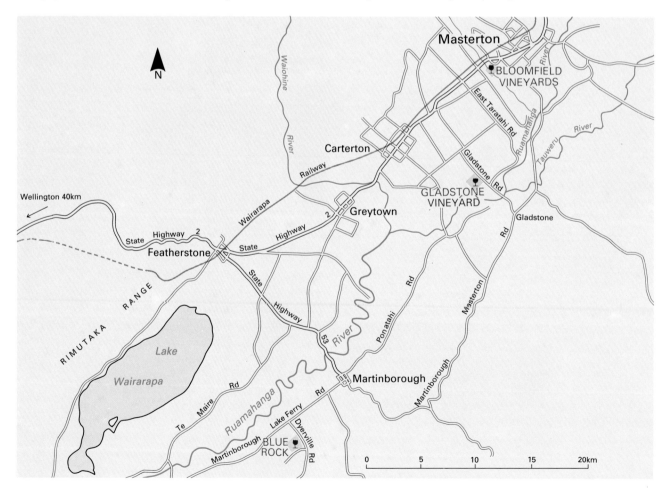

side, being a PhD, and I knew how to use tanks and pumps and pH measuring equipment, because of my laboratory experience, but none of us had technical training in winemaking. We learnt from our mistakes.'

A turning-point was the 1986 arrival of the first experienced winemaker, Larry McKenna of Martinborough Vineyard. In the eyes of Richard Riddiford, of Palliser Estate, 'Larry put Martinborough on the map.'

Several — although not all — of Martinborough's winemakers share a conviction that their district is a sort of 'southern Burgundy'. According to an early brochure of the Martinborough Winemakers Association, their study of relevant climatological and soil data indicated that 'conditions were similar to those in Burgundy'.

Martinborough and Burgundy have roughly similar total rainfall figures (Burgundy 650–700mm per annum, Martinborough 750mm). The two winegrowing areas are also believed to accumulate comparable heat readings (1150 degree days Celsius) during the vines' growing season. (This heat summation figure has been disputed by some Martinborough winemakers currently enjoying success with Cabernet Sauvignon-based reds, including Bill Benfield, who claims readings taken by the local meteorological service confirm a significantly higher figure of 1395 degree days Celsius.) The Martinborough grapegrowers have planted both Burgundy and Bordeaux grapes: Pinot Noir, Chardonnay and Cabernet Sauvignon are (in that order) by far the most heavily planted varieties.

Despite its location in the North Island, Martinborough's viticultural climate in fact resembles Marlborough's more closely than Hawke's Bay's. Martinborough's spring bud burst, for example, is typically two weeks behind Hawke's Bay's. Cold rainy conditions in spring can be a problem for Martinborough's viticulturists, causing poor flowering and fruit 'set'.

Martinborough is the driest area in the North Island. 'Being further south than Gisborne or Hawke's Bay,' says Derek Milne, 'Martinborough escapes the worst of the late summer and autumn rains that result from tropical depressions that travel down the eastern side of the North Island in some years.' Martinborough's and Marlborough's autumns are on average markedly drier than Hawke's Bay's.

Droughts are a definite threat in summer, leading some of the wineries to install fixed irrigation systems in their vineyards. This relative dryness, however, also encourages the winemakers to hang their grapes late on the vines to achieve full fruit ripeness, without facing any undue risk of disease.

In March and April, the vital months leading into the harvest, Martinborough's average daily temperature of 14.7°C is more akin to Marlborough's 14.3 degrees than Hawke's Bay's 15.8°C. The cooler autumns enable the development of intense flavour in the berries without any pronounced loss of acidity.

The chief weather drawback here is the wind: the shape of Martinborough's trees tells the story. Martinborough is

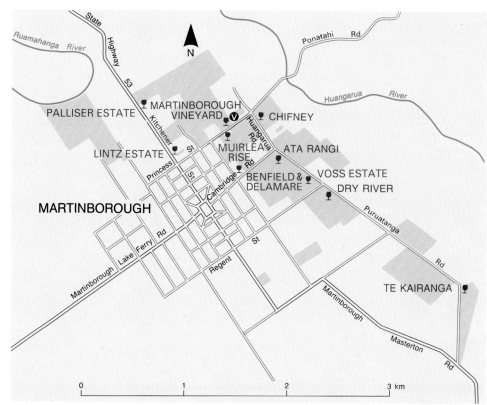

not only pummelled by strong southerlies from Cook Strait, but is also exposed to regular northwesterly gales. The Merlot variety is especially vulnerable to the winds' onslaught, suffering severe shoot damage and loss. Shelter belts are a necessity.

Martinborough's three basic soil types are not all equally well suited to quality viticulture. Neither the areas of poorly drained clay loam overlying silt pans, nor the fertile silt loams well supplied with water, are regarded as ideal for grapes. Most highly sought-after are the pockets of friable, gravelly silt loams overlying free-draining alluvial gravels.

An area of high-quality, free-draining soils on the northeast side of the town, bounded by the Ruamahunga and Huangarua Rivers, was early delineated by the Winemakers Association, laying the foundation for an appellation system designed to protect and promote Martinborough's vinous reputation. Since 1986 the wines produced from grapes grown within the confines of this area, and also processed there, have been eligible to carry a handsome black and gold seal, guaranteeing their regional authenticity.

The seal has not been entirely free of controversy, with some recent arrivals on the Martinborough wine scene arguing it gives the winemakers whose vineyards are in the defined area an unfair marketing advantage. According to Dr Neil McCallum of Dry River, however, 'the appellation within Martinborough (now called "The Martinborough Terrace" to avoid confusion) is being presented as a homogeneous viticultural area typified by free-draining alluvial gravels and very low rainfall. It is not being promoted as better or worse than any other part of Martinborough which contains vineyards.'

Land prices have soared in Martinborough in the wake of the wines' success. In the late seventies, blocks of stony land fetched $2500 per hectare; today $20,000 per hectare and more is common. With large tracts of land not generally available — the land around Martinborough is mainly divided into four to eight-hectare farmlets — the larger wine companies have not yet been attracted to the district.

Winslow is a new label. Grown in Ross Turner's Martinborough vineyard and produced at the Chifney winery, Winslow Cabernet Sauvignon/Franc 1991 scored a silver medal at the 1992 Air New Zealand Wine Awards.

Other areas not far from the township are now being planted in vineyards, including Ponatahi, three kilometres to the north of Martinborough, and Te Muna, three kilometres to the south-east. Gladstone, between Masterton and Martinborough, is also showing promise.

Few other vineyards are found in the southern part of the North Island. Pierre is a tiny winery at Waikanae, sixty kilometres north of Wellington on the west coast. Owner Peter Heginbotham, a Wellington optometrist, planted his three-hectare vineyard of Pinot Noir, Cabernet Sauvignon, Sauvignon Blanc and Sémillon in gravelly river silts in 1960. His debut commercial vintage was 1966. The Pierre range is produced in an underground cellar and almost entirely sold in Wellington. 'Cabernet Sauvignon is my best wine,' says Heginbotham.

The Grape Republic, on the state highway at Te Horo, south of Otaki, produced its first grape wines in 1990. Alistair Pain (54), a pioneer of viticulture in the Horowhenua, is a bearded, affable ex-Methodist minister, who has made fruit wines on the same property under the Parsonage Hill label since 1984.

In The Grape Republic's boulder-strewn two-hectare vineyard, Chardonnay, Riesling, Gewürztraminer, Cabernet Sauvignon, Merlot, Pinot Noir, Malbec and Cabernet Franc are planted. Fruit is also purchased from Hawke's Bay.

Pain's wines — produced in a modest winery converted from a haybarn — have, he says, been 'studiously ignored by some, misunderstood by others, while a third group has taken to the best wines with alacrity.' The key wines in The Grape Republic's range are a dry non-wooded Sauvignon Blanc, two Chardonnays based on Te Horo and Hawke's Bay fruit, and a Cabernet Sauvignon. These are sound wines but very high-priced.

Ata Rangi

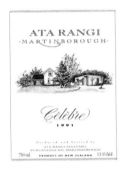

Ata Rangi Vineyard
Puruatanga Road, Martinborough

Owners: Clive Paton and Phyllis Pattie

Key Wines: Pinot Noir, Célèbre, Chardonnay

Clive Paton (44) set up his Ata Rangi ('new beginning') winery to specialise in red wines. His deliciously perfumed and supple yet firm-structured Pinot Noir now rivals Martinborough Vineyard's as the finest in the country.

After taking a diploma in dairying from Massey University, Paton went sharemilking before buying his Purua-

Ata Rangi is an increasingly celebrated 'boutique' winery. Clive Paton and Phyllis Pattie produce only three wines — Célèbre, Pinot Noir and Chardonnay — but they have an extraordinary track-record in show judgings.

tanga Road property in Martinborough in 1980. He worked the 1981 vintage with Malcolm Abel and the 1982 vintage at Delegat's before erecting his own small wooden winery in 1987. Phyllis Pattie (39), formerly a Montana wine-maker at Blenheim — in charge of day-to-day table wine production — then moved north to join him.

'Clive has the long-term ambitions and strategy, but is largely self-taught, relying on flair — feel, taste and smell,' says Pattie. 'I'm more technical, giving attention to detail and worrying about today.' Pattie, a Massey University graduate in food technology, oversees Ata Rangi's white wine production, administration and marketing, while her partner concentrates on viticulture and red wine production. 'But there's a fair bit of crossover,' says Pattie.

Ata Rangi's three-hectare vineyard of Cabernet Sauvignon, Merlot, Pinot Noir, Syrah and Chardonnay is planted in free-draining gravelly soils beneath a shallow layer of top soil; Gewürztraminer was recently uprooted. A nearby two-hectare vineyard owned by Clive's sister, Alison, planted in Pinot Noir, Cabernet Sauvignon and Cabernet Franc, is also a key source of fruit. Most of the winery's Chardonnay grapes are drawn from Kelvin Bremner's Craighall vineyard, across the road from Ata Rangi.

Paton, who first sprang to prominence when his 1986 Pinot Noir collected a gold medal, sees Pinot Noir as his 'biggest challenge'. He travels to Burgundy to study Pinot Noir production: 'I want to see what their grapes are like at harvest and how different it is to our own fruit quality,' he says. 'I want to find out whether top Burgundy is made in the vineyard or in the winery.'

Paton attributes the recent evolution of his Pinot Noir from a light, supple, pleasantly raspberryish wine to a more absorbing, complex and richer style to the advancing age of his vines, a wider selection of clones, some whole-bunch fermentation and longer oak maturation — twelve to fifteen months in Burgundy oak barriques. This is a graceful, mouthfilling Pinot Noir with a delicious concentration of strong, ripe, almost sweet cherryish flavours, savoury, succulent and sustained.

Ata Rangi produces only three wines. 'I want to be known as a specialist,' says Paton. 'A fault of many wineries is that they go for all sorts of things.' Ata Rangi's second specialty is Célèbre, a Cabernet Sauvignon/Syrah/Merlot/ Cabernet Franc blend in which the typical presence of fifteen percent Syrah makes its presence well felt. 'I wanted to create a wine distinctively mine,' says Paton. 'Cabernet is a bit boring on its own and I enjoy Rhone reds' weight and power. Syrah fills the palate up, taking the flavour to all corners of the mouth.' With its brilliant, deep red hue, intensely smoky and spicy bouquet, and rich, peppery, firm flavour, this is a very robust and distinctive red.

Ata Rangi's first Chardonnay was fermented in 1988 in a single oak barrel. Pattie, intent on producing a 'food style' of Chardonnay, adds complexity by giving the wine a partial malolactic fermentation and maturing it on its lees in new oak barriques. This is a weighty, stylish and deep-flavoured wine, delicious in its youth yet built to last.

Ata Rangi's annual production is small — only 3000 cases. Paton sells his wine by mail order or at the vineyard shop during summer. 'The rest of the year, you see people turning into the gate and turning out again. It's hard work — seven days a week and at night. But I'm looking forward to the day I can make wine off forty-year-old vines.'

Benfield and Delamare Vineyards
Cambridge Road, Martinborough

Owners: Bill Benfield and Sue Delamare

Key Wine: Benfield and Delamare

Benfield and Delamare is an extremely rare red-wine label. The first 'commercial' release from the 1990 vintage amounted to only forty-two cases, and even in the mid-1990s, when the company's vineyards are full-cropping, only 800 cases will be marketed. To judge from the early vintages, this is a top-flight red on the rise.

Bill Benfield (53) was born in Christchurch and until recently practised architecture in Wellington. His first contact with wine, he recalls, was at Auckland University, 'going out to Western Vineyards to buy wine for the architects' ball.' His partner, English-born Sue Delamare (45), is a librarian who commutes each day to Wellington.

Benfield and Delamare started planting in Martinborough in 1987. Their non-irrigated, close-spaced vines are in three blocks: a half-hectare vineyard of predominantly Cabernet Sauvignon and Cabernet Franc, planted in

shallow, gravelly soils adjacent to the winery; a tiny, 0.2-hectare block in New York Street; and a 1.6-hectare, sandy clay vineyard a few hundred metres away in Oxford Street, principally planted in Merlot.

Benfield sees viticulture as 'very important. We lay down only one cane per vine, trained low to the soil, to get below

Benfield and Delamare

Bill Benfield, Sue Delamare and friend (a guard against feathered vandals) specialise in the production of a single, claret-style red. Its eye-catching quality and New Zealand and North American gold medals have elevated the reputation of the Wairarapa's Cabernet Sauvignon-based reds.

the wind and allow the fruit to ripen in the heat reflected from the ground.'

Why the emphasis on Cabernet Sauvignon and Merlot, in a district far better known for its Pinot Noirs? Benfield has written that Martinborough's 'rainfall distribution, length of season, heat summation and frost incidence are all positive pointers towards Cabernet and Merlot grape varieties.' Benfield initially set out to produce a Cabernet Sauvignon-based red, but now Merlot is 'figuring larger.

We're moving towards a Merlot-predominant blend.'

In Benfield's small winery, a converted barn — 'It's big enough for what we need', he says — he produces a single, claret-style red. The experimental 1989, debut 1990 and 1991 vintage display impressive depth of colour, strong, vibrant, spicy, concentrated fruit flavours and a taut tannin backbone. Benfield and Delamare is one of the few truly distinguished non Pinot Noir-based reds yet produced in Martinborough.

Bloomfield

David Bloomfield and his former viticulturist, Janine Petersen, in the boulder-strewn estate vineyard at Masterton. Bloomfield wines are rare but full of character.

Bloomfield Vineyards
119 Solway Crescent, Masterton

Owners: The Bloomfield family

Key Wines: Bloomfield Cabernet Sauvignon/ Merlot/Cabernet Franc, Pinot Noir, Sauvignon Blanc; Solway

No road sign beckons visitors to the Bloomfield winery in Solway Crescent, just off the highway south of Masterton. Wine is sold directly to the public, but there are no routine tastings: 'There'd be nothing left,' says winemaker David Bloomfield.

Bloomfield Vineyards is a small family partnership, involving David Bloomfield, his wife, Janet, and his parents Eric and Pamela. After planting their first vines in 1981 — and having experimental wines made at the Te Kauwhata research station — the family processed their first commercial vintage in 1986.

Masterton-born David Bloomfield (36) worked as an architectural draughtsman in Wellington until in 1986 he launched into his full-time career in wine. 'Our family has

always drunk wine,' he recalls, 'and in the late 1970s Dad [a Masterton optometrist] and I were looking for something we could do together. We knew there'd been vineyards here previously, looked at the climatic data and chose wine.' Bloomfield is an engaging personality, affable, lively and intelligent: books crowd his office and opera fills the air in his small concrete winery.

Bloomfield's eight-hectare estate vineyard is planted in deep river shingles, making cultivation extremely difficult. Cabernet Sauvignon, Merlot and Cabernet Franc are the principal varieties, with smaller plots of Pinot Noir, Sauvignon Blanc and Sémillon. This is a very close-planted vineyard, its high density modelled on Bordeaux.

Bloomfield also draws fruit from local growers, including the Lyndor Vineyard, where 1.5 hectares of Merlot, Cabernet Franc and Cabernet Sauvignon are cultivated on a north-facing clay and limestone slope in the hills to the east of the town. David Bloomfield hopes 'the differences between these two vineyards . . . will play a significant role in producing a multi-dimensional claret-style wine.'

The winery's output is currently tiny — about 1000 cases per year. Bloomfield has deliberately not pursued a high early profile. 'At the moment we're more interested in getting the vineyard right. We're in it for the long haul. There's no point in saying: "This is it." It's not "it". Our wines are still changing.'

David Bloomfield's wines are full-flavoured and distinctive. Bloomfield allows his Sauvignon Blanc 'to age on the vine long enough to pass the stage where green-pepper, gunshot characters predominate, to the stage where tropical-fruit characters begin to show'. Lees-aged but not barrel-matured, this is a weighty wine with restrained herbal flavours — a style rather reminiscent of white Bordeaux.

Bloomfield Pinot Noir is produced in tiny quantities; only two barriques — equivalent to 600 bottles — are made each vintage. 'This is a Pinot from a non-serious Pinot Noir maker,' says Bloomfield. A rich red with impressive depth of cherry and coffee-like flavours, this is an absorbing wine with complexity and length.

The Bloomfield Cabernet Sauvignon/Merlot/Cabernet Franc is full bodied, dark, spicy, chocolatey and herbaceous, with plenty of flavour and a firm tannin grip. David Bloomfield's recent introduction of Solway, a second-tier claret-style label, should enhance the already impressive quality of the top blended red.

Blue Rock Vineyard
Dyerville Road, Martinborough

Owners: The Clark family

Key Wines: Chardonnay, Sauvignon Blanc, Oaked Sauvignon, Rhine Riesling, Pinot Noir, Cabernet, Magenta

Blue Rock

On an elevated, wind-assaulted site eight kilometres from Martinborough township, Wairarapa farmer Nelson Clark and his family have recently produced the first vintages at Blue Rock. The vineyard's name, says Clark, is derived from 'the area called Blue Rock up the road, where the stones in the river are blue.'

The Clarks planted their first vines in late 1986 and made their first, tiny vintage in 1990: 'We drank it ourselves.' Blue Rock's first 'commercial' wines were produced in 1991. This is a family affair: Clark's wife, Beverley, concentrates on viticulture; their daughter, Priscilla (32), oversees the marketing; and another daughter, Jenny (31), is the winemaker.

Nelson Clark (58) has farmed in the Wairarapa for most of his life. 'Farming was shot, so we chose to diversify. With other types of horticulture we're defeated by the [westerly] wind but with grapes it's not such a problem.' Why aren't other farmers in the locality planting vineyards? 'We're a very conservative region. You try and sell a new brand of tractor. They all wait until someone else buys it. They're all watching us with grapes.'

In their north-facing vineyard, principally clay with a band of shingle running through the centre, the Clarks have planted six hectares of Chardonnay, Sauvignon Blanc, Riesling, Pinot Noir, Cabernet Sauvignon, Cabernet Franc and Meunier (for a bottle-fermented sparkling). Nearby trees are so wind-buffeted they have grown limbs only on one side. In summer the winds' blasts can damage the vines' canes, but the Clarks are experimenting with unconventional training systems in a bid to reduce the problem.

Blue Rock's 1990–91 vintages were produced by Phyllis Pattie and Clive Paton of Ata Rangi. Jenny Clark, who worked the 1989 vintage at Amity Vineyards in Oregon, and in 1990 gained a postgraduate Diploma in Horticultural Science from Lincoln University, made the 1992 vintage 'under the stars'. Blue Rock's new winery rose in time for the 1993 vintage.

When I tasted the 1991 wines in their infancy, they all displayed the copybook, clear, penetrating flavours and appetising acidity typical of Martinborough wines.

Jenny ('Jen') Clark is the winemaker at Blue Rock, which boasts one of the Wairarapa's loveliest vineyard settings. The Clark family run Blue Rock in conjunction with their 200-hectare sheep and cattle farm.

Chifney Wines
Huangarua Road, Martinborough

Owners: Stan and Rosemary Chifney

Key Wines: Cabernet Sauvignon, Garden of Eden Red, Rosé, Garden of Eden White, Chenin Blanc, Gewürztraminer, Chardonnay

Chifney

Stan Chifney, the snowy-bearded, gentlemanly, popular seventy-three-year-old owner of Chifney Wines in Huangarua Road, has yet to be seduced by Pinot Noir. 'Thirteen years ago, I wanted to just plant Cabernet Sauvignon and what goes with it, but we hedged our bets and planted other varieties too. Now it turns out we should have planted all Cabernet. That's my favourite wine.'

London-born Chifney arrived in this country twenty-one years ago, after a career spent in vaccine manufacture in the Middle East and Nigeria. He and his wife, Rosemary, having made their own fruit wines, thought winemaking would be an ideal retirement hobby.

In 1983 they erected their concrete-based winery with its partly subterranean cellar. Their 4.5-hectare vineyard of Cabernet Sauvignon, Chardonnay, Chenin Blanc and Gewürztraminer, planted in loamy surface soils overlying stony subsoils, yielded its first commercial crop in 1984. Chifney recently added a plot of close-planted Cabernet Sauvignon vines: 'It's my "retirement block", for when I'm too old and doddery to look after the rest'.

Only about 1000 cases of wine flow each year from the Chifney winery. Chifney has achieved greater success with red than white wines. Those vintages I have tasted of the

Stan Chifney recommends that his Cabernet Sauvignon be served 'with Beethoven, Wagner, Strauss (Richard, of course), and possibly Mahler ...' To match his lighter Garden of Eden Red, Chifney suggests Schubert or Mendelssohn.

Chenin Blanc and Chardonnay have been of variable quality.

The Cabernet Sauvignon is undoubtedly the highlight of the range. 'A few bunches of Merlot go in,' says Chifney, but essentially this has been a 'straight', unblended Cabernet Sauvignon. The 1986 won a gold medal — Martinborough's first for a Cabernet-based red. Chifney's 1985 Cabernet Sauvignon (a vintage he sees as 'particularly good') is still in lovely condition: densely coloured, complex and crammed with sweet, softening blackcurrant and spicy flavours. Only five bottles of this, his first Cabernet

Sauvignon, survive at the winery; Chifney opens one each year.

Garden of Eden Red, a single-vineyard wine made from bought-in grapes, is a drink-young style blended from such varieties as Pinot Noir, Cabernet Sauvignon, Merlot and Cabernet Franc. Fresh and supple, with 'minimal' oak influence, this is fruity, flavoursome and soft.

Stan Chifney doesn't like the idea of winding down. 'I'll have to be wound down,' he smiles. 'I'll need some help in the winery in a year or two. But I'll keep pottering around for as long as I can.'

Dry River

Dry River Wines
Puruatanga Road, Martinborough

Owners: Dr Neil and Dawn McCallum

Key Wines: Pinot Gris, Gewürztraminer, Chardonnay, Sauvignon Blanc, Riesling, Riesling Botrytis Selection, Pinot Noir

With their spare, understated labels and fleeting presence on retail shelves, Dry River wines are easy to overlook. Amongst keener members of the wine-drinking fraternity, however, Dr Neil McCallum's immaculate, robust, slowly evolving Martinborough wines are much sought-after.

Dry River's tiny output fascinates wine buffs for several reasons. Most of the wines are cultivated on the estate vineyard in Puruatanga Road or at the nearby Craighall vineyard, giving them precise 'single vineyard' origins. They demonstrate the outstanding quality of currently less fashionable grape varieties like Gewürztraminer, Riesling and Pinot Gris, and in their infancy they are shy and cry out for a lengthy spell in the cellar. These are classic wine styles for the 'serious' collector.

The emergence of Neil McCallum as a gifted, individualistic winemaker has been greatly assisted by his previous career as a DSIR scientist. Born in Auckland in 1943, he capped his high-flying academic record with an Oxford doctorate, awarded for his dissertation on penicillin substitutes. At one memorable Oxford dinner he was 'bowled over' by a Hochheimer Riesling — and launched on his love affair with wine.

McCallum and his wife, Dawn, planted the first vines on their shingly, free-draining block at Martinborough in 1979. The vineyard, now covering four hectares, is close-planted in Gewürztraminer, Pinot Gris, Sauvignon Blanc, Chardonnay and Pinot Noir vines, trained on the Scott-Henry system and not irrigated. Riesling, Chardonnay and Pinot Noir are also drawn from the Craighall vineyard 200 metres down the road.

McCallum's corrugated-iron winery is a traditional Wairarapa barn and he describes his approach to winemaking as 'low-tech, involving minimum processing and placing an emphasis on cellaring qualities rather than short-term attractiveness for early drinking.'

'Cellaring' is a popular word in the McCallum vocabulary. His ability as an organic chemist to precisely control each stage of the winemaking process lies at the heart of the slow-maturing style he has evolved. The 'oxidative' approach to winemaking, which involves deliberately exposing grape-juice or wine to oxygen in a bid to enhance its early drinking appeal, finds no favour with him. McCallum wants to produce wines capable of maturing over the long haul and unfolding the subtleties of old age. One-year-old Dry River Pinot Gris only hints at the wealth of savoury, earthy, stone-fruit flavours it will later unleash.

McCallum is chiefly interested in producing white wines. His annual output is low at about 2500 cases.

Dry River wines are stylish, with plenty of extract, and are usually — although not always — bone dry. The Gewürztraminer has precise peppery/spicy varietal characteristics, without the pungent seasoning of the more ebullient Gisborne-grown examples of this variety. This is a 'fine-grained' (to borrow McCallum's adjective) dry wine of excellent weight and unusual flavour depth.

The Dry River label is inextricably linked in most wine lovers' eyes with the Pinot Gris variety; McCallum's is the finest in the country. His Pinot Gris vines, sourced from Mission Vineyards, are probably an old Alsace clone called Tokay à petit grain (small berry Pinot Gris) known for its low yields. Dry River Pinot Gris is mouthfilling and savoury with a subtle bouquet and concentrated, slow-building peachy/earthy varietal flavour. At a retrospective tasting in early 1992, the highlights were the 1986 — still lively, firm and lingering — and 1990 vintages. The 1992 looks even better. Here is a much-needed Chardonnay alternative.

Dry River Sauvignon Blanc is a vigorously crisp, dry, non-wooded style, bursting with flavour. Its notably ripe fruit flavours — described by McCallum as 'stone-fruit flavours and hints of capsicum' — are achieved by employing a platoon of local schoolchildren to remove all the leaves around the ripening bunches.

Dry River Chardonnay is a bold wine with strong grape-fruit and stone-fruit flavours, savoury oak and fresh, steely acidity. The Riesling, with its lovely outpouring of botrytis-enriched scents, concentrated lime/lemon fruit flavours, hint of sweetness (except in very ripe vintages, when the sugar level soars) and tense acidity, is a classic. Dry River

Neil McCallum was in 1979 the first winemaker to plant vines in Martinborough, and from the start his wines have been among the district's finest. Dry River Pinot Gris is the country's top example of that variety, but the same could be said of the Gewürztraminer.

also produces high-scented, fragile, delectably botrytised sweet Rieslings labelled — in ascending order of fruit ripeness — as 'Selection', 'Bunch Selection' and 'Berry Selection'.

Dry River Pinot Noir is a floral, cherryish red, full and supple. Its quality has risen sharply since the 1989 vintage, with longer oak handling enhancing the wine's structure and complexity. This now takes its place as one of the district's most stylish Pinot Noirs.

In ten years' time McCallum will be content if his wines are being accepted as 'amongst the best of their type' and he has retained his faithful clientele. 'I didn't have a crystal ball in the beginning. The statistics suggested Martinborough could do it, but until it happens you're holding your breath. Now we've been here over a decade and the vines are really getting into their stride.'

Gladstone Vineyard
Gladstone Road, Gladstone

Owner: Dennis Roberts

Key Wines: Riesling, Sauvignon Blanc, Cabernet Sauvignon/Franc/Merlot

Out on a limb from the principal cluster of Wairarapa wineries at Martinborough, Gladstone Vineyard's initial releases are nevertheless of such promise they look sure to spur others to explore the area's viticultural potential.

Gladstone lies between Martinborough and Masterton, about ten kilometres from Carterton. The climate is 'more Bordeaux than Burgundy-orientated', says Dennis Roberts (47), an Australian who has lived in New Zealand since 1974. Roberts, a university-trained chemist and former Wellington veterinarian, describes himself as a 'post mid-life crisis professional'.

Why did he plant his vineyard in Gladstone? 'I used to drive through here and thought the eastern hills looked promising. Then when I decided to "retire", I put an advertisement in the paper specifying my land requirements and got a response from Gladstone. It's less windy here than in Martinborough and probably a bit hotter.'

On old alluvial terraces which once formed the bed of the nearby Ruamahanga River, Roberts has planted three hectares of Cabernet Sauvignon, Merlot, Cabernet Franc, Sauvignon Blanc and Riesling. He focuses on making the wine, while an employee tends the vineyard. Roberts' handsome cream and red-brown winery, set amidst landscaped grounds and thousands of trees, was erected in time for the second 1991 vintage.

Gladstone Vineyard's output is small — only about 1500 cases. The dry Riesling is a classy, floral wine with intense lemon/lime flavours and tense acidity. Packed with sweet-tasting, supple fruit, the Cabernet Sauvignon/Franc/Merlot is soft and very easy-drinking in style. 'I guess we have kept a pretty low profile,' says Roberts — but the impressive quality of his wines has swiftly put Gladstone Vineyard on the map.

Gladstone

Gladstone
RIESLING
—— 1992 ——

Lintz

Lintz Estate
Kitchener Street, Martinborough

Owners: The Lintz family

Key Wines: Spicy Traminer, Sauvignon Blanc,
Chardonnay, Noble Selection Optima, Pinot Noir,
Cabernet/Merlot

Hand-pickers combed the vineyard nine times to secure the shrivelled, botrytis-infected berries which formed the foundation of Lintz Estate's first gold medal wine, the honey-sweet Noble Selection Optima 1991. 'Good stickies are one of my major aims,' says winemaker Chris Lintz.

The Lintz Estate winery, with its conspicuous tower and New Zealand flag, lies on the main road into Martinborough, not far from Palliser. The tower conceals an elevated, gravity-fed crushing and destemming system. During vintage the grapes are sucked up by a pneumatic system and then fed down through the crusher into joint

The tower atop the Lintz winery — housing a crusher and drainer tanks — is visible from afar in the pancake-flat town of Martinborough. Lintz has tasted early success with his rich, golden, honey-sweet Noble Selection Optima.

drainer and fermenter tanks. The winery's first vintage was in 1991.

Chris Lintz's parents, Harold and Uni, came to New Zealand from Germany after the Second World War. Lintz (32), a zoology graduate from Victoria University, has worked on a wine estate in the Saar once owned by the Lintz family, and in 1988 gained a diploma in viticulture and oenology from the famous Geisenheim Institute. After returning to New Zealand, he worked at Montana and Brookfields, and then in 1989 planted his first vines in Martinborough.

The tiny plot of Riesling vines adjacent to the winery has been planted for 'image' reasons. Across the road from the Ata Rangi winery, the two-hectare Vitesse vineyard has been planted in Cabernet Sauvignon and Gewürztraminer. Another eight-hectare Moy Hall block between Dry River and Te Kairanga, only partly established, has been planted in Merlot, Cabernet Franc, Cabernet Sauvignon, Pinot Noir, Gewürztraminer, Chardonnay and Optima.

Lintz Estate's output is currently small — about 2000 cases — but Lintz is aiming for a peak production level of 8000 cases by 1997. The lemony, crisp 1991 Chardonnay (based on Hawke's Bay fruit) and the powerful, strongly herbal, oak-influenced 1991 Sauvignon Blanc made very solid debuts. The 1992 Spicy Traminer is weighty and full-flavoured.

With their floral perfume and mouthfilling, citric-fruit and honey flavours, the 1991 and 1992 vintage Noble Selections made from the Optima variety — an early-ripening cross of Müller-Thurgau with another crossing of Riesling and Sylvaner — are an eye-catching initial feather in Lintz's cap. As part of his 'strategy to do something different', Lintz has also produced a Riesling-based bottle-fermented sparkling and a red bottle-fermented sparkling from Pinot Noir.

Martinborough Vineyard

Martinborough Vineyard
Princess Street, Martinborough

Owners: Derek and Duncan Milne, Claire Campbell,
Russell and Sue Schultz, Larry McKenna

Key Wines: Pinot Noir, Chardonnay, Riesling,
Late Harvest Riesling, Gewürztraminer,
Sauvignon Blanc

Martinborough Vineyard is the highest profile winery in the Wairarapa. Winemaker Larry McKenna — often dubbed the country's 'king' and, more poetically, 'prince' of Pinot Noir — won the top Pinot Noir trophy three years in succession (1988–90) at the Air New Zealand Wine Awards. During the same period, Martinborough Vineyard also carried off trophies for the champion Chardonnay, Müller-Thurgau and Riesling. Such a formidable show track-record indicates Martinborough Vineyard is on course to achieve its 'ultimate goal of becoming an internationally

rated Pinot Noir, Chardonnay and Riesling producer'.

McKenna, as Martinborough Vineyard's winemaker and general manager, is often in the limelight. The company's success also reflects the talents of its founding partners: Dr Derek Milne, formerly a soil scientist with the DSIR and now a consultant; his brother Duncan and his wife, Claire Campbell; and businessman and pharmacist Russell Schultz and his wife, Sue.

The company planted its first vines in Martinborough's deep alluvial gravels in 1980. The first, 1984, vintage based on a mere two tonnes of grapes, yielded two outstanding wines: both the Pinot Noir — only 150 bottles were made — and Sauvignon Blanc unleashed enormous body and flavour length. The next challenge was to reproduce these standards in larger-volume wines.

For the first commercial-scale crop in 1985, the coolstore winery's roof was erected only days before the harvest. McKenna arrived in early 1986, the first experienced winemaker in the district.

'I don't live in Martinborough because I like the area,'

says McKenna. 'I'm here because I believe we can grow good fruit.' McKenna is an Australian, now 41, who graduated with a diploma in agriculture from Roseworthy College, and then crossed the Tasman in 1980 to work under John Hancock — an old friend from his boarding-school days in Adelaide — at Delegat's. Following Hancock's departure for Morton Estate, McKenna headed Delegat's winemaking team for three years, until in early 1986 he was lured south by Martinborough Vineyard.

Burly and soft-spoken, McKenna loves outdoor pursuits — climbing, tramping, trout fishing. But he came to Martinborough for other reasons. 'I thought the district had a lot of potential, which was confirmed after I had a look at all the wines . . . I wanted to be fully involved in a winery in all aspects from viticulture to marketing . . . I also had the possibility to have a share in a vineyard and winery — and that is the sort of commitment I wanted to make.' Martinborough Vineyard crushes about 130 tonnes of fruit, giving it a decidedly small output (only about 9000 cases) for such a well-known label.

McKenna's Pinot Noir is the most multidimensional in the country. Ata Rangi, Palliser and Dry River have recently emerged as extremely stiff competitors in the district, but only Martinborough Vineyard can point to a string of top-flight wines back to the 1986 vintage (and before that, to the flawed 1985 and arrestingly good experimental 1984). Their key achievement has been to transcend the light, simple, shallow style of Pinot Noir that was previously the norm in New Zealand.

What characters is McKenna pursuing in his Pinot Noir? 'Berry fruit flavours, with something more than strawberry ripeness, underpinned by quality oak and complexity — that earthy, mushroomy character that is almost indefinable.' McKenna works predominantly with the 10/5 clone of Pinot Noir, selected for its 'good fruitiness', but has an increasing proportion of 'Pommard' coming on stream, a clone which yields 'full bodied, tannic, structured wine'.

In his incessant search for higher quality, McKenna is also exploring what he terms the 'grey areas' of Pinot Noir vinification. For the 1992 Pinot Noir, for instance, he macerated the skins in the juice before the fermentation; included a proportion of the normally discarded stalks in the ferment; and whole-bunch fermented (rather than crushed) one-third of the crop. These fermentation techniques are all designed to enhance the wine's complexity and structure. This is a hugely satisfying red. Medium-full cherry-red in hue, it is very fine and complex on the palate, with the delectable, rich beetroot/oak/mushroom characters of quality Pinot Noir.

Martinborough Vineyard Chardonnay is almost as widely acclaimed as the Pinot Noir. Bold, peachy-ripe fruit and strong, savoury oak are hallmarks of the winery's Chardonnay style. Powerful and chewy, with long, complex flavours from its barrel fermentation, partial malolactic fermentation and lengthy lees-aging, this is a firm, mealy, buttery, very stylish wine. Both the Chardonnay and Pinot Noir have been exported to the United Kingdom and Australia.

Larry McKenna's superb string of show successes in the late 1980s earned Martinborough Vineyard a formidable reputation. McKenna's Pinot Noir is the most complex in the country and his Chardonnay is equally absorbing.

Martinborough Vineyard — The Site

The bold, meaty fruit characters typical of his Pinot Noir are in Larry McKenna's eyes a reflection of the vineyard's 'terroir' — its key influences of climate, terrain and soil.

The majority of Martinborough Vineyard's grapes are drawn from the six-hectare vineyard which surrounds the winery — planted in equal plots of Chardonnay, Pinot Noir, Riesling, Sauvignon Blanc and Gewürztraminer — and the new, four-hectare company vineyard across the road, entirely devoted to Chardonnay and Pinot Noir. Fruit is also drawn from Graham and Jill Cleland's nearby two-hectare vineyard, and Dr Jack McCreanor's neighbouring four-hectare vineyard, established in Pinot Noir, Chardonnay and Pinot Gris.

For McKenna, the crucial attraction of these shingly vineyard sites — previously used for potato-growing, and before that for grazing sheep — is their free-draining nature. 'It devigorates. Our vines don't tap a permanent source of water. When the ground dries out, the vines stop growing. You don't get fruit-shading problems, and the vines are freed-up to put their energy into ripening the grapes.'

The soil is 'fairly low' in fertility, with a natural lack of nitrogen and phosphorus, and uniform in structure throughout the company's vineyards.

Wind can be a curse in Martinborough. Scorching northwesterlies during summer can damage the vines' leaves and canes, retarding photosynthesis and fruit-ripening. Winter brings freezing southerly blasts: 'We're open all the way to the South Pole,' says McKenna. The flatness of the vineyards also has two drawbacks: the land is not north-sloping for maximum exposure to the sun, and the lack of air drainage increases the frost risk. But the district's light rainfall in late summer and autumn creates excellent fruit-ripening conditions (most years it is too dry for the grapes to be infected by 'noble rot'.) Temperatures 'are ideal for the long, slow ripening of earlier-maturing grape varieties, especially Chardonnay and Pinot Noir.'

As the vines mature, they are producing earlier-ripening grapes with lower pH (which enhances the wine's longevity) and higher levels of sugar, acidity and colour. Martinborough Vineyard's finest wines are undoubtedly yet to come.

Few vineyards in the country can boast of a lineup of wines as uniformly excellent as Martinborough Vineyard's. The winery has fine tuned its range over the years, focusing on the varieties the partners consider best suited to Martinborough; one notable casualty was the vibrantly fruity but light Cabernet Franc/Cabernet Sauvignon. Overshadowed by the reputation of its Pinot Noir and Chardonnay stablemates, Riesling is nevertheless the third jewel in the winery's crown, with a lovely outpouring of citric/marmalade-like scents and strong, lingering, lemony flavours.

The Gewürztraminer is a strapping, dry style with deep citric and spicy flavours, needing two or three years' cellaring to break into full stride. (In favourable vintages McKenna has also produced a botrytised sweet Riesling and Gewürztraminer under a Late Harvest label.) The Sauvignon Blanc is a fresh, non-wooded style with restrained herbal flavours. The Müller-Thurgau was delightfully floral and crisp, with an impressive depth of citric/lemony fruit flavours — quite Riesling-like — and a gentle touch of sweetness; the 1992 vintage, however, was the last.

Seven years after his arrival in Martinborough, is McKenna happy he made the right move? 'I'm more than happy. I never dreamt we would be so successful.' What of the future? 'Our show results probably climaxed at the 1989 competition, and others in the district are catching up with us. The key challenge now is to market our rising output.'

Muirlea Rise

Muirlea Rise
Princess Street, Martinborough

Owners: Willie and Lea Brown

Key Wines: Pinot Noir, Après

'Controlling the grape is the closest I'll ever get to God,' says Willie Brown, owner of one of Martinborough's tiniest wineries. 'It all comes back to yields. It's fruit concentration we're after in our Pinot Noir.'

Brown (56) and his wife, Lea, planted the first vines on their Princess Street site, just across the road from Martinborough Vineyard, in 1988. Brown, who was raised in Auckland and spent the early part of his working life as an instrument maker in the air force and as a dairy owner, later became absorbed in wine while working in Wellington's retail liquor trade. After spending ten years at the giant distribution company, New Zealand Wines and Spirits, and three years at Brown and Garvey — a smaller wine distribution company he co-founded — Brown headed for Martinborough.

Muirlea Rise, the estate's name, is a conjunction of Brown's mother's maiden name, Muir, and his wife's name, Lea.

'I went into winemaking because I wanted to get away from working seven days a week,' says Brown. 'Running a winery is in some ways the stupidest occupation a man could undertake; you still have to be a workaholic — but you love it. Pinot Noir is the big challenge; to me a bottle of Chambolle-Musigny is pure buttercups and blue skies.'

Brown's two-hectare vineyard is planted in a mix of clay and gravel soils. Close-planted Pinot Noir vines — spaced one metre apart in rows, with 1.5 metres between the rows — take up eighty percent of the space, with smaller plots of Cabernet Sauvignon, Merlot, Cabernet Franc and Syrah.

Brown has built a miniature winery to match the hundred-year-old house on the property. The first 1991 Muirlea Rise Pinot Noir — of which only 100 cases were bottled — was made 'in the garage', says Brown. A new barrel hall will be erected 'out of profit' in the next couple of years, but Muirlea Rise will always be a tiny company, producing only 800 cases even when at full throttle.

Brown plans to produce three wines under the Muirlea Rise label — the Pinot Noir, a barrel-matured 'port' (or 'fortified red', as Brown prefers to call it) called Après, and a quaffing red from 'harvest extras'. The Pinot Noir 1991 is soft and supple, with lovely, floral fruit aromas and strong, vibrant, raspberryish fruit flavours — a good debut. Après, a Cabernet/Syrah blend, is also satisfying: spicy, rich and drier than most dessert wines.

From their little plot of vines, Willie and Lea Brown produce a floral, vibrantly fruity and supple Pinot Noir. Brown's long-handled hoe keeps his back straight and the under-row weeds in retreat.

Palliser Estate Wines
Kitchener Street, Martinborough

Owner: Palliser Estate Wines of Martinborough Limited

Key Wines: Chardonnay, Sauvignon Blanc, Rhine Riesling, Riesling Botrytis Late Harvest, Pinot Noir

Palliser Estate is one of the newer, largest and best wineries on the bustling Martinborough wine scene. 'We're aiming to produce about 20,000 cases by the mid-1990s,' says managing director Richard Riddiford. 'I genuinely believe these will be some of the best wines in New Zealand.' The top-flight quality of the winery's initial releases lends credibility to Palliser's high ambition.

Palliser's handsome cream and green-coloured colonial-style winery stands on the main highway into Martinborough. 'Wyatt Creech, now an MP and Minister, was the driving-force in the early days,' says Riddiford. 'He established the vineyards and is still a shareholder.' Creech

planted the first vines in his Om Santi vineyard in 1984 and four years later formed an unlisted public company to take over his vineyard and build a winery. Om Santi, the name first proposed for the winery, was dropped in favour of Palliser Estate — a reference to Cape Palliser, the southern-most tip of the North Island.

Palliser Estate is owned by about 130 shareholders. 'It's a captive market,' grins Riddiford, 'and they all drink a lot of wine and make the right noises in trendy outlets.' The shares are well spread, with the largest individual share-holder owning only five percent of the company.

Palliser's extensive vineyards are all within close prox-imity of the winery. The three-hectare block adjacent to the winery is planted in Riesling and Sauvignon Blanc. Two further vineyards totalling twenty hectares have been estab-lished in Chardonnay, Riesling, Sauvignon Blanc and Pinot Noir. A small proportion of Palliser's annual grape intake is purchased from local growers. The large scale of these vineyards — by the district's standards — is reflected in the fact that Palliser is currently the only Martinborough winery to employ a mechanical harvester.

Palliser's first 1989 vintage was produced by Larry

Palliser

Allan Johnson made wine in Hawke's Bay, Western Australia and Marlborough before he came to the Wairarapa in 1990. Johnson's uniformly well-crafted, intensely flavoured wines have made Palliser one of the region's high-fliers.

McKenna at Martinborough Vineyard, and the 1990 by Australian winemaker Rob Bowen. Allan Johnson (36) took over the winemaking reins in late 1990. While growing up in Hawke's Bay, Johnson worked in vineyards during the school holidays, and in 1980 joined McWilliam's as a cellarhand.

After graduating from Roseworthy College in 1984, Johnson became winemaker at Capel Vale in Western Australia. He was there for the 1985–89 vintages, but says he 'yearned to get back to New Zealand. Our fruit has more concentration of flavour, more power in the middle and end palate.'

Palliser has swiftly established itself as one of the leading up-and-comers of the New Zealand wine industry. The Chardonnay is full with rich, vibrant stone-fruit flavours and a buttery, mealy complexity. Barrique-fermented and lees-aged for eight months, this is a stylish and well-structured wine with early-drinking appeal yet plenty of cellaring potential.

The non-wooded Sauvignon Blanc abounds with lush, ripe tropical-fruit aromas and flavours in an appetisingly fresh and vibrant style. The equally distinguished Rhine Riesling, which is only fractionally off-dry, displays piercing lemon/lime fruit flavours and tense acidity. The delicate, delectable Riesling Botrytis Late Harvest is much sweeter.

The style of Palliser Estate's Pinot Noir reflects Allan Johnson's fondness for 'rich Pinot Noir with the roast coffee aromas of ripe fruit and good structure'. The wine is fermented at warm temperatures, including about fifteen percent whole-bunch fermentation, and matured in oak casks (about twenty-five percent new) for twelve months. Soft and plump, its bold cherryish fruit flavours underlaid by charry, smoky wood, this is an impressively complex, savoury and supple red.

Te Kairanga

Te Kairanga Vineyard
Martins Road, Martinborough

Owners: Tom and Robin Draper and shareholders

Key Wines: Chardonnay, Reserve Chardonnay, Sauvignon Blanc, Pinot Noir, Reserve Pinot Noir, Cabernet Sauvignon

Te Kairanga ('the place where the soil is rich and the food plentiful') winery rests on a stunning site above the Huangarua River, against a backdrop of sunlit green hills.

Tom Draper (60) — a former building contractor and co-founder of one of Wellington's winetasting groups, the Magnum Society — and his wife Robin in 1983 bought Alister Taylor's vineyard, then in a run-down condition. After they brought in partners, Te Kairanga and — across the road — East Plain vineyards spread out over thirty-two hectares of free-draining stony river terraces. Chardonnay, Pinot Noir and Cabernet Sauvignon are the principal varieties, with smaller plantings of Sauvignon Blanc, Merlot, Cabernet Franc and Durif. An irrigation system has been installed: 'I know how dry it can get here,' says Draper.

Tom Draper retired in 1993. Te Kairanga's new general manager is Andrew Shackleton (34), formerly Wellington area manager for Villa Maria. Robin Draper remains involved in the sales arena.

The Drapers quietly released their first wine, Te Kairanga Chardonnay 1986, in 1987. Stan Chifney and Heiko Tutt made the 1986–89 vintages of Te Kairanga wines, which overall proved disappointing. A concrete-walled, iron-roofed winery was erected for the 1988 vintage, and a 125-year-old, pit-sawn timber cottage removed from Martinborough township to the vineyard site. Originally built by the founder of Martinborough, John Martin, for a farm worker and his family, it now enjoys a new lease of life as Te Kairanga's sales and administration facility.

The appointment of Te Kairanga's first full-time wine-maker, Chris Buring, in late 1989 has been followed by a marked upswing in wine quality. Buring (47), a scion of the famous Australian winemaking family, holds a bachelor's degree in fermentation science from the University of California, Davis. After twenty-three years at Lindemans, where he rose to the post of production manager, in 1986 Buring left to pursue a career as a wine consultant, columnist and vineyard tour operator. He came to New Zealand because he 'wanted to get back into production and it was an opportunity to build something almost from the ground to a reasonable size'. Soon after, Buring set the seal on his involvement at Te Kairanga by marrying one of the partners.

With its current annual output of about 10,000 cases, Te Kairanga is one of the district's most widely seen labels. Two Chardonnays are produced: a non-wooded style with mouthfilling body and flinty, green-appley, crisp fruit flavours; and the barrel-fermented Reserve label, a bold, peachy-ripe, creamy wine underpinned by the 'fresh, racy

Chris Buring, a member of the famous Australian winemaking family, has markedly elevated Te Kairanga's reputation since his arrival in 1989. Te Kairanga's robust, strong-flavoured Reserve Pinot Noir and rich, creamy-smooth Reserve Chardonnay are his key achievements to date.

acidity' Buring favours. There have also been two Sauvignon Blancs: a non-wooded style with fresh, crisp, vigorously herbal fruit flavours; and an Oak Fermented Sauvignon Blanc to which the wood adds depth and structure.

Four reds are marketed, a diverse array by the district's standards. Te Kairanga Pinot Noir shows strong raspberry fruit flavours and oak fullness in a supple, straightforward style. The Reserve Pinot Noir, made from 'specially ripe fruit' matured longer in oak, is more subtle and markedly richer.

Te Kairanga Durif is an almost opaque, chunky, high-acid red. The Cabernet Sauvignon is much more enjoyable: full-flavoured and crisp with fresh, strong, vibrant fruit.

Voss Estate
Puruatanga Road, Martinborough

Owners: Gary Voss and Annette Atkins

Key Wines: Pinot Noir, Cabernet Sauvignon/Merlot, Chardonnay

'When I tasted Clive's [Paton of Ata Rangi] '86 Pinot Noir I was excited; I thought it was the best Pinot Noir made in New Zealand or Australia,' recalls Gary Voss. 'Then I tasted his Cabernet Sauvignon-based Célèbre and liked that too.' So in 1987 Voss and his partner, Annette Atkins, bought land next to Ata Rangi and a year later planted both red wine varieties.

Voss (36), who has a BSc in zoology, is a former Fisheries Research diver. After studying oenology for a year in Australia, he worked vintages at deRedcliffe and Ata Rangi before plunging full-time into wine under his own label. Annette Atkins (30), a Fisheries Research kahawai specialist,

is also the winery's general manager. 'For lifestyle reasons,' she says, 'we intend to keep small.'

The 2.5-hectare estate vineyard has been planted in one-third Pinot Noir, one-third Cabernet Sauvignon and Merlot, and one-third Chardonnay. Gary's brother, Murray (32), who has worked in Swiss vineyards, oversees the viticultural operations.

The early vintages were made on an outdoor concrete pad. For the first 1991 vintage, Hawke's Bay grapes were used to make about 150 cases of a fresh, buoyant, supple Merlot ideal for 'soft summer drinking'. The 1992 vintage brought a Merlot and Chardonnay, both grown in Hawke's Bay and both silver medal winners. The first estate-grown wines flowed in 1993. When production is in full swing, each vintage should produce about 2000 cases of Voss Estate wines.

'Our technical background is strong,' says Gary Voss. 'In future we hope to make a Pinot Noir, Chardonnay and a Cabernet-based red as good as, or better than, anyone's in the district.'

Voss

NELSON

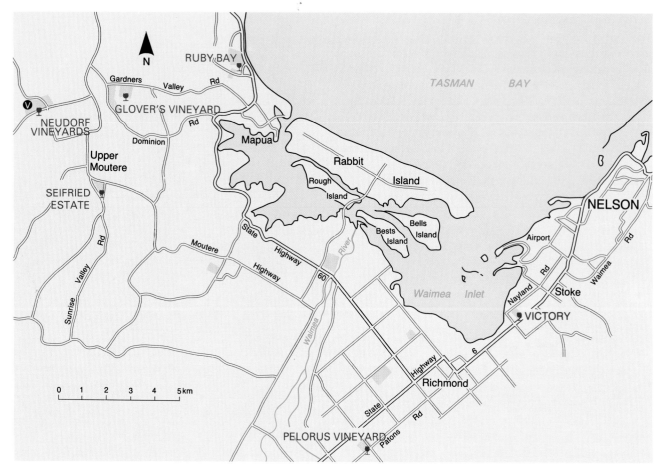

New life is stirring in the Nelson wine region, for so long largely the preserve of two couples — the indefatigable Hermann and Agnes Seifried and Tim and Judy Finn of Neudorf. Responding to the urges of their overseas agents, the Seifrieds abandoned their former, Austrian-sounding winery name, Weingut Seifried, in favour of Seifried Estate.

The urge to change winery names proved infectious: David and Christine Moore, who in 1989 purchased Korepo Wines, renamed it Ruby Bay; and Andrew Greenhough and Jenny Wheeler, who bought Ranzau in 1991, now call it Pelorus. More important than these superficial name changes are the energy and new ideas both couples are bringing to two small wineries that have previously struggled to make any real impact.

As a group, Nelson's wineries do not enjoy the high profile of those in the four principal growing regions: Marlborough, Hawke's Bay, Gisborne and West Auckland. When wine talk turns to new and exciting areas, the Wairarapa and Central Otago invariably come up. Nelson is often overlooked.

The reasons are not hard to find. Early German winemakers who landed at Nelson in 1843 and 1844 looked askance at the steep, bush-clothed hills and departed for South Australia. Other problems have included a shortage of large holdings suitable for viticulture and the region's distance from principal transport routes.

The undoubted climatic advantages for viticulture — warm summers and high sunshine hours — are slightly reduced by the risk of damaging autumn rains as harvest approaches. In this respect Nelson parallels most North Island wine districts more closely than other South Island regions. Nelson winemakers thus sometimes struggle to match the sugar levels achieved over the hills in the relatively dry Marlborough climate.

Overshadowed by fruit, tobacco and hop growing, viticulture has always had only a modest foothold in Nelson province. Only about eighty hectares are under vines, just over one percent of the national vineyard area; the region is equally recognised for its cider and fruit wines.

Austrian-born Seifried, who established his vineyard in the hills at Upper Moutere in 1974, has so far produced the most wines. But others preceded him. In the 1890s, F.H.M. Ellis and Sons were 'substantial' winemakers at Motupipi, near Takaka, according to historian Dick Scott. Established in 1868, making wine from cherries and wild blackberries as well as grapes, the Ellis winery stayed in production for more than seventy years, until it was converted into a woolshed in 1939.

Later, Viggo du Fresne, of French Huguenot descent, from 1967 to 1976 made dry red wine at a tiny, half-hectare vineyard planted in deep gravel on the coast at Ruby Bay. The vineyard, dating back to 1918, was originally established with Black Hamburgh table grapes; du Fresne took over in 1948 and waged a long, unsuccessful struggle to establish classical vines. After his Chardonnay, Sémillon and Meunier vines all failed — probably due to viruses —

he produced dark and gutsy reds from the hybrid Seibel 5437 and 5455 varieties.

Victory Grape Wines is a pocket-size vineyard on the main road south at Stoke, near Nelson city. Irish-born Rod Neill (63), who named his winery in remembrance of Lord Nelson's ship, made his first trial plantings in 1967 and began hobbyist winemaking in 1972. His loam clay soils are planted in one hectare of Breidecker, Seibel 5455, 'Gamay Beaujolais' (Pinot Noir) and Cabernet Sauvignon.

Victory rose fleetingly to prominence at the 1980 National Wine Competition by scoring a silver medal for its 1978 Gamay Beaujolais — a light, pale and fruity red. Another silver at the 1987 Air New Zealand Wine Awards for its Seibel 5455 red proved a measure of ongoing success but, with production amounting to only a few hundred cases per year, Victory wines are rarely seen beyond Nelson. 'It's my hobby — a paying hobby,' says Neill.

Most wine lovers know that the Seifrieds produce a wide range of sharply priced wines, with Riesling — in several styles ranging from dry to sweet — the focal point; and that

Neudorf produces a tight range of exceptionally fine wines, the most significant being its powerful, mealy Chardonnay.

What surprised me during a recent visit to Nelson was a quartet of rewardingly ripe and meaty Cabernet Sauvignons. Seifried Estate is plummy and flavourful, with soft, easy tannins; Ruby Bay has a strong, concentrated palate with ripe, almost sweet fruit flavours. The Neudorf is well-structured, with firm tannin and deep blackcurrant-evoking flavours; Glover's is gutsy and packed with rich, brambly fruit. Almost gone are the green, leafy characteristics that a few vintages ago detracted from the quality of Nelson's reds.

In the past decade, the efforts of a small knot of enthusiasts — most prominently Hermann and Agnes Seifried and Tim and Judy Finn — have put Nelson firmly on the New Zealand wine map. Tourists responding with ever-increasing enthusiasm to promotion of the Nelson wine trail are able to explore wines varying in quality from plain to world class in some of the most stunning vineyard settings in the country.

Glover's Vineyard
Gardner Valley Road, Upper Moutere

Owners: David and Penny Glover

Key Wines: Pinot Noir, Cabernet Sauvignon, Riesling, Sauvignon Blanc

David Glover, like Michael Erceg of Pacific Vineyards, holds a doctorate in mathematics. Unlike Erceg, who has channelled his energies into large-scale liquor enterprises, Glover is focusing on the production of a rivulet of Upper Moutere wine, highlighted by a rich, fleshy Pinot Noir.

Glover's Vineyard lies in pretty, undulating countryside at Upper Moutere, between the inland and coastal Nelson–

Glover's

Dr David Glover traded his high-flying career in the Australian Defence Department for winemaking in the Upper Moutere hills. Glover's reds are full of character — bold and beefy with taut tannins.

Motueka highways. Glover (48) and his wife, Penny, planted their first vines in these soft, blue-green hills in 1984.

David Glover is a former Wellingtonian who spent sixteen years in Australia studying algebra and working for the Defence Department. 'We became keen wine drinkers over there. I'd always thought the Upper Moutere climate was ideal for winemaking, and after tasting Tim Finn's '82 Cabernet Sauvignon, I knew the potential was there. I used to go to Neudorf for days on end and sit there drinking a bottle of the '82.'

The two-hectare estate vineyard lies on a gentle, north-facing slope. In low-fertility clay threaded with decomposing rock, the Glovers have planted Pinot Noir, Cabernet Sauvignon and Sauvignon Blanc. 'Something funny is happening,' says Glover. 'I thought New Zealand's advantage lay with Pinot Noir and Sauvignon Blanc, but Cabernet Sauvignon is also performing well.' Riesling grapes are also bought from a local grower.

Glover's annual output is currently very small — less than 1000 cases. Since the first 1989 vintage, the Pinot Noir has attracted the most attention. This is a robust, meaty style of Pinot Noir, emerging from its six months' maturation in one- and two-year-old French oak barrels with strong raspberry/beetroot aromas and satisfying flavour depth.

Glover's Cabernet Sauvignon is a muscular red, dark-hued and chunky with assertive tannin. The white wines to date have impressed me more with their body and fruit intensity than their delicacy: the Sauvignon Blanc is weighty and tangy, the Riesling bursting with strong lemon-and-grapefruit flavours in a crisp, bone-dry style.

'I guess I'm in wine for the lifestyle,' says Glover. 'But I'm also aiming to be known for the quality of my wines.'

Neudorf

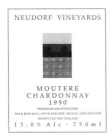

Neudorf Vineyards
Neudorf Road, Upper Moutere

Owners: Tim and Judy Finn

Key Wines: Chardonnay, Riesling, Sémillon, Sauvignon Blanc, Pinot Noir

Opulent, with layers of lush fruit and a powerful surge of buttery, mealy flavour, Neudorf Chardonnay is the most eye-catching achievement of winemaker Tim Finn; the 1991 won the trophy for the champion high-priced white at the 1993 Liquorland Royal Easter Wine Show. Neudorf, which derives its name (pronounced Noy-dorf) from the surrounding district settled by Germans last century, lies

Tim and Judy Finn's vineyard setting is ravishingly beautiful; so too are their wines. Neudorf Chardonnay is one of the finest in the land — with a cluster of top trophies to prove it.

not far from Seifried Estate just off the Nelson–Motueka inland highway. This is one of New Zealand's finest small wineries.

Tim Finn (46) was born in India and brought up in Wellington; he is an MSc graduate and a former dairying advisory officer with the Ministry of Agriculture and Fisheries. His wife, Judy, a former rural reporter for radio, is involved in bottling, labelling, administration and sales.

For their first vintage in 1981, the Finns used old stables on the property as a temporary winery. For 1982, Tim Finn built a handsome macrocarpa winery, pitching its roof high to accommodate his fermentation and storage tanks inside. Today during summer visitors picnic in the winery garden, which Judy Finn calls her 'bring your own food' restaurant.

Neudorf Chardonnay is the Nelson region's greatest wine. A robust, creamy style fermented in fifty percent new French oak barriques and given lengthy lees contact, the 1989 vintage won a major trophy at the 1991 Sydney International Winemakers' Competition. According to Australian wine writer James Halliday, the judges singled it out for its 'fantastic oak handling and extraordinary richness of fruit', comparing it 'to Bâtard-Montrachet or Meursault'.

Neudorf's distinguished white wine range also features a floral, vibrantly fruity, flavour-packed Riesling with racy acidity; a zingy, non-wooded, Sémillon; and a Sauvignon Blanc as fresh, aromatic and penetrating as Marlborough's.

The Cabernet Sauvignon, barrique-matured for a year, is a very claret-like red with strong, moderately ripe cassis-like flavours and the tightness to mature well.

Neudorf used to be famous for its (now deleted) Young Nick's Red, a floral and fragrant luncheon-style red made from 'Gamay Beaujolais' (Pinot Noir) grapes. Its success encouraged Tim Finn to attempt a 'serious' Pinot Noir — with impressive results. Neudorf Pinot Noir displays supple cherry and plum-like fruit flavours, French oak complexity and a sweet, sustained, firm finish.

In pursuit of greater knowledge of Pinot Noir vinification techniques, Finn worked the 1991 vintage in Mercurey, in the south of Burgundy. According to Judy, 'He returned in favour of the two-hour lunch, hot chocolate at dawn and hard work only when absolutely necessary.'

Neudorf — The Site

'Our real emphasis is in the vineyard,' says winemaker Tim Finn. 'That's where your real fruit flavours come from.'

If you ask Finn for the key reasons behind the consistently outstanding quality of his wines, the soil is the first thing he points to. 'I like heavier soils. Light soils give lightness and clays give depth of flavour; I don't know why it is. I base that observation on Chardonnay; mine has more depth than Marlborough's, yet the two regions' climates are pretty similar.'

The five-hectare, non-irrigated vineyard is planted on Moutere clays, threaded with layers of gravel. Having experimented with numerous varieties, the Finns are now concentrating on Chardonnay, Riesling, Sauvignon Blanc, and Pinot Noir, although they still grow Sémillon and Cabernet Sauvignon.

'Aspect is very important in a cool climate,' says Finn. 'We're on a gentle north-facing slope and the rows of vines run north–south to maximise the fruit's exposure to sunlight.'

To devigorate Neudorf's vines, Finn is gradually uprooting every second row and replacing it with two new rows — thereby reducing the spacing between the rows of vines from three to two metres. 'With fifty percent more rows,' says Finn, 'we can reduce each vine's vigour yet still increase the total grape crop.' 'Grassing down', selection of low-vigour rootstocks, and pruning the roots with a ripper also reduce the vines' vigour — and channel more of their energy into ripening their fruit.

During winter pruning, Finn is laying down fewer and shorter canes to reduce bud numbers and hence the size of each vine's eventual crop. The entire vineyard is trained on the Scott-Henry system, to open up the vines' canopies and expose the bunches for selective hand-picking. Cabernet Sauvignon is extensively leaf-plucked, but Sauvignon Blanc less so, lest the fruit lose its desired herbaceousness. 'The land is no longer such a supreme influence as it once was,' says Finn. 'We're learning how to compensate.'

Finn is ruthless about uprooting grape varieties that don't perform well on his site. 'You must match the grape to the region. You have to ask yourself what does consistently well and make the hard decisions.' Chardonnay, Pinot Noir, Riesling and Sauvignon Blanc are all secure in the Neudorf vineyard, but the two latest ripeners, Sémillon and Cabernet Sauvignon, are still question-marked: 'I won't give them up without a fight,' says Finn.

Pelorus

Pelorus Vineyard
Patons Road, Hope, Richmond

Owners: Andrew Greenhough and Jenny Wheeler

Key Wines: Chardonnay, Riesling, Müller-Thurgau, Pinot Noir, Cabernet Sauvignon

'We're going to drive ourselves hard for the next three or four years to produce top quality wines and get on the map,' says Andrew Greenhough. 'We also want to create an environment people will wander around in and enjoy.'

Andrew Greenhough and Jenny Wheeler are MA graduates who, after purchasing the low-profile Ranzau winery, now produce a rivulet of Nelson wine under the Pelorus label.

Greenhough and his partner, Jenny Wheeler, bought Pelorus Vineyard, then called Ranzau Wines, from its founder, Trevor Lewis, in early 1991. Lewis, a medical technologist who tended his vines and wines in spare hours in the evenings and at weekends, planted his vineyard at Hope, south of Richmond, in 1980. In his tiny, concrete-block winery, from 1983 Lewis produced a rivulet of wine rarely seen beyond Nelson.

Greenhough (34), a former Aucklander, holds an MA degree in art history. 'Dad's always drunk wine, which encouraged us to buy and cellar wine. Wine's always been a hobby.' Greenhough worked as a cellarhand at Villa Maria in 1990, where he started to gain confidence and experience, and says he initially got 'a lot of help from Villa Maria staff and Saralinda McMillan when she was at Seifried Estate'. Jenny Wheeler (32) also runs a company supplying children's books to libraries and schools. 'We're lucky to have the alternative income,' says Greenhough.

Pelorus Vineyard takes its name from the Pelorus River, which rises in the Richmond hills. Chardonnay, Riesling and Pinot Noir are the three varieties Greenhough will concentrate on under the Pelorus label: 'They've already demonstrated their potential in the area and we like them.'

The three-hectare irrigated estate vineyard, planted in silty loams overlying river gravels, was established by Trevor Lewis in Riesling, Müller-Thurgau, Pinot Noir and Cabernet Sauvignon vines. Greenhough is extending the Pinot Noir, planting Chardonnay and cutting back shelter belts to eliminate their shading of the vines. During summer the vineyard is covered with netting: 'Otherwise the birds are devastating.'

Pelorus's annual output is only about 900 cases of wine, although this will rise when the new plantings come on stream. An array of uniformly clean and promising wines has flowed from Greenhough's first vintages, highlighted by a full, savoury Chardonnay and a floral, off-dry Riesling with delicate lemony flavours and incisive acidity.

Ruby Bay

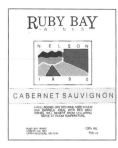

Ruby Bay Wines
Korepo Road, Ruby Bay

Owners: David and Christine Moore

Key Wines: Cabernet Sauvignon, Pinot Noir, Ruby Bay Red, Chardonnay, Sauvignon Blanc, Riesling

The Korepo winery, founded by Craig Gass on a stunning site overlooking Ruby Bay, is now enjoying a new lease of life with new owners, David and Christine Moore, and a new name — Ruby Bay.

Gass planted his 3.5-hectare vineyard on a north-facing, clay and gravel slope in 1976. From 1979 on there flowed a stream of solid although generally unspectacular wines; although the botrytised Tröckenbeerenauslese 1987 won a gold medal, Korepo never achieved a high profile.

David Moore (45) has an MSc from Lincoln University, and for sixteen years lectured in biochemistry at Christchurch Polytechnic, where he also taught courses in wine appreciation. After gaining a Graduate Diploma in Wine from Roseworthy College in South Australia, Moore made the 1989 vintage at Torlesse Wines in Canterbury. The Moores bought Korepo in late 1989: 'I wanted to make wine more than anything else,' says Moore, 'and Nelson appealed for the lifestyle.'

The estate vineyard, terraced for ease of cultivation, is planted in Pinot Noir, Meunier, Cabernet Sauvignon, Riesling, Sauvignon Blanc, Sémillon and Chardonnay. The fruit is supplemented by Chardonnay purchased from local growers. The site is warm and sunny, says Moore, with little wind. 'Our fruit is picked a good two or three weeks ahead of the other vineyards in the area. One result of this is the decreased likelihood of rain at harvest.'

Moore's principal interest is reds. 'White winemaking is

more predictable; with good ripe fruit, handled non-oxidatively and properly made, you know what you're going to get. Reds aren't that straightforward.' Pinot Noir is a special challenge, 'because it's so difficult to make. I know this area can produce the fruit.'

Ruby Bay's annual output is low at about 1500 cases of wine. My favourite is the Cabernet Sauvignon, a deep-hued, American oak-aged red with plenty of stuffing and strong, sweet fruit flavours. The Pinot Noir is lighter and raspberryish; Moore favours a 'softer, more fragrant' style than Gass's darker, more tannic Pinot Noirs.

Ruby Bay's lineup of whites includes a Sauvignon Blanc with a Sémillon-like grassiness; a Riesling with strong, ripe grapefruit-and-lemon flavours; and a full-bodied, American oak-aged Chardonnay with good depth of peachy-ripe flavours. (French oak will in future play a bigger role in Moore's wines.)

During summer visitors throng into the vineyard restaurant, devouring an array of local seafood and cold meats washed down — of course — with Ruby Bay wines. 'I'm committed to producing a limited volume of high-quality wine with regional character,' says Moore. 'I'm also committed to complementing our wines with appropriate food in a beautiful and relaxed environment.'

Ruby Bay's north-facing coastal vineyard site is one of the most stunning in the country. The chunky, flavour-packed Cabernet Sauvignon is a highlight of the range.

Seifried Estate
Main Road, Upper Moutere

Owners: Hermann and Agnes Seifried

Key Wines: Rhine Riesling Reserve Dry, Rhine Riesling Oak Aged, Rhine Riesling, Rhine Riesling Late Harvest, Rhine Riesling Beerenauslese, Gewürztraminer, Gewürztraminer Ice Wine, Chardonnay, Sauvignon Blanc, Müller-Thurgau, Old Coach Road Dry White, Pinot Noir, Refosca, Cabernet Sauvignon

At a winemakers' Pinot Noir conference, Hermann Seifried addressed the audience on his Roto-fermenters — horizontal cylinders which extract the skins' colour by rotating. Compared with traditional extraction techniques, Seifried claimed his Roto-fermenters were more efficient. 'We don't like to work too hard,' he added. The assembly erupted with laughter.

Hermann Seifried's appetite for work is legendary. He works six days per week, from 7.30 a.m. to 6.00 p.m., and on Sundays 'checks' the vineyards. According to his former winemaker Saralinda McMillan, 'He likes his staff to work hard, too.' The result: Seifried Estate is by far the largest Nelson winery and the only one with its labels on shelves throughout the country.

The Seifried label, proudly adorned with the Austrian eagle, early won respect when, from the first vintage, the Sylvaner 1976 won a silver medal. Hermann Seifried (46)

graduated in wine technology in Germany, and made wine in Europe and South Africa before arriving in New Zealand in 1971, as winemaker for the ill-fated venture by the Apple and Pear Board into apple-wine production. In spring 1974 Seifried planted his own vineyard in the clay soils of the Upper Moutere. A year later his wife Agnes, a South-lander, resigned her teaching job to join him in the winery. Today she oversees the company's administration, exports and public relations, but can still be found behind the vineyard counter on Saturdays.

The Seifried winery lies on the outskirts of the tranquil village of Upper Moutere. The ten-hectare Sunrise Valley estate vineyard is planted in Chardonnay, Riesling, Sylvaner and Pinot Noir. The Seifrieds also own a twenty-five-hectare vineyard in the Redwood Valley and at a third, twenty-four-hectare vineyard on the coast at Rabbit Island, a restaurant will open in late 1993. A small percentage of the annual crush is also bought from growers over the hills in Marlborough.

Saralinda McMillan, an Australian graduate of Rose-worthy College, was appointed to a full-time winemaking job in 1988. Previously Seifried had employed a stream of Austrian and German winemakers, but for only a few months at a time. Between 1988 and 1992, the winery's annual output soared from 20,000 to 35,000 cases. McMillan oversaw the cellar operations, taking the wine from the juice stage to its preparation for bottling; Hermann Seifried controlled the packaging operation and had 'the final say on questions of style'. Following McMillan's departure in 1992, Jane Cooper, who holds a graduate diploma in viticulture and oenology from Lincoln Univer-

Seifried

The indefatigable Hermann Seifried pioneered commercial viticulture in Nelson in the mid-1970s, and today owns the region's only large winery. Riesling is the variety indivisibly associated with the Seifried name, but his Gewürztraminers, Sauvignon Blancs and Chardonnays can also be rewarding.

sity, was appointed assistant winemaker.

The standard of Seifried wines has improved in recent vintages — the whites have greater delicacy and consistency; the reds are notably riper and richer. Confirmation of the winery's ascending quality came at the 1992 Liquorland Royal Easter Wine Awards, where Seifried Estate Sauvignon Blanc 1991, Gewürztraminer Ice Wine 1991 and Rhine Riesling Beerenauslese 1990 all collected gold medals.

Most wine lovers still think of Seifried in terms of Riesling. A quintet of Rieslings flow under the Seifried label, ranging in style from dry to sweet. My favourite is the Rhine Riesling Reserve Dry, a steely, austere style with strong lemon-and-grapefruit-like flavours and spine-tingling acidity. The Rhine Riesling Oak Aged is an absorbing style departure — a fresh, tangy, bone-dry wine with a smoky undercurrent of wood. The wine simply labelled as Rhine Riesling is an easy-drinking, slightly sweet style. The freeze-concentrated, medium-sweet Rhine Riesling Late Harvest was praised by Michael Broadbent, director of Christie's wine department, as 'almost literally a lollipop. It should have been on a stick, to be licked.' Sweeter again is the freeze-concentrated, briefly oak-matured Rhine Riesling Beerenauslese, an oily, strong-flavoured, slightly honeyish beauty.

Seifried Müller-Thurgau is consistently appealing: fresh, fruity and delicate, with a touch of sweetness and lively acidity. Both the Gewürztraminer Reserve Dry and the mainstream, medium wine labelled simply as Gewürz-

traminer display appealing floral/spicy bouquets and full, well-spiced palates. The lusciously fruity Gewürztraminer Ice Wine is another freeze-concentrated, sweet style: 'Every second person who calls at the cellar requests a taste,' says Agnes Seifried.

Chardonnay, Sauvignon Blanc and Cabernet Sauvignon are also coming to the fore in the Seifried range. The Chardonnay — both stainless steel and oak-fermented — is citric-flavoured and crisp; the occasional riper vintage really shines. The non-wooded Sauvignon Blanc, a weighty, ripe, moderately herbal style, can achieve real distinction in favourable vintages. The latest Cabernet Sauvignons are much darker, weightier and riper in flavour than earlier vintages.

Both the Pinot Noir and Refosca — an Italian variety also known in north-east France as Mondeuse — are light, raspberryish, soft, simple reds.

Seifried Estate wines often deliver irresistible value for money — a key factor in the winery's recent sales surge. The Rieslings and Gewürztraminers offer particularly fine value. Like all substantial New Zealand wineries, Seifried is eager to penetrate overseas markets. Seifried Estate was among the first New Zealand wines stocked in the duty-free stores at Heathrow and Gatwick airports in England. Virgin Atlantic Airlines started pouring Seifried Estate Sauvignon Blanc on its trans-Atlantic routes in 1992 — it was the first New Zealand wine served by the airline.

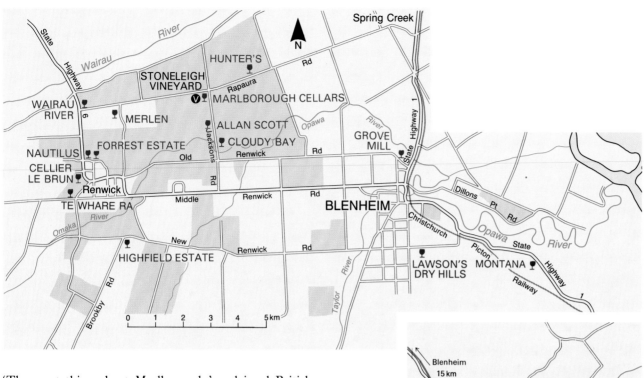

'The great thing about Marlborough,' acclaimed British wine writer Oz Clarke said during a Radio New Zealand interview, 'is that it produced for the first time since the War, maybe this century, a flavour which no-one's ever found before. Marlborough flavour is unbelievably strong, unbelievably memorable . . . The Sauvignon is absolutely, stabbingly strong fruit: it's a mixture of apricots and asparagus and grassiness which is terribly exciting because the Sauvignon grape in Europe, where they've been growing it for donkey's years, has got drier and drier and leaner and leaner . . . Suddenly, out comes Marlborough with this flavour which is so strong.'

The emergence of Marlborough as the country's most heavily planted wine region rates among the most crucial developments of the last twenty years. From its endless rows of vines marching across the pebbly, pancake-flat Wairau Plains have flowed the deep-flavoured, scented white wines which — more than any rival region's — have awakened the world to the beauty of New Zealand wines.

Marlborough, the north-eastern edge of the South Island, contains the inland Kaikoura Ranges, which reach an elevation approaching 3000 metres. The Wairau River, draining the ranges of silt and gravel, descends from the back country to the Wairau Plains; it is on the plains, formed by massive alluvial deposits from the river, that Montana, Corbans and others have planted their vines.

Sheep inhabited the Wairau Plains as early as the 1840s. Later, small- and medium-scale mixed farming established a stronghold and recent decades have witnessed developing interest in peas, grass seed, lucerne, garlic and cherries. Not until 1973, however, did viticulture stake a significant commercial claim.

When Montana planted its first vine in the province on 24 August 1973, it triggered the modern era of Marl-

borough viticulture. The region's first wines, however, had flowed almost a century earlier. David Herd's Auntsfield vineyard, in the hills to the south of Fairhall and Brancott, produced its first commercial harvest around 1875. Auntsfield's sweet red wine was made from red Muscatel grapes, 'crushed with a machine made from the wheels of an old flax mill stripper, then pressed in a barrel . . . [and] matured in oak brandy casks'. (Cynthia Brooks, *Marlborough Wines and Vines*, 1991). Only about eight hundred litres were produced each vintage, but the trickle of Auntsfield wine survived Herd's death in 1905; his son-in-law Bill Paynter carried on the family tradition until 1931.

At Mount Pleasant Wine Vaults, just south of Picton, in 1880 George Freeth started making wine from a wide array of fruits, including grapes. Yet no surge of vine plantings in Marlborough followed the 1895 publication of Romeo Bragato's 'Report on the Prospects of Viticulture in New Zealand'; he was more impressed with Nelson's potential. In the first half of this century, in the heart of Blenheim, Harry Patchett and Monsoor Peters grew grapes and sold a trickle of wine. Patchett lived until 1974 — just long enough to witness the Montana-led revival of Marlborough wine.

Marlborough is one of the few South Island regions that is sufficiently warm for viticulture on a commercial scale.

The heat summation figure, 1150–1250 degree days Celsius, is higher than at Geisenheim on the Rhine, in Germany, which has 1050–1250 degree days Celsius. Blenheim frequently records the highest total sunshine hours in the country, and this plentiful, although not intense sunshine affords the grapes a long, slow period of ripening. According to Montana, Marlborough's heat and sunshine are usually sufficient for 'good sugar levels to be attained in white grapes and adequate to good levels in red grapes'.

The risk posed by heavy autumn rains is lower than across the hills in Nelson. March is usually the driest month of the year; April rainfall, averaging sixty-one millimetres, compares favourably with the average seventy-two millimetres in Bordeaux during the harvest month.

The warm, dry northwesterly winds that 'come out of the Kaituna Valley like a freight-train', according to Richard Bowling of Vavasour, can pose drought problems, dehydrating the vines and severely reducing crop sizes. Most vineyards have installed a trickle-irrigation system, feeding water to the vines and greatly enhancing grape yields. Irrigation is most important during the vines' early years, before they have had the opportunity to develop an extensive root system. Late frosts also pose a risk — most are insufficiently intense to cause real damage but October, when three ground frosts strike on average, is a danger period; should temperatures drop below around minus 0.6°C, the vine shoots and flowers can die. A heavy frost in the autumn of 1990 killed the vines' leaves, preventing the full ripening of Riesling, Sauvignon Blanc and Cabernet Sauvignon: 'The whole valley turned black overnight,' recalls Jane Hunter.

Owing to the relatively dry summers and low humidity in Marlborough, during the ripening season the vines are sprayed less frequently than in northern regions. Such dry weather diseases as powdery mildew pose more of a threat than botrytis and downy mildew, associated with wet climates. During the harvest month of April the average temperature is quite low, which by slowing the spread of disease allows the grapes to be left late on the vines to ripen fully.

Not all the various soil types found on the plains adapt well to viticulture. Large areas of deep silt loams are fertile, with a high water storage capacity. The preferred sites are of lower fertility, with a noticeably stony, sandy loam topsoil overlying deep layers of free-draining shingle with sand infilling. These shallow, stony soils promote a moderately vigorous growth of the vine. 'The key benefit of the stones is that they reduce the soil's fertility,' says John Belsham of Vintech.

One problem facing viticulturists here is that soil types often vary enormously even within individual vineyards. The Wairau Valley is a flood plain, with braids of soil of varying fertility running in an east–west direction, crossed by north–south-facing vine rows. According to Montana, 'it is common to find vigorous vines with a dense canopy and heavy crop in the same row as can be found stressed weak vines with little or no canopy. With such variability,

harvest decisions become a compromise.' Vines planted in Marlborough's more fertile areas thus share the problem of too-dense foliage canopies found further north. Improved canopy control promises to bring a further upgrading of the quality of the region's fruit.

Some of Marlborough's winemakers are convinced that climatic and soil differences between parts of the Wairau Valley are having a significant impact on wine styles. The south side of the valley, where Montana's vineyards are concentrated, has less rainfall and a lower water table, and thus drier soils. The grapes, planted in heavier soils than the warmer, stonier soils on the north side of the valley, ripen about ten days later with more intensely herbaceous flavours.

Marlborough's winemakers are now starting to exploit these differences in soil and climate within the Wairau Valley. 'If you want to make a very herbaceous, greener style of Sauvignon Blanc,' says Belsham, 'you choose a site with medium-high fertility and good soil moisture retention. For a lusher, less aggressive Sauvignon Blanc, you choose a stony site which matures its fruit two to three weeks earlier.'

Vavasour in the mid–late 1980s pioneered viticulture in the Awatere Valley, situated east and over the Wither hills from the more sweeping Wairau Valley, where all Marlborough's vineyards had hitherto been concentrated. According to Richard Bowling of Vavasour, 'climatically the Awatere Valley is akin to the Wairau, the Dashwood having a tendency to be drier than Rapaura . . . The braided river pattern resulting in very uneven [soil] profiles in the Wairau Valley is also much less pronounced . . .' The hill country on the south side of the Wairau Valley is still unexplored for quality table-winemaking, but has obvious potential.

From the start, Montana has established a commanding presence and its pioneering move into Marlborough and Marlborough regional wines are discussed under the Montana entry on pages 75–9. Wayne Thomas's 1973 report on the region's viticultural potential (see page 75) was clearly a decisive factor in Montana's move into Marlborough, but earlier others had speculated about the possibilities of Marlborough wine. 'One undeveloped area with distinct possibilities for viticulture is behind Blenheim, among the northern foothills of the Kaikouras', John Buck wrote in his book *Take a Little Wine*, in 1969. Laurie Millener concluded in 1972, after analysing New Zealand's regional climates, that 'Nelson, and especially Blenheim, offer great promise. It should be possible to make specialist table wines there, both reds and "Mosels".'·

Marlborough was often cited by politicians of that era as a typical example of a 'forgotten' region needing diversification. 'Nothing more vividly recalls the sudden realisation of what wine could do for Marlborough,' Terry Dunleavy, Montana's sales manager in the early 1970s, has written, 'than the stunned reaction of Lucas Bros. when faced with a [Montana] order for twenty-six tractors.' Yet cropping farmers then reliant on hormone sprays for weed control

Stoneleigh Vineyard

When builders dug the foundations of the Marlborough Cellars winery, adjacent to the Stoneleigh Vineyard, 'they went down six metres and it was all stones', says Corbans winemaker Alan McCorkindale. This famously stony vineyard is the source of four wines under the Stoneleigh label — Sauvignon Blanc, Rhine Riesling, Chardonnay and Cabernet Sauvignon. Some of the best fruit from older vines is also earmarked for Corbans' top Private Bin selection.

Sheep grazed on the land before the first vines sank root in 1980. The vineyard is braided with different soil types, with the stonier parts marking the old course of the Wairau River. The structure of Stoneleigh Vineyard is 'very representative of the north side of the valley', says McCorkindale.

The sweeping, one hundred-hectares vineyard, split into four blocks in the Jacksons Road/Rapaura Road area, has been planted in Sauvignon Blanc, Sémillon, Riesling, Chardonnay, Gewürztraminer, Pinot Noir, Cabernet Sauvignon and Merlot. So impressed are Corbans with Merlot's performance, some Cabernet Sauvignon vines are being replaced with the earlier-ripening variety.

Frost and wind killed some of the young vines planted between 1980 and 1989. Phylloxera is another problem, since forty percent of the vines are not grafted onto phylloxera-resistant rootstocks.

The vines are trained on standard vertical trellises with moveable foliage wires. Spacing is conventional: 2.7 metres between rows, 1.8 metres between vines. The vines are summer-pruned ('hedged') about six times each season and leaf-plucked to aid the penetration of light into the canopy.

When the hot northwesterly wind blows, the vines respond by shutting down their leaves to reduce water loss, thereby extending their ripening season. The wind also helps to combat the threat of wet rot. 'Noble rot' is common: between 1986 and 1993, in every vintage except 1987 and 1992 Corbans produced a sweet, botrytised wine at Stoneleigh.

and fearful they would be banned from using them in the proximity of vineyards, vigorously opposed the planting of the early vineyards; Philip and Chris Rose's plans to plant the first vineyard in the central Rapaura area in 1978 ran a gauntlet of fifty-six objections.

Penfolds and Corbans established vineyards here much later than Montana. Penfolds' first contract vineyards were planted in the winter of 1979 and subsequently the company arranged contracts amounting to about 400 hectares. Its plans for a Blenheim winery, to be operational by 1983, were ultimately shelved and, until Montana purchased Penfolds, Penfolds' Marlborough grapes were trucked to the North Island.

Corbans commenced its vineyard development in the region in 1980 and unlike Penfolds planted its own company vineyards, notably Stoneleigh. Unlike Montana, which had established its vineyards on the southern margins of the plains, where the less gravelly soil was thought likely to be kinder to machinery, Corbans planted its vineyards in the stony Rapaura district.

Since 1985, when Australian capital financed the erection of Cloudy Bay's handsome concrete winery in the heart of the Wairau Valley, overseas investment has streamed into the Marlborough wine scene. The Marlborough Cellars winery just along Jacksons Road from Cloudy Bay, now wholly Corbans-owned, was initially funded by Corbans and the Australian company Wolf Blass (now Mildara Blass).

Links between Marlborough and the great houses of Champagne are also mounting. Deutz has lent technical assistance and its own name to Montana's bottle-fermented sparkling, Deutz Marlborough Cuvée. Veuve Clicquot Ponsardin is now the majority shareholder in Cloudy Bay. Two of the three shareholders in the Highfield winery are Japanese, also involved in the Champagne house of Drappier. And Moët & Chandon's Australian subsidiary, Domaine Chandon, recently unveiled its first New Zealand bottle-fermented sparkling, produced at Hunter's winery.

Vintech, set up in 1991, is a contract winery whose principal shareholders are winemaker John Belsham — formerly of Matua Valley and Hunter's — and engineer Geoff Taylor. Vintech's clients are principally companies which draw fruit from Marlborough but have their headquarters in other regions. Previously, many wineries trucked Marlborough grapes for twenty-four hours to Auckland, giving the juice extended, uncontrolled skin contact. Now Vintech can de-juice the grapes and ferment them to dryness, before railing the wine to its clients in large stainless steel tanks. 'Each wine doesn't taste the same,' Belsham points out. 'The wine companies themselves choose the vineyards, timing of harvest and so on.'

With 2071 hectares of vines in 1992, Marlborough has emerged as the country's most heavily planted wine region, clearly ahead of Hawke's Bay and Poverty Bay. Sauvignon Blanc (especially) and Chardonnay have proliferated,

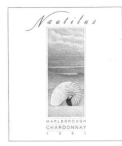

accounting for one-half of the vines, but numerous other varieties are well-established, notably Müller-Thurgau, Pinot Noir, Cabernet Sauvignon, Riesling, Sémillon and Merlot.

That Sauvignon Blanc and Riesling thrive in Marlborough's cool ripening conditions has been demonstrated by Montana for more than a decade. Lately the standard of Marlborough's Chardonnays has soared. In the past they often lacked the weight and flavour richness of those from the North Island; this is no longer so. The power and subtlety of Vavasour, Corbans Private Bin, Montana Renwick Estate, Cloudy Bay, Villa Maria Reserve Marlborough, Hunter's and others is proof that Marlborough has now emerged as a formidable rival to Hawke's Bay and Gisborne in the Chardonnay quality stakes.

Pinot Noir has not shone as a red-wine variety in Marlborough — although recent releases from Hunter's and Robard & Butler are promising — but is proving to be an ideal base for bottle-fermented sparkling wines. Marlborough's Cabernet Sauvignons often display lightness and strong leafy-green flavours; the variety has traditionally flourished in a warmer climate. 'With careful site selection you can produce good Cabernet-based reds,' says John Belsham, 'but you don't get enough warm years to consistently produce good wines.'

Marlborough's Merlots are another story. Corbans Private Bin and Highfield are both dark-hued, spicy and complex. Merlot ripens two to three weeks earlier than Cabernet Sauvignon in Marlborough, giving the grapes a much higher chance of achieving optimal ripeness. Will it be Merlot, rather than Cabernet Sauvignon, that will finally allow Marlborough to shed its 'stunning white wines, pity about the reds' reputation?

Jackson Estate is one of the newest wine companies on the block. Warwick and John Stichbury own the thirty-six-hectare Jackson Estate vineyard in Jacksons Road — but no winery. It's more economic to process their wine at Vintech. The first 1991 vintage of their Sauvignon Blanc picked up a gold medal at the 1992 Liquorland Royal Easter Wine Awards. Jackson Estate's range also includes a promising Chardonnay and Riesling and a bottle-fermented sparkling is planned for the future.

The highly regarded Nautilus Chardonnay, Sauvignon Blanc, Cuvée Marlborough and Cabernet/Merlot, distributed by Negociants, are also sourced from Marlborough. Produced in Marlborough by Alan Hoey, a leading Australian winemaker, the Chardonnay is a partly barrel-fermented wine of eye-catching elegance and length.

The newest names on the Marlborough wine scene include Phil and Chris Rose's Wairau River, source of an excellent Sauvignon Blanc and Chardonnay; Murray Brown's Cairnbrae Wines; Craig Gass's promising Conders Bend range, and grapegrower Ross Lawson's auspicious releases under the Lawson's Dry Hills label.

Allan Scott

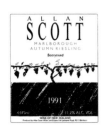

Allan Scott Wines and Estates
Jacksons Road, Blenheim

Owners: Allan and Catherine Scott

Key Wines: Sauvignon Blanc, Riesling, Autumn Riesling, Chardonnay

Allan Scott is a vastly experienced viticulturist who recently plunged into winemaking under his own label. His deep-scented, incisively flavoured Riesling ranks among the region's finest.

Allan Scott, formerly Corbans' national vineyards manager, launched his own label with a 1990 Sauvignon Blanc. After years of 'wheeling and dealing in land and having faith in the district's wine future', Scott and his wife, Cathy, now own 'about as much of the stony Jacksons Road area [where Corbans and Cloudy Bay have their wineries] as anyone'. Their atmospheric, rammed-earth winery was built in early 1992, just across Jacksons Road from Cloudy Bay. Chardonnay, Riesling and — inevitably — Sauvignon Blanc will be the mainstays of the Scotts' range.

Scott (45) has long been a key figure in the development of Marlborough's vineyards. He was born on a North Canterbury farm; Cathy is from Blenheim. When Montana arrived in Marlborough in 1973, Scott got a job as a vineyard labourer, tearing down fences and planting vines. Within a month he was appointed vineyard foreman, and later supervisor of Montana's Fairhall vineyard.

In 1980, Scott moved to Corbans to oversee the establishment of their Marlborough vineyards — notably Stoneleigh. By 1982 he was Corbans' national vineyards manager, in which role he undertook the 'difficult' task (in human terms) of dismantling the company's Auckland vineyards. After resigning from Corbans in 1989 he set up as a viticultural consultant, but has recently been absorbed in developing the Allan Scott range of wines.

The Scotts' vineyards encircle the winery, spreading over twenty-four hectares on both sides of Jacksons Road. In future Scott is planning to make his own wines on site, but the early vintages have been crushed and fermented at Vintech. The Scotts' output has grown rapidly to reach 10,000 cases in 1993.

Visitors to the Allan Scott winery and vineyard restaurant enjoy a tight range of well-crafted wines. The Sauvignon Blanc is a non-wooded style, brimming with lush tropical-fruit and moderately herbal flavours. The Chardonnay, about one-half barrel-fermented, is weighty and savoury. The Riesling displays the strong, ripe citric-fruit flavours typical of Marlborough and a light touch of sweetness. Autumn Riesling is a late-harvested style, perfumed and honey-sweet.

Scott is confident of his new winery's future. 'I won't be giving away our hard work in the vineyards by selling inferior wine.' In mid-1993, however, the Scotts were negotiating the possible sale of their winery to Appellation Vineyards.

Cellier Le Brun
Terrace Road, Renwick

Owner: Appellation Vineyards Limited

Key Wines: Daniel Le Brun Méthode Champenoise Brut, Blanc de Blancs, Méthode Champenoise Rosé, Vintage Méthode Champenoise Brut, Blanc de Noirs, Cuvée Adele, Chardonnay, Pinot Noir

Daniel Le Brun's Blanc de Blancs 1988 won a gold medal and trophy for the champion sparkling wine at the Air New Zealand Wine Awards in late 1991. He upstaged that effort at the 1992 Liquorland Royal Easter Wine Awards: of the three gold medals awarded in the bottle-fermented sparkling class, one went to his non-vintage Méthode Champenoise Brut, the second to his Blancs de Noirs 1989 and the third to his Vintage Méthode Champenoise 1989. I wonder how he celebrated?

Daniel Le Brun (48) is a Champenois, the scion of a family of French Champagne makers stretching back over twelve generations to 1648. In search of new horizons, Le Brun came to New Zealand and at Renwick, near Blenheim, he discovered the combination of soil and climate he wanted. His ambition: to fashion a bottle-fermented sparkling wine in the antipodes able to challenge the quality of Champagne itself.

Le Brun speaks with a thick French accent. He was born at Monthelon, only a few kilometres south of Épernay and recalls, 'the only thing for me to do was to carry on the family tradition.' However, after graduating from the École de Viticulture et Oenologie at Avize, he grew more and more frustrated by the very tight restrictions placed on the size of individual landholdings in Champagne.

After visiting New Zealand in 1975, he emigrated here the same year, and three years later met his future wife, Adele, in Rotorua. By 1980 they had purchased land just outside Renwick and had begun establishing their vineyard.

Cellier Le Brun

The locals called Daniel Le Brun 'the mad Frenchman' during his early days in Marlborough. With his beloved, arrestingly rich, toasty, flavour-packed bottle-fermented sparkling wines, however, Le Brun has had the last laugh.

Labelled 'the mad Frenchman' by the locals after they got wind of his unorthodox winery plans, Le Brun set out to duplicate the cool subterranean storage conditions of Champagne by burrowing twelve metres into his Renwick hillside, to form steel-lined caves under four metres of earth. In these cool caves, varying only a couple of degrees in temperature between summer and winter, the Le Brun bottle-fermented sparklings age after bottling.

The classic varieties of Champagne are naturally featured in the Le Brun vineyards (two hectares adjacent to the winery, twelve hectares down the road): six hectares of Chardonnay, five hectares of Pinot Noir and three hectares of Meunier. The vines, planted in river gravels, are densely spaced as in Champagne. Few contract-grown grapes are used.

Le Brun is intent on 'carrying on the old techniques — everything is handled according to Champagne tradition'. To make his beloved Méthode Champenoise Brut, he blends the base wine from the three varieties and across vintages. After adding yeasts and sugar for the second fermentation, the wine is then bottled and rests on its lees for two years.

The bottles, rather than being laboriously hand-turned, are then loaded into heavy metal-framed riddling machines called gyropalettes. They stay there for five days, being automatically shaken every four hours. Then each bottle is disgorged, topped up, corked and wired.

And the wine? Daniel Le Brun Méthode Champenoise Brut — which constitutes two-thirds of the winery's output — is a distinctively bold, full-flavoured style, reflecting Le Brun's liking for 'gutsy' sparkling wines: 'I'm a fan of Bollinger rather than Taittinger,' he says. Blended from sixty percent Pinot Noir, thirty percent Chardonnay and ten percent Meunier, this is a mouthfilling wine with plenty of yeast-derived complexity and impressive flavour richness. Some early batches proved disappointing; Le Brun concedes that his recent releases have 'achieved a consistency of style that wasn't there before'.

Described as their 'crème de la crème', the Le Brun's purely Chardonnay-based Blanc de Blancs is stunningly bold for a style usually cast in the 'light and fresh' mould. Le Brun himself views New Zealand's bottle-fermented sparklings as 'bigger and fruitier' than their Champagne counterparts, principally due to their more advanced fruit ripeness.

Pushed to reveal his favourite among his own wines, Le Brun says he prefers 'the vintages — they're better structured'. His Vintage Méthode Champenoise Brut, a blend of Pinot Noir and Chardonnay held on its yeast lees six months longer than the non-vintage Brut, is an exceptionally lively, deep-flavoured, buttery and complex wine. This is my pick of the bunch too.

The Le Brun range also includes a robust, salmon-pink Méthode Champenoise Rosé; the new, high-priced Cuvée Adele, a solid but plain Pinot Noir; and a still Chardonnay, matured in Argonne oak 'foudres' (5000L casks), which displays a pungent yeastiness of taste reminding one of the sparklings.

Regal Salmon Limited gained a controlling interest in Cellier Le Brun in 1987 for $700,000, but Daniel Le Brun stayed on as the winemaker and a director, and Adele Le Brun still oversees sales. The Le Bruns and other original investors retained minority shareholdings. In mid-1993, the shareholders of Cellier Le Brun agreed to sell their shares to Appellation Vineyards Limited for shares in Appellation, in which Regal Salmon is the largest shareholder.

'The Courtyard Café' at the winery serves platters of smoked beef, eel, salmon, croissants and bagels. Le Brun sparklings are sold in a range of seven different bottle sizes, culminating in a fifteen-litre Nebuchadnezzar at $900.

Le Brun returns to Champagne every year. At a major wine trade fair he meets the suppliers of his bottles, yeasts, crown seals, corks, capsules and wire — all shipped from Champagne to Renwick. 'When I told my brother how much land I had, all in the same place and all excellent for growing grapes, he was so jealous.'

Cloudy Bay

Cloudy Bay Winery
Jacksons Road, Blenheim

Owner: Cape Mentelle (NZ)

Key Wines: Sauvignon Blanc, Chardonnay, Pelorus, Cabernet/Merlot

When Kevin Judd arrived in Marlborough in 1985 to take up his new job as Cloudy Bay's winemaker, no fanfare greeted him. The wine industry, on the verge of the massive vinepull scheme, was sunk in glut-induced despondency. 'We had no vineyards, no winery, no equipment, nothing,' recalls Judd. 'Things could only get better.' Explosively flavoured and stunningly packaged, Cloudy Bay's 1985 Sauvignon Blanc swiftly sent a ripple through the international wine world.

In the conviction that the quality of New Zealand-grown Sauvignon Blancs could not be equalled in Australia, David Hohnen, part-owner of Cape Mentelle, a prestigious Western Australian winery, had crossed the Tasman to set up a second winery in Marlborough. Cape Mentelle's reputation was based on having captured the Jimmy Watson Trophy — awarded to the top one-year-old Australian red — in successive years, with its 1982 and 1983 Cabernet Sauvignons. Cloudy Bay, the name of the nearby bay whose waters, when the Wairau River fills them with silt, turn cloudy, was finally chosen as the name of the new venture, but not before the name of a prominent local cape had been entertained but swifty rejected — Farewell Spit.

Construction of the handsome, concrete-slab winery in Jacksons Road, Rapaura, began in August 1985 under the

Kevin Judd — also a talented photographer — heads the production of New Zealand's most internationally prestigious wine — Cloudy Bay Sauvignon Blanc. His strapping, savoury Chardonnay also ranks among New Zealand's finest.

direction of Australian-born Kevin Judd, a Roseworthy College graduate who elevated the standard of Selaks' wines between 1983 and 1985. Hohnen was first introduced to and impressed by Judd at a tasting following the 1984 National Wine Competition: 'He wasn't making a social event of it, but seriously working his way through the medal-winning wines.'

Judd, a lean, greying 34-year-old, now has responsibility for the day-to-day running of Cloudy Bay. 'I've seen it come from a paddock to what it is today,' says Judd. 'I'm the manager and winemaker. I'm not in the laboratory day in and day out, and not much involved with finance, but I'm involved in everything else.' David Hohnen sees Judd as having 'an appreciation of quality, a very acute palate — and a competitive edge'. Hohnen crosses the Tasman about five times each year for crucial blending decisions, but 'the final decisions are Kevin's,' says Hohnen.

In 1990 Veuve Clicquot Ponsardin, the illustrious Champagne house, purchased a majority interest in Cape Mentelle — and thus Cloudy Bay. In Judd's eyes, the link with Veuve Clicquot has brought two key benefits: 'Financial stability and greater access to overseas markets.' Cloudy Bay wines are now sold in twenty-five countries including Australia — the biggest export market — Hong Kong, the United States, the United Kingdom, France, Italy and Denmark.

On flat land with stony, well-drained soils adjacent to the winery, the fifty-hectare Cloudy Bay vineyard is planted in Sauvignon Blanc, Chardonnay and Sémillon, with small plots of Pinot Blanc, Merlot and Malbec. Grapegrowers on long-term contracts, supported by Cloudy Bay's fulltime viticulturist, Ivan Sutherland — himself a grower — supply over half the winery's annual fruit intake.

Why was the first 1985 Sauvignon Blanc (based on grapes bought from Corbans and made by Judd at Corbans' winery) such a roaring success? 'It all came together,' says Judd. 'We had a simple name, attractive label and a distribution system in place in Australia. The flavour obviously had wide appeal. When I was at Selaks I used to say: "I shouldn't be here making this. I should be in Australia selling it — they'll freak when they see it."'

Judd seeks to produce Sauvignon Blanc having 'a lively gooseberries and lychees — rather than green peas — fruit character and a touch of oak complexity'. His adroitly structured blend includes about ten percent wood-fermented wine, a varying but low proportion of lower-acid Sémillon and also a hint (up to five grams per litre) of sugar which is barely perceptible. The style is bracing and firm, with mouth-filling body and a powerful surge of herbal flavours that makes it a truly exciting mouthful.

From the start, Hohnen planned to extend the Cloudy Bay range beyond Sauvignon Blanc, although the Australian wine trade initially opposed the idea. Cloudy Bay's output, 50,000 cases per year, makes it a medium-sized winery by New Zealand standards. Chardonnay, Cabernet/Merlot and Pelorus, the new bottle-fermented sparkling, are the three other key wines in the Cloudy Bay lineup.

A strapping, high alcohol, oaky, citric-flavoured wine,

David Hohnen

When a party of four Kiwi wine-makers visited his Cape Mentelle winery in Western Australia in 1983, David Hohnen tasted his first New Zealand wine. 'I had my '82 Sémillon/Sauvignon Blanc in barrels, and said: "Get a load of this". They said: "If you think that's herbaceous, see what we've got in the car". Penfolds 1983 Sauvignon Blanc from Marlborough just blew me away. It was a bit sweet, but it had fruit characters that we would never get in Australia.'

A fit, tanned, grey-bearded forty-four-year-old, Hohnen was born in New Guinea, where 'there was always wine on the table'. After his father, a mining engineer, took the family to Perth in 1964, the Hohnens bought a block of land in the Margaret River, adjoining the Vasse Felix vineyard. His interest sparked, Hohnen worked for a year as a cellarhand at Stonyfell in the Barossa Valley, and then studied oenology at the University of California, Fresno. He topped the wine classes, but 'because I didn't do the courses in American history, basket weaving and baseball, I didn't graduate, only getting a diploma.'

The first vines were planted at Cape Mentelle in 1970. Hohnen spent the early–mid 1970s helping to set up the Taltarni winery in Victoria, before returning to the Margaret River for the 1977 vintage. Double Jimmy Watson Trophy triumphs with his 1982 and 1983 Cabernet Sauvignons soon put Cape Mentelle firmly on the Australian wine map.

When Hohnen plotted to establish a new winery in Marlborough, his accountants advised him to stay out of the New Zealand industry, then on the verge of the cut-throat price war of 1985–86. 'It was a terrific gamble. I just had this gut feeling that told me it was the right thing to do. New Zealand Sauvignon Blanc simply hadn't been discovered and seemed to me to have a great future.' Before finally committing himself, Hohnen flew his distributors across the Tasman, and asked them if they could sell New Zealand Sauvignon Blanc. The answer: 'Yes.'

Hohnen and his brother, Mark, then 'borrowed a whack of dough [$NZ1.5 million], put up the winery, negotiated a three-year grape supply contract with Corbans — and Cloudy Bay was born.'

The elegant and uncluttered Cloudy Bay label was also a Hohnen inspiration. 'Looking out the window of an Air New Zealand Friendship as we were landing in Blenheim, I saw the three-tiered silhouette of the mountains in the distance and got an instant vision of the Cloudy Bay label.'

Today forty percent of Cloudy Bay's Sauvignon Blanc is sold in Australia. 'It's the perfect consumer wine,' says Hohnen, 'a real leap-out-of-the-glass style.' Cloudy Bay's problem in 1985, however, was persuading Australians to even taste a New Zealand wine. So Hohnen sent key wine retailers a pack containing a bottle, a glass and a jar of New Zealand mussels, carrying the message: 'Before you open this box, get a corkscrew and some fresh crusty bread.' Bottles were broached and almost overnight Cloudy Bay was a smash hit.

Cloudy Bay is today a larger winery than Cape Mentelle. Veuve Clicquot purchased a majority share in both wineries in 1990, but David Hohnen has retained a twenty percent shareholding in the Australian parent company. 'If I keep some of my money in the business, I'll retain my enthusiasm,' says Hohnen.

Hohnen's fundamental approach to wine is clear. 'The French have always understood that quality is what you build your reputation and wealth on, and won't allow any compromises. Cloudy Bay is seen as a benchmark wine, and we are under acute scrutiny with each new release.'

For an exceptionally successful businessman, Hohnen is not very interested in finance. 'So long as the cashflow is there, and I understand balance sheets, it doesn't interest me. I've got a good nose and an instinct for trends in the market. Another strength is recognising good people and being able to get the best out of them. The acclaim for Cloudy Bay is a real ego boost, but my greatest pleasure comes from the people I work with.'

Cloudy Bay Chardonnay is predominantly barrel-fermented and held on its yeast lees for a full year. A rich, high impact style, it has proven its ability to cellar well: the first 1986 vintage is still in splendid condition.

The Cabernet/Merlot is robust, spicy, complex and firm, with a touch of the leafy greenness that often is a characteristic of cool-climate reds. The first 1986 vintage was an unblended Cabernet Sauvignon; in 1987 Merlot was introduced; in 1988 came Cabernet Franc. Hohnen is frank: 'If I'd come to New Zealand to make reds, I'd have gone to Hawke's Bay. But so long as we're not intimidated by Hawke's Bay, and don't try to make a Hawke's Bay style, we'll make great reds.' This is one of Marlborough's — although not New Zealand's — finest reds.

Pelorus, launched in 1992, is a bottle-fermented sparkling of high class. The first 1987 vintage was blended predominantly from Pinot Noir and Chardonnay, with fifteen percent Pinot Blanc, by Harold Osborne, a Californian sparkling wine specialist. Matured on its yeast lees for three years, this is a stylish wine with rich green-appley flavours, strong yeast autolysis and a firm, sustained finish.

The Cloudy Bay range also includes small volumes of a promisingly supple and savoury Pinot Noir and a perfumed, ravishingly nectareous Late Harvest Riesling.

Forrest Estate
Blicks Road, Renwick

Owners: John and Brigid Forrest

Key Wines: Barrel Fermented Sémillon, Chardonnay, Select Late Harvest Sauvignon Blanc, Cabernet Rosé, Merlot, Gibsons Creek

John Forrest is a relative newcomer to the Marlborough wine scene. The owner of a cosy, unpretentious little winery near Cellier Le Brun, Forrest produces notably scented and full-flavoured reds. His white wine focus will be on the unfashionable but noble Sémillon variety, both in dry and sweet styles.

From the Forrest winery, visitors enjoy a stunning view over vine-swept plains to the Richmond Range on the valley's northern flanks. Forrest, a former biochemist at the DSIR in Palmerston North who quit because 'there was no money to do the work I wanted to do', gained his practical winemaking experience by working as a cellar hand at Marlborough Cellars in 1989 and Grove Mill in 1990.

Sémillon, Sauvignon Blanc, Chardonnay, Merlot, Cabernet Franc and Cabernet Sauvignon are the principal varieties in the twelve-hectare estate vineyard, where the first vines were planted in 1989. The early vintages have been made at Grove Mill and Vintech, although some of the range is then barrel-matured at the Forrest winery.

With its current output of about 3000 cases, which is planned to peak at over double that level, Forrest will always be a small winery. The Chardonnay is a very lightly wooded, green-appley, crisp style with fresh, direct fruit flavours. Why is it so sparingly oak-influenced? 'I want Sémillon to be my main white,' says Forrest, who is looking

Forrest

John and Brigid Forrest first made an impact with their trophy-winning Cabernet Rosé 1990. Forrest Estate's early releases of Sémillon, Chardonnay, Merlot and Cabernet-based reds have also been auspicious.

for a 'less herbaceous' style. The Barrel Fermented Sémillon bursts with tangy, green-edged flavour.

John Forrest is also part-owner of a red wine vineyard in Hawke's Bay's famously stony Gimblett Road. In future, Forrest envisages a red blended from Marlborough Merlot — which he says has 'better backbone than Hawke's Bay Merlot' — and Gimblett Road Cabernet Sauvignon. Currently his flagship red is a Marlborough-grown Merlot. Gibsons Creek, the medium-priced label blended from Cabernet Sauvignon, Merlot and Cabernet Franc, is unexpectedly dark, rich and minty.

Forrest's name first achieved prominence when his debut Cabernet Rosé 1990 won the trophy for the champion rosé at the 1990 Air New Zealand Wine Awards. This is a salmon-pink, raspberryish wine, charmingly soft and fresh. There is also a Late Harvest Sauvignon Blanc with a lovely outpouring of passionfruit and citric-like aromas and a gentle, medium-sweet palate.

Grove Mill Wine Company
1 Dodson Street, Blenheim

Owners: Private shareholders

Key Wines: Lansdowne Chardonnay, Marlborough Chardonnay, Riesling, Gewürztraminer, Sauvignon Blanc, Drylands, Blackbirch

One of the most applauded wines on the judging circuit in 1990 was Grove Mill Lansdowne Chardonnay 1989. Yet before its show-stealing successes, many wine lovers had never heard of the Grove Mill winery.

A fat and succulently soft wine deliberately fashioned by winemaker David Pearce to 'win immediate recognition', Lansdowne Chardonnay 1989 served its purpose brilliantly, putting the fledgling winery well and truly on the Marlborough wine map.

Grove Mill, only a few minutes' walk from the heart of Blenheim, is ensconced in part of the former Wairau Brewery, founded by Henry Dodson at Blenheim in 1858.

The 'Old Malt House', as it has long been known by locals, later served as the Coker and Mills ice-cream factory, until in 1988 the brick-walled building was converted into a small but atmospheric winery and restaurant complex.

Grove Mill was funded by twenty-three shareholders. Terry Gillan, a director of the company, is an English property developer who 'retired' to New Zealand a few years ago; from the start he's been a key force behind the rise of Grove Mill. Toni Gillan — Terry's wife — was also previously involved as marketing manager. The general manager, and winemaker from the first vintage in 1988, is David Pearce.

Only thirty-three, Pearce oozes confidence, which no doubt springs from the many vintages already under his belt. Christchurch-born, as a child he helped his father make cider for home consumption, and by the time he left school he had decided to be a winemaker. Corbans hired him for a year as a cellarhand and laboratory technician; then he did a food technology degree at Massey University. After graduating, he returned to the Corbans fold, working his way up at the giant Gisborne winery from trainee to

Grove Mill

Winemaker David Pearce is justifiably proud of his wines under the Grove Mill label. Pearce's aromatic, intense, appetisingly crisp white wines and dark-hued, full-flavoured reds are consistently successful on the show circuit.

assistant winemaker to the top post — winemaker.

Pearce abandoned his flourishing career at Corbans because he became intrigued by the potential of Marlborough: the early vintages of Corbans' Marlborough wines were all fermented at the Gisborne winery and each vintage, he recalls, 'nine times out of ten, irrespective of the variety, the Marlborough fruit was the best.' At Grove Mill he also had an opportunity to design the winemaking process and equipment: 'It was exciting building everything from scratch.'

'Lansdowne', contrary to popular belief, is not a premium vineyard site, but the name of the Marlborough rugby ground adjoining the winery. Lansdowne Chardonnay is a blend of fruit from various vineyards, but single-vineyard wines do feature in the Grove Mill range. From Rex and Paula Brooke-Taylor's Framingham vineyard, Pearce

draws 'very, very good Riesling', and Grove Mill's flagship red is grown at the Brooke-Taylors' Blackbirch vineyard.

David Pearce has spent much of his energy pursuing a great Marlborough Chardonnay. 'I've hung my hat on Marlborough Chardonnay. Sauvignon Blanc is lovely, but there's a limit to how much of that people will drink.'

Steely acidity, which gives a fresh, mouth-watering crispness, and grapefruit and lemon flavours are both distinctive features of Marlborough's slow-maturing Chardonnays. The gold medal 1989 Lansdowne Chardonnay, however, was a blend of lower-acid Gisborne and Marlborough fruit, softened by a secondary, malolactic fermentation, with the sweet perfume of lashings of American oak. A 'big softie', you either loved or hated it. I found it a bold, peachy-ripe but slightly flabby and cloying wine.

Recent vintages of Lansdowne Chardonnay are much more subtle. Based entirely on Marlborough fruit, and aged in the more restrained French, rather than American, oak casks, it has a richness and barrel-ferment complexity that promises a great future. The mid-priced Marlborough Chardonnay is savoury, lemony and crisp, in a slightly leaner style than its Lansdowne stablemate.

Lifted, lime-juice aromas and racy, mouth-watering acidity are features of Grove Mill's stylish Riesling. The partly barrel-fermented Gewürztraminer is a bold, full-flavoured, well-spiced wine. Bursting with fresh, ripe, invigoratingly crisp herbal flavours, the Sauvignon Blanc is straight out of the classic Marlborough mould; Drylands, the botrytised Sauvignon Blanc, fermented in new French oak barriques, is tangy and limey-sweet.

Le Clairet is a deliciously buoyant, refreshing rosé. The Cabernet/Pinotage displays vibrant, raspberryish fruit flavours, fresh and supple. Grove Mill's flagship red is Blackbirch, a dark-hued, Cabernet Sauvignon-predominant wine with a wealth of minty, herbal and plummy flavours underpinned by strong oak and taut tannins.

Highfield

Highfield Estate
Dog Point Road, Blenheim

Owners: Neil Buchanan, Shin Yokoi and Tom Tenuwera

Key Wines: Sauvignon Blanc, Chardonnay, Rhine Riesling, Noble Late Harvest, Merlot

Highfield, a relatively new winery on a knoll overlooking the Omaka Valley, on the south side of the Wairau, is the source of a rewardingly dark, spicy and complex Merlot. Its future, however, will principally revolve around the export of bottle-fermented sparkling wines to Japan.

Highfield was founded in 1989 by the Walsh family, who have owned the surrounding land for sixty years. Bill Walsh, one of the first Marlborough farmers to diversify into grape-growing in the mid-1970s, had 'always had an

ambition to build a winery', says his son, Philip. Philip's brother, Gerald, was also active in the venture.

Highfield's first 1989 vintage, processed at Hunter's, yielded two wines, a Grey Riesling and Walsh Surprise, which commanded little respect. The 1990 vintage, for which the company's own winery was (barely) in operation, produced a range of classic varietals of varying quality. In 1991 Highfield slid into receivership.

The company is now owned by Neil Buchanan, a Wellington businessman, and two Japanese investors, United Kingdom-based Tom Tenuwera and Japan-based Shin Yokoi, who have connections with the French Champagne house, Drappier. By tapping Drappier's expertise, the partners plan to produce a bottle-fermented sparkling wine in Marlborough, which will be distributed in Japan by Yokoi's Champagne-importing company.

Highfield's new owners are initially restricting its range to four wines — Sauvignon Blanc, Chardonnay, Rhine

The Highfield winery stands on a knoll overlooking the south side of the Wairau Valley. After an ill-planned start and a change of ownership, Highfield now produces a tight array of wines — notably an appealingly plummy and spicy Merlot.

Riesling and Merlot. Tony Hooper, a lanky South Australian who is a graduate of Roseworthy College, and was previously winemaker at Yarra Burn in Victoria and Esk Valley in Hawke's Bay, processed his first Highfield vintage in 1991.

The Sauvignon Blanc is fresh and zingy, with the penetrating herbal flavour typical of Marlborough. The Rhine Riesling is tangy, limey and flavourful. With Chardonnay, Hooper is aiming for a 'fruit uppermost' style; both tank and barrel-fermented, it is citrusy and crisp.

A delectably perfumed and honeyish Müller-Thurgau, labelled Noble Late Harvest 1990, earned Highfield's first gold medal at the 1991 Air New Zealand Wine Awards. The other highlight of the range to date is the robust Merlot, with its generous coffee and chocolate-like flavours and strong early-drinking appeal.

Hunter's Wines
Rapaura Road, Blenheim

Owners: E.C. Hunter Trust and M.J. Hunter

Key Wines: Sauvignon Blanc, Sauvignon Blanc Oak Aged, Chardonnay, Riesling, Gewürztraminer, Pinot Noir, Cabernet Sauvignon

Jane Hunter, in many wine lovers' eyes, *is* Hunter's Wines. Her smile, which lights her entire face, appears frequently in the media and the winery's newsletters, and she often travels to overseas markets to wave the flag on Hunter's behalf. 'It wasn't me who decided to make it a "person" winery,' she says. 'Ernie established that.'

Ernie Hunter, his spectacular career tragically cut short at the age of only thirty-eight, at the time of his June 1987 death in a motor accident, had started to savour worldwide applause for the Marlborough wines he toiled so tenaciously and with such charisma to promote.

An ebullient Ulsterman, Hunter joined the retail liquor trade in Christchurch, started a hotel wine club, then bought twenty-five hectares of land at Blenheim to grow grapes for Penfolds. After meeting Almuth Lorenz — a young German winemaker here on a working holiday — at a New Year's Eve party in 1981, at her suggestion he elected to plunge into commercial winemaking.

With Lorenz as winemaker the first, 1982, vintage of Hunter's wines was made under primitive conditions using borrowed gear at an old Christchurch cider factory. Observers were soon startled when, after entering six wines in that year's National Wine Competition, the fledgling company emerged with six medals, including three silvers. The promise in that performance was not fulfilled, however, in the 1983 vintage wines.

Suffering severe financial and marketing problems in 1984, Hunter turned his formidable energy to export — and shipped thousands of cases of wine to the United Kingdom, the United States and Australia. In his widow Jane's words: 'Ernie didn't just sell Hunter's wines. When he was in New Zealand he always talked about Marlborough wines and when he was overseas he talked about New Zealand wines.' His most publicised successes came at the 1986 and 1987 *Sunday Times* Wine Club Festivals in London when the public voted his 1985 Fumé Blanc and 1986 Chardonnay as the most popular wines of the shows.

After Ernie Hunter's premature death, his wife Jane stepped in as managing director. Born in 1954 into a South Australian grapegrowing family, she graduated in agricultural science from Adelaide University, majoring in viticulture and plant pathology. After a two-year stint running a restaurant at Waikanae — 'I came to New Zealand to get away from the wine industry,' she says — she came to Marlborough in 1983 to take up a new post as Montana's

chief viticulturist.

Following her husband's death, Jane thought about leaving the region. 'But then I thought, what's the point? We worked so hard to build it up. It would have been a waste if I'd walked away.'

The winemaker is Gary Duke (40), a tall, quiet, mustachioed Australian who worked at the Tisdall and Hanging Rock wineries in Victoria before joining Hunter's in 1991. 'It was a chance to do something different,' says Duke. 'New Zealand wines are going places. Australians can only dream of the intensity of fruit and natural acidity in Marlborough.' Dr Tony Jordan, an eminent Australian oenologist, has been a consultant since the 1986 vintage.

The Hunter's winery in Rapaura Road, erected from coolstore materials in 1983, has since 'grown like Topsy' to handle its current output of 23,000 cases. Fruit is drawn from the sixteen-hectare estate vineyard — Chardonnay, Sauvignon Blanc, Gewürztraminer, Pinot Noir and Cabernet Sauvignon in silty loams overlying riverstones — and from twenty-four hectares of contract vineyards.

Hunter's wines have deserved their acclaim. Although the two Sauvignon Blancs and the Chardonnay are the most notable wines, these are also supported by a fine Gewürztraminer and Riesling. Deep-scented, elegant white wines of appetisingly crisp, sustained flavour are the Hunter's hallmark.

The Sauvignon Blanc Oak Aged, of which only thirty percent is handled in casks, abounds with vibrant Sauvignon fruit, crisp and fresh, barely toned down by the background hint of wood. Duke sees it as 'a winemaker's wine, riper and less herbaceous' than the equally outstanding, non-oaked Sauvignon Blanc, which is packed with lush, ripe gooseberry and herbal flavours. Hunter's Sauvignon Blanc 1991 won the Marquis de Goulaine Trophy for the champion Sauvignon Blanc at the 1992 International Wine and Spirit Competition in London. Both wines are Marlborough classics.

The Chardonnay is weighty, lemony and savoury, with fresh, lively acidity. 'We want obvious fruit,' says Duke, who ferments the wine in a mix of new and older oak casks; a small percentage also undergoes a softening malolactic fermentation. It performs strongly in the cellar, peaking at about five years old.

Hunter's Gewürztraminer is a satisfying dinner wine: not too pungently spicy but flavoursome and with excellent palate weight. The Riesling is crisp, fragrant and brimming with fresh lime and citric flavours.

Hunter's reds are enjoyable but lack the distinction of the whites. 'We can make good reds in a distinctive Marlborough style in, say, two years out of five,' says Jane Hunter, who was awarded an OBE in December 1992. Duke is confident he can improve the light, strawberryish straightforward Pinot Noir (and in 1991 he did). The plummy and herbaceous Cabernet Sauvignon-based red is improving from vintage to vintage, reflecting the growing presence in the wine of Merlot and Cabernet Franc.

Merlen Wines
Rapaura Road, Renwick

Owner: Merlen Wines Limited

Key Wines: Chardonnay, Riesling, Riesling Oak
Matured, Magic Riesling, Gewürztraminer,
Sauvignon Blanc, Fumé Blanc, Sémillon, Müller-
Thurgau, Müller-Thurgau Dry, Morio-Muskat

The conversation of Almuth Lorenz, a tall, charismatic
35-year-old with a sunny face, is liberally sprinkled with
'Ja's. Raised in the Rheinhessen, as a child she worked in her
parents' eight-hectare vineyard and winery, and then
studied the production of alcoholic beverages for four years
at the acclaimed Geisenheim Institute. After arriving in
New Zealand in 1981, Lorenz built her early reputation as
Hunter's winemaker between the 1982 and 1986 vintages.

Lorenz left Hunter's in 1986, and briefly marketed a
handful of wines under her own Lorenz label. Jeremy
Cooper, a Blenheim accountant and business developer,
then organised a group of investors to back her by funding
the erection of the Merlen winery. Lorenz says the winery's
name is 'derived from the ancient name for Marlborough
in England. The area was originally called Merlborough
from the legend that Merlin the magician practised his arts
there.' After the first wine under the Merlen label, the
superb (one is tempted to say spellbinding) 1987 Chardonnay was barrel-fermented in a rented refrigerated container, the small concrete-slab winery near Renwick opened
in 1988.

Chardonnay, Gewürztraminer, Sauvignon Blanc, Sémillon and Morio-Muskat are the featured varieties in the
sandy, shingly five-hectare estate vineyard. Growers supply
half of the annual crush.

Lorenz's favourite wines are Chardonnay and Riesling.
Chardonnay is 'no headache' in the vineyard, soaring to
very high sugar levels — 26 degrees brix in 1991. Lorenz
ferments her Chardonnay in predominantly new oak barrels
and puts twenty percent through a malolactic fermentation
to soften its vigorous Marlborough acidity. This is a bold,
fat, ripe-tasting wine, packed with strong stone-fruit
flavours and lees-aging complexity.

For her beloved Riesling, Lorenz aims for 'the equivalent
of a kabinett halbtröcken — light, with a touch of botrytis,
and closer in style to a Rheingau than Mosel.' The 'traditional' wine, labelled Riesling, is a fractionally sweet
style of varying quality. The Riesling Oak Matured subdues
the variety's floral characters with a seasoning of wood.

Merlen Magic Riesling is a ravishingly beautiful, rich,
honey-sweet dessert wine. For her gold medal 1991 vintage, Lorenz harvested her fruit in early June, when the
shrivelled, botrytis-infected berries had soared to an extraordinarily high (45 degrees brix) sugar level. The grape
juice, a dark, chocolate-coloured sludge, was so thick that
it blocked the winery's pumps. Once fermented and clari-

fied, however, it yielded one of the finest sweet whites ever
made in this country.

Two styles of Sauvignon Blanc are produced. For the
non-wooded version, labelled Sauvignon Blanc, Lorenz
aims for a 'young, fresh and aggressive' style. The Fumé
Blanc is a blend of sixty percent lusher, tank-fermented
Sauvignon Blanc and forty percent barrel-fermented, more
stridently grassy Sémillon.

A dry and delicate Gewürztraminer, a pair of Müller-
Thurgaus (off-dry and medium), and a seductively aromatic,
slightly sweet Morio-Muskat flesh out the Merlen range. 'I
don't like reds,' says Lorenz. 'I haven't grown up with
them. I tell my customers to go down the road if they want
a red.'

The majority of Merlen's annual output of 6000 cases is
sold by mail-order or at the winery, where light, German-
style food is sold in the cosy 'weingarten'. 'Try our magic,'
invites the sign in the winery shop. Lorenz herself is often
behind the vineyard counter: 'Our mail-order customers
like to see me and I like talking about what I'm doing.'

Lorenz has sunk deep roots in Marlborough. 'It's hard
work — a battle in the marketplace. I have to go out and
see my customers and meet new customers. I'm settling
here because Marlborough's going to be great.'

*Tall, extroverted, charming
— Almuth Lorenz is one of
the great personalities of the
Marlborough wine scene. Not
surprisingly in view of her
Rheinhessen origin, Lorenz is
a white wine specialist.*

Te Whare Ra

Allen and Joyce Hogan produce a trickle of wine, ranging in style from strapping, bone-dry whites to fragile, sweet 'stickies'. Rampantly botrytised sweet Rieslings and Sauvignon Blancs are the jewels in the Te Whare Ra crown.

Te Whare Ra Wines
Anglesea Street, Renwick

Owners: Allen and Joyce Hogan

Key Wines: Duke of Marlborough Chardonnay, Gewürztraminer, Riesling; Sarah Jennings Cabernet Sauvignon/Merlot/Cabernet Franc; Fumé Blanc Botrytis Bunch/Berry Selection Riesling, Sauvignon Blanc

Allen Hogan is one of the fierce individualists of the Marlborough wine scene. Describing himself as a 'self-made, bootstraps winemaker' who is 'suspicious of received book wisdom on winemaking techniques', under his Te Whare Ra label he fashions some of this country's most rampantly botrytised sweet white wines. His strapping, richly alcoholic dry wines are also full of character.

Bay of Plenty-born Hogan (50) gathered his initial winemaking experience at a small Perth winery, then spent a couple of vintages with Montana at Marlborough and another vintage at the Te Kauwhata research station. He and his wife Joyce planted their first vines in Anglesea Street, Renwick in 1979.

It 'felt right' when he first saw the land over which their future vineyard would spread. Today the Hogans have four hectares of Gewürztraminer, Riesling, Sémillon, Chardonnay, Cabernet Sauvignon, Merlot and Cabernet Franc vines planted in variable loam and gravel soils. Growers supply all of the Sauvignon Blanc, a small amount of Chardonnay and some of the botrytised fruit.

The Te Whare Ra (House In The Sun) winery, begun in 1982, has been built from earth bricks with a timbered exterior. A beautiful leadlight glass panel in the entrance to the tasting room features the sun, vine leaves, a bunch of ripe grapes and a glass of red wine, accompanied by the motto 'Na Te Ra Nga Mamahi' (By Sun And Hard Work). Joyce Hogan oversees the vineyard, while Allen concentrates on making and promoting the wine.

Te Whare Ra's varietal white wines are marketed under the Duke of Marlborough label (in memory of the crushing defeat inflicted by John Churchill, Duke of Marlborough, on the French in 1704 at the battle of Blenheim). Hogan makes distinctively 'big, high-alcohol styles as is to be expected with the late-harvesting techniques currently employed'. He likes 'high-flavoured wines, and hand in hand with that goes high alcohol, unless you make the wine sweet. James Halliday, the Australian wine writer, tasted my 14.2 percent alcohol Chardonnay, and told me his Australian clients favour a lighter, more elegant style. I asked my mail-order clients their opinion and overwhelmingly they said: "Don't change the meaty style."'

The barrel-fermented and lees-aged Chardonnay is a blockbuster wine drunk by heroes. Its style is well caught by Hogan's description of the 1991 vintage: 'Alcohol right at the top of the scale and flavours to match. Very sweet fruit character, therefore confusing to the uninitiated.' This is a succulent, fat, slightly fiery Chardonnay.

Other dry-to-medium whites include a bold, powerfully spiced, slightly honeyish Gewürztraminer, a rich, dryish Riesling, and a very full-flavoured Fumé Blanc, blended from Sauvignon Blanc and Sémillon. Hogan has no interest in producing a non-wooded Sauvignon Blanc: 'Everyone else does it.'

Hogan says he is not persuaded by the 'myth of New Zealand, and especially Marlborough, not being the place to make red wine'. By cropping his vines lightly and harvesting the fruit late in the season, he is achieving red wines of impressive quality. Te Whare Ra Sarah Jennings Cabernet Sauvignon/Merlot/Cabernet Franc (named after the Duke of Marlborough's wife) is stocky and deep-flavoured, standing out as markedly riper and more robust than most Marlborough Cabernet-based reds have been.

'It's unfortunate my fame hinges on the sweet whites,' says Hogan. 'Writers focus on our sweet whites because they are different; everyone has a Chardonnay. But I think our Chardonnay and reds are as interesting as our sweet whites.'

The top Te Whare Ra wines, however, are indeed the stunning botrytised whites. In 1979 Hogan noticed others ignoring the nobly rotten grapes in their Marlborough vineyards and determined to remedy this. Working with Rainer

Eschenbruch at the Te Kauwhata research station later enhanced his knowledge of how to handle botrytised fruit.

Marlborough's cold autumn nights, followed by a heavy dew and high humidity in the morning, drying out later in the day, are tailor-made for the spread of favourable *Botrytis cinerea*, says Hogan. The wines, usually based on Riesling or Sauvignon Blanc, are stop-fermented, leaving them with relatively low alcohol and soaring sugar levels.

The Botrytis Bunch Selection Riesling, made from grapes with less than sixty percent botrytis infection, is 'approximately equal in grape sweetness to a German beerenauslese,' says Hogan. This is a poised, honeyish, delectably botrytised wine, brimming with residual sugar, yet displaying a perfect alcohol/sweetness/acid balance. Even more stunning is the Botrytis Berry Selection Riesling, made not from individual berry pickings, but heavily raisined bunches with sixty to ninety percent botrytis infection. Harbouring up to 240 grams per litre of residual sugar (equivalent to a German tröckenbeerenaus-lese), intensely botrytised and nectareous, this wine stands at the very forefront of this country's sweet white wines.

Te Whare Ra Botrytis Bunch Selection Sauvignon Blanc represents Hogan's bid to fashion a sweet wine 'as near as we can to a true French Sauternes, taking into consideration this is Marlborough not France'. With a bouquet of honey and straw, this is a markedly more robust style of sweet wine than its Riesling stablemates.

More than most other winemakers, Hogan has put his relationship with mail-order customers on a personal footing by three times each year hosting Te Whare Ra tastings in cities from Whangarei to Invercargill. Hogan's personality enlivens his newsletters: 'According to some of the writers and oracles of the popular press, winter arrived in January [1992]. Bunkum, Baldadash [sic] and Hokum. Sure, the average temperatures were down, sure we had a wetter than normal summer, but boy, did we have a dry autumn and autumn's when we harvest the grapes.' He is determined to stay small: 'I like the hands-on feel and never had any visions of becoming big like Hunter's or Cloudy Bay.' Only about 3000 cases of Te Whare Ra wines are produced each year — but that's enough to supply Hogan's cult following.

Vavasour Wines
Redwood Pass Road, Dashwood

Owner: Vavasour Wines Limited and Company

Key Wines: Vavasour Reserve Sauvignon Blanc, Chardonnay, Cabernet Sauvignon; Dashwood Sauvignon Blanc, Chardonnay, Cabernet Sauvignon; Stafford Brook Cabernet Franc/Merlot

Vavasour is a relatively new name worth noting for three reasons: its pioneering of winemaking in the Awatere Valley; the consistently outstanding quality of its wines; and finally its glorious site on terraces bordering the Awatere River.

Peter Vavasour (42) is an entrepreneurial Awatere Valley farmer who owns 'The Favourite', part of the Ugbrooke Estate purchased by the Vavasour family in the 1890s. (One of his ancestors served as a cup-bearer for William the Conqueror, thus early setting the family on its wine-tasting path.) Until a few years ago his 350-hectare property of stone-and-tussock was devoted exclusively to sheep and beef.

The second prime mover in getting Vavasour off the ground was Richard Bowling (38), a burly viticultural expert. Bowling served a seven year stint with Corbans, at Taupaki and as second-in-charge of their Marlborough vineyards, before branching out as an independent viticultural consultant, based in Marlborough. Bowling was early convinced of the Awatere Valley's wine potential; in Peter Vavasour he found a natural partner, one equally convinced of the district's viticultural future, but commanding the financial resources and skills to do something about it.

To fund the $1.3 million venture, Peter Vavasour formed a special partnership in 1986 with carefully chosen shareholders — wine merchants, advertising executives, merchant bankers and accountants.

Although every winemaker is adamant that his vineyard enjoys a mesoclimate superior to his neighbour's, Vavasour and Bowling researched their project well. The Wairau, says Bowling, developed vineyards first because of the Awatere's perceived labour problem and distance from the Blenheim wineries. But the Awatere Valley has at least equal suitability for viticulture.

The vineyard site itself is memorable. Mt Tapuaenuku's 2,900-metre peak rears to the south; to the north lies the sea. The vines are planted on the terraced banks of the Awatere River. The river's banks reveal an ideal soil structure for vines: a one-metre-deep surface layer of alluvial silt, over two metres of gravelly, silty substrata, down to a base strata of papa (mudstone), rich in calcium and iron. 'It's a low-vigour site,' says winemaker Glenn Thomas, 'with good drainage and hard, bony soils.' The twelve-hectare vineyard, established in 1986, is principally planted in Sauvignon Blanc, Chardonnay, Cabernet Sauvignon and Cabernet Franc, with smaller plots of Sémillon, Pinot Noir, Merlot, Malbec and Syrah.

The handsome concrete and cedar winery was erected in 1988. English-born Thomas (36) entered the wine industry in Australia, working for Orlando Wines 'packing pallets with casks. It was better for your body than your mind.' After graduating from Roseworthy College in 1979, Thomas worked at the Ryecroft, Kaiser Stuhl and Normans wineries, before arriving in New Zealand in 1985.

Appointed winemaker at Corbans' Gisborne winery from 1986 to 1988, from the start Thomas was also involved in handling Marlborough fruit, by making the early vintages

Vavasour

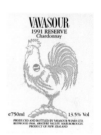

of the Stoneleigh Vineyard range. He joined Vavasour 'for the challenge of setting up a small winery in a new area'.

Vavasour's finest achievement has been its white wines, above all the Reserve Sauvignon Blanc and Chardonnay. At the 1992 Liquorland Royal Easter Wine Awards, for instance, Vavasour Chardonnay 1991 won the trophy for champion wine, and Thomas the overall award for wine-maker of the show.

Yet Vavasour is equally committed to producing premium reds in Marlborough. Bowling has predicted that Vavasour will make 'a blended, claret-style red equal to any in the country'. That target has yet to be hit but the early signs are auspicious: Vavasour Reserve 1989 topped the class for one-year-old Cabernet Sauvignon-predominant reds at the 1990 Canberra Wine Show.

The premium Vavasour label is exclusively reserved for Awatere Valley-grown wines; those carrying the mid-priced Dashwood label may be based on either Awatere Valley or Wairau Valley fruit. In Thomas's eyes Dashwood is 'not a second label, just a different style of wine, with greater complexity in the Vavasour wines, and the Dashwood wines more for early drinking.' The cockerel which dominates the Vavasour label is drawn from an ancient Vavasour family crest.

Vavasour's current annual output is about 7000 cases. 'We'll move to about 12,000 cases,' says Thomas, 'but no more. Our philosophy is to be "hands-on".'

The wines are immaculate. The Sauvignon Blancs are awash with lush, vibrant, penetrating herbal flavours: the Dashwood a fresh, direct style, the Vavasour Reserve (of which sixty percent of the final blend is barrel-fermented) riper and fleshed out by its deft oak handling. The Vavasour Reserve, especially, ranks among Marlborough's finest Sauvignon Blancs.

The top, cockerel-labelled Chardonnay is impressively taut and deep-flavoured. 'Chardonnay is not a blank canvas; the fruit is very important,' says Thomas, who is aiming for a 'complex style, needing bottle-age'. Fermented in French oak barriques and lees-matured for ten months, this is a very stylish wine with intense melon-like fruit flavours, a malolactic-derived butteriness and a savoury, firm, lingering finish. The Dashwood Chardonnay is a lightly wooded style with fresh, tangy, lemony fruit flavours.

Vavasour Reserve, the winery's top red, is a blend of Cabernet Sauvignon and Cabernet Franc, matured for eighteen months in French oak barriques. With its vibrant, spicy, herbaceous, crisp fruit flavours, this is a distinctively Marlborough red wine style with good weight, colour and flavour depth. It is partnered by the Dashwood Cabernet Sauvignon, a less complex, fruitier and softer style. A trickle of light, slightly peppery Shiraz (with an 'undecided' future in the Vavasour vineyard) and an easy-drinking Stafford Brook Cabernet Franc/Merlot have also been marketed.

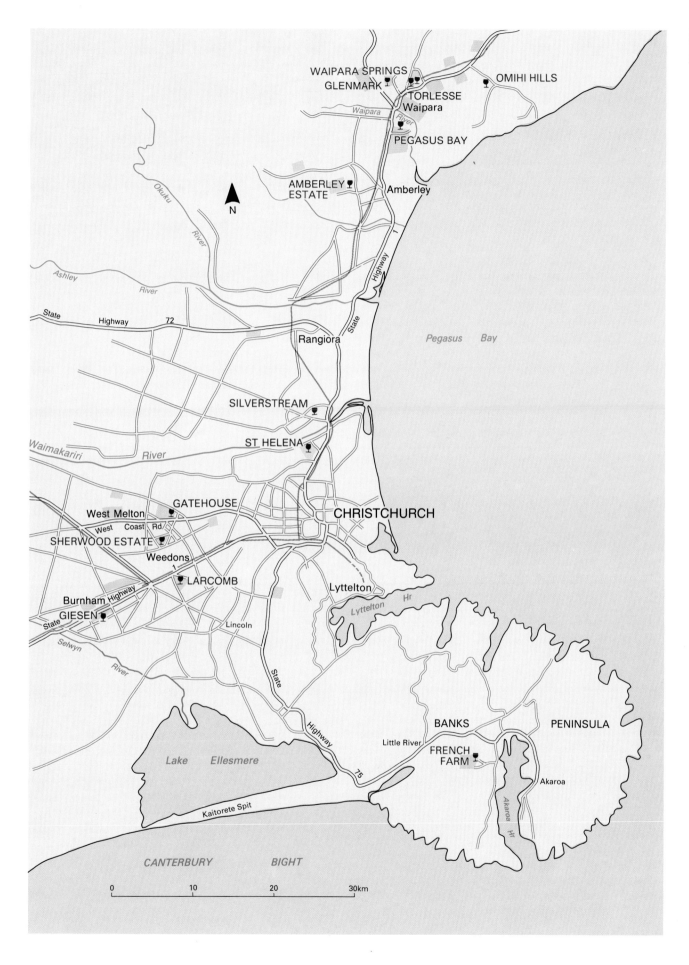

CANTERBURY

WAIPARA SPRINGS
GLENMARK
TORLESSE
Waipara
OMIHI HILLS

Waipara

PEGASUS BAY

River

AMBERLEY
ESTATE
Amberley

Okuku

River

N

Ashley

River

State   Highway   72

Pegasus   Bay

State

Rangiora

Highway

SILVERSTREAM

Waimakariri   *River*

ST HELENA

GATEHOUSE
West Melton
CHRISTCHURCH
West   Coast   Rd
SHERWOOD ESTATE
Weedons
1
LARCOMB
Lyttelton

Lyttelton   *Hr*

Burnham   Highway
GIESEN
State
Lincoln

Selwyn

River

State

BANKS

PENINSULA

Little River
FRENCH
FARM

Highway

Lake   Ellesmere

75

Akaroa

Akaroa   *Hr*

Kaitorete Spit

CANTERBURY   BIGHT

0   10   20   30km

Only fifteen years after the planting of the province's first commercial vineyard, Canterbury has emerged as New Zealand's fifth most heavily planted wine region. Christchurch wine lovers enjoy the luxury of numerous wineries on their doorstep and a handful of top-flight Pinot Noirs, Rieslings and Chardonnays has underlined the region's potential.

The Canterbury wine scene is dominated by the two 'giants' of the region, St Helena and Giesen. There's a natural rivalry here between the Mundy brothers at the St Helena winery, who pioneered commercial winemaking in Canterbury and won the region's first gold medals with their 1982 and 1984 Pinot Noirs, and the Giesen brothers, more recent arrivals from Germany, who have swiftly built Giesen into the region's largest winery and are now winning international recognition for their dry and rampantly botrytised sweet Rieslings.

The majority of the companies are small and content to sell much of their output directly to the public from the winery. The Larcomb winery at Rolleston, which produces consistently fine Pinot Gris, Riesling, Gewürztraminer and Pinot Noir, is a model of the family-owned winery serving a local rather than a national clientele. I was strongly reminded here of the family-run wine estates on the Rhine, where the locals gather to sip wine, eat and chat in much the same way as the British socialise in their local pubs.

The climatic hazards for viticulture in Canterbury are more severe than for districts further north. Canterbury, although nearer the equator than many European wine regions, in cooler years can fail to accumulate the heat readings necessary to fully ripen grapes. 'This is a region of very marked vintage swings, from very good to poor,' says St Helena winemaker, Petter Evans. In this respect Canterbury parallels parts of Germany.

October spring frosts are a risk (causing problems 'about one year in five', according to Mark Rattray of Mark Rattray Vineyard) and April frosts can retard ripening. Canterbury is also subject to strong winds — both hot dry northwesterlies and cool easterly sea breezes — making shelter-belts a necessity.

Canterbury, however, like Marlborough, enjoys one vital advantage over most North Island wine growing regions — low rainfall. During Canterbury's long dry autumns, the warm days and cool nights enable the fruit to ripen slowly, with high levels of acidity and extract. Müller-Thurgau is usually harvested in mid-April — a month or more later than in the North Island — and Riesling and Pinot Noir hang on the vines until May (in extreme cases June). 'A dry vintage period is something we have come to rely on,' says John Thom of Larcomb. 'It's probably the most important feature of Canterbury viticulture.'

The fortunate viticultural combination of low rainfall and low-to-moderate soil fertility means that excessive vine foliage growth is not a problem here. Open vine canopies and dry weather also reduce disease problems and can encourage the development of 'noble rot'.

Most of the vineyards are located in two areas: southwest of Christchurch and at Waipara, Omihi and Amberley in North Canterbury. The soils in both districts are typically silty loams — shallower in the south — overlying river gravels, free-draining and in most seasons needing to be irrigated. Much excitement centres on North Canterbury, where the vineyards, protected from the cooling easterly breezes by the hills to the east, bask in hot summer temperatures regularly climbing over 30°C.

French peasants who landed in 1840 at Akaroa on Banks Peninsula carried vine cuttings, from which wine soon flowed for their domestic consumption. A century after their arrival, W.H. Meyers built a small winery, Villa Nova, in the Heathcote Valley. By 1945 he had a tiny vineyard of about 0.8 hectares planted in Verdelho — a Portuguese variety — Pinot Gris, Muscat and other grapes. Although wine was made, Meyers' vines were uprooted around 1949 after they failed to flourish.

The current resurgence of interest in Canterbury wine stems from research conducted at Lincoln College (now Lincoln University) under the direction of Dr David Jackson. When the first trials commenced in 1973, research focused on identifying the most suitable varieties for Canterbury's cool climate. After losing seventy percent of his vines to a late frost, Jackson began 'wondering if I really was making a mistake.'

Trial plantings of more than sixty varieties later demonstrated, according to the university, that Canterbury produces grapes of high acidity and high sugar levels. Jackson sees Canterbury as 'borderline' for such mid-to-late season ripeners as Sauvignon Blanc and such late-season ripeners as Cabernet Sauvignon, but Pinot Noir and Chardonnay are 'particularly promising'.

Riesling is also flourishing in Canterbury. These are steely Rieslings, in cooler vintages lean and austere, but at their best packed with spine-tingling acidity and intense citric-fruit flavours. The 1992 vineyard survey reveals the three principal varieties planted in Canterbury are (in order): Pinot Noir, Chardonnay and Riesling.

You'd never starve on the Canterbury wine trail. Several wineries offer fresh, wholesome lunches featuring an array of local specialties, such as eel and salmon, pâtés and cheeses. Vineyard restaurants are more common here than in any other wine region in the country.

At the three-hectare Silverstream Vineyard at Clarkville, Kaiapoi, Peter and Gesina ('Zeke') Todd produce a Chardonnay and a fleshy, soft Pinot Noir.

The Pegasus Bay winery recently joined the burgeoning Canterbury wine scene. Ivan Donaldson, an associate professor of medicine at Christchurch Hospital and a wine columnist, and his wife, Christine, planted their twenty-hectare Waipara vineyard in the mid-1980s. The first stage of the winery was completed for the 1992 vintage, processed by the Donaldsons' son, Matthew, a Roseworthy College oenology graduate. The first releases under the Pegasus Bay label were red wine from 1991 and whites from 1992.

Amberley Estate Vineyard
Reserve Road, Amberley

Owners: Jeremy and Lee Prater

Key Wines: Riesling, Müller-Thurgau, Gewürztraminer, Chardonnay, Pinot Noir

For Jeremy and Lee Prater, owners of the small Amberley winery in North Canterbury, the past decade has been 'a long, hard road. The hours we're working today are as long, if not longer, than when we started.'

Amberley lies fifty kilometres north of Christchurch and two kilometres inland from Amberley township. The Praters planted their first vines here in 1979. Today, their gently undulating vineyard on north-facing Waipara loam-clay slopes includes eight hectares of Riesling, Müller-Thurgau, Gewürztraminer, Chardonnay and Pinot Noir. Other grapes are drawn from growers in Marlborough.

Prater (44), who was born in England, arrived in New Zealand in 1970 and later graduated with a BA from Canterbury University. He then spent four years in Switzerland — gaining diplomas in viticulture and wine-making from the Swiss Federal College, near Geneva — Germany and France, learning cool-climate winemaking, and more than a year with Montana at Marlborough, before processing his first Amberley vintage in 1984.

In his insulated timber winery, Prater makes about 2000 cases of wine per year. Much of this is consumed in the cosy vineyard restaurant, which opens daily from spring until autumn, serving up to 130 diners on Sundays.

'Pinot Noir is my favourite wine,' says Prater, 'but it's so difficult to make well. You must get the right clone. You want to get close to the style of Burgundy, but there are a hundred schools of thought on how to handle the grapes.' Prater's Pinot Noir, based on Marlborough fruit, is a 'light, Beaujolais-style wine', briefly barrel-aged, designed to be

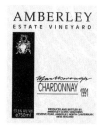

Jeremy and Lee Prater produce Canterbury estate-grown and Marlborough wines, much of which is consumed in the Praters' humming vineyard restaurant.

consumed during the summer following the vintage. His first estate-grown Pinot Noir is due in the mid-1990s.

Sauvignon Blanc failed to fully ripen in the Amberley vineyard; the 1990 vintage was the last. The future of Gewürztraminer is also question-marked: 'It has a powerful floral character, but lacks spice,' says Prater. Amberley Estate Gewürztraminer displays a full-bloomed scent, subdued varietal spiciness and a very soft finish. The 1990 Hawke's Bay and 1991 Marlborough Chardonnays are solid.

Fragrant, with a pleasing depth of lemony, slightly honeyish flavours in a firm, bone-dry style, the Marlborough-grown Riesling is one of the high points of the Amberley range. 'People told us our dry Rieslings were too austere,' says Prater, 'so in 1989 I gave the wine a hint of sweetness. Its sales were no higher, so I've gone back to the fully dry style.'

With the myriad tasks of a vineyard and winery to attend to, and their restaurant proving popular, the Praters work seven days a week. In mid-1993, however, they put Amberley Estate up for sale.

French Farm Vineyards
French Farm Valley Road, Banks Peninsula

Owners: John Ullrich, John Hibbard and Tony Bish

Key Wines: Oak Aged Sauvignon Blanc, Riesling, Rosé, Reserve Cabernet/Cabernet Franc

'About 100,000 cars come over the Banks Peninsula hills to Akaroa every year, and there's not a lot for them to do,' says John Hibbard, co-owner of French Farm. The vineyard and winery were established at French Farm Bay, on the western shores of the Akaroa Harbour, 'both for the area's climatic and tourist potential'.

The strikingly handsome brown concrete winery, erected in 1991, with its arched windows and wooden shutters, is intended to exude a traditional French character; the roof is clad in Marseilles tiles. French Farm is jointly owned by

Hibbard, a Christchurch accountant; John Ullrich, a Canterbury fencing manufacturer; and winemaker Tony Bish. The French Farm label should eventually be widely seen: the first 1991 vintage yielded 3500 cases of wine, and the partners plan to gradually raise the winery's output to around 10,000 cases.

French Farm buys grapes from Marlborough, and Sauvignon Blanc and Cabernet Sauvignon from a Waipara, North Canterbury vineyard which the partners are confident will 'rival Marlborough in producing world-class Sauvignon Blanc'. Intense interest also centres on the wines to flow from the three-hectare estate vineyard planted in Pinot Noir and Chardonnay in 1990 on a north-facing, loam clay slope. Here the autumn rainfall is higher than on the Canterbury plains, but the partners believe the heat during the growing season is 'above the Canterbury average and similar to Marlborough'. The first crop from the home vineyard was harvested in 1993.

With Tony Bish, an experienced winemaker, at the production helm, an eye-catching building and a captive Banks Peninsula tourist market, the French Farm winery looks sure to flourish.

Winemaker Tony Bish has previously worked for Corbans, Vidal, Martinborough Vineyard and Brown Brothers in Victoria, and from 1986 to 1989 was the winemaker at Rippon Vineyard at Wanaka. In 1990 Bish graduated with a Bachelor of Applied Science degree in wine science from Charles Sturt University at Wagga Wagga in Australia.

French Farm's early releases have been solid. The Oak Aged Sauvignon Blanc is robust and full-flavoured; the Riesling displays a vigorous surge of fresh, dry citric and lime juice flavours and racy acidity. The Reserve Cabernet/Cabernet Franc is a gutsy red, fruity and green-edged. The Rosé, which according to the winery newsletter has a 'remarkably good ability to kick-start people into party mode', is fresh, light and dry — ideal as a thirst-quencher in summer.

Diners in the vineyard bar can feast on seafood and game, seasoned with fresh herbs, and local produce such as Barry's Bay cheeses and Akaroa bread — washed down, of course, with French Farm's enjoyable wines.

Gatehouse

Gatehouse Wines
Jowers Road, West Melton

Owners: The Gatehouse family

Key Wines: Riesling, Chardonnay, Gewürztraminer, Merlot/Cabernet, Pinot Noir

A vibrantly fresh, plummy, full-flavoured Merlot/Cabernet, one of Canterbury's rare successes with these traditional Bordeaux varieties, is the chief claim to fame of the fledgling Gatehouse winery in West Melton.

Peter Gatehouse marketed his early releases under the Makariri brand, but now the name Gatehouse is coming to the fore in the winery's marketing.

Peter Gatehouse (44), proprietor of Gatehouse Wines, is a ball of energy. Raised in Christchurch, he graduated with a BSc in geology and now manages an audio-visual unit, while tutoring in wine science, at Lincoln University. Somehow he also finds the time to grow his grapes and make and sell his annual output of about 1000 cases of wine.

Gatehouse started experimenting with fruit wines in the late 1960s. 'Some turned out not too jolly bad, so I thought I'd give winemaking a crack.' He planted his first vines in Jowers Road, West Melton, in the early 1980s. Today the four-hectare vineyard, planted in fine sandy loams overlying shingle and sand, is planted in Riesling, Gewürztraminer, Chardonnay, Breidecker, Pinot Noir, Cabernet Sauvignon, Merlot and Malbec.

Gatehouse produced his first, experimental grape wines in 1985. In 1987 and 1988 his grapes were sold to the Torlesse winery, in which Gatehouse was a shareholder. The year 1989 brought the first commercial release of wines under Gatehouse's Makariri label.

In his spacious West Melton winery — formerly Torlesse's — Gatehouse produces a very solid lineup of wines. The Chardonnay, Gewürztraminer and Riesling all display plenty of weight and fresh, mouthwatering acidity. So impressively deep-coloured was the Cabernet/Merlot 1989, it 'aroused suspicions the grapes had come from elsewhere,' recalls Gatehouse. Now a Merlot-predominant blend, this is a strong-flavoured, American oak-aged, very crisp wine with little of the leafy-green character which can tether the quality of cool-climate reds.

Peter Gatehouse believes Gatehouse's future will 'very much revolve around tourism. We'll sell most of our wines on-site as part of a vineyard and winery experience. We'll provide winery tours and displays and slide-shows.'

Giesen Wine Estate
Burnham School Road, Burnham

Owners: Theo, Alex and Marcel Giesen

Key Wines: Riesling, Riesling Dry, Riesling Late Harvest, Botrytised Riesling, Marlborough School Road Chardonnay, Burnham School Road Reserve Chardonnay, Müller-Thurgau, Botrytised Müller-Thurgau, Gewürztraminer, Ehrenfelser, Sauvignon Blanc, Pinot Noir, Cabernet/Merlot

Marcel Giesen, twenty-eight-year-old winemaker at Canterbury's largest winery, says 'Canterbury wines need a far stronger image outside their own province.' With his recent headline-grabbing competition successes, Giesen himself is doing more than anyone to generate wider awareness of Canterbury's often delightful Rieslings.

The Giesen Wine Estate lies at Burnham, eighteen kilometres south of Christchurch. Its Botrytised Riesling 1990 was the overall champion at the 1991 Royal Easter Wine Show in Auckland; its Riesling Dry Reserve 1989 won a gold medal at the 1991 International Wine and Spirit Competition in London.

Marcel Giesen, although one of the country's youngest winemakers, already has ten vintages under his belt. Raised in Germany, at Neustadt in the wine-growing Rheinpfalz region, Giesen started tasting wine as a four-year-old. Granite quarrying, construction and masonry were the family's chief occupations, but the Giesens — like thousands of other German families — also owned a one-hectare plot of grapevines and made wine for their private consumption.

Why did they uproot themselves to start a new life in New Zealand? 'For opportunity, space, clean air and freedom,' says Marcel. Winemaking was not part of their early plans in the new country; Marcel's brothers, Theo (35) and Alexander (34), were contracted to work for a Canterbury construction company specialising in natural stone.

'One weekend we went to Warners, then *the* bottle-store in Christchurch, and met Ernie Hunter, who founded Hunter's Wines. He loaded our trolley with New Zealand wines. We were confused by the number of Müller-Thurgaus labelled as 'Riesling', and astonished by the lack of dry Rieslings. We thought: "Why not make some wine in the style we had produced at home?"'

The Giesens bought land at Burnham — but not before looking at other potential vineyard sites in Auckland, Hawke's Bay, the Wairarapa and Marlborough. 'We particularly wanted to grow Riesling, which flourishes in a cool climate, so in the end it was obvious we should go south.' The brothers' parents, Kurt and Gudrun, helped to fund the new winery.

The Giesen estate vineyard now covers eighteen hectares. Riesling, Müller-Thurgau, Gewürztraminer, Sauvignon Blanc, Ehrenfelser (a cross of Riesling and Sylvaner),

Chardonnay, Pinot Noir and Cabernet Sauvignon are the major varieties planted in the stony, free-draining soils. The Giesens also own two other vineyards, at Burnham and West Melton, and buy grapes from growers in both Canterbury and Marlborough.

To keep this fast-growing, 25,000 cases per year company on the rails, the Giesen brothers have divided between them the myriad tasks of a modern winery. Both Theo and Alexander are immersed in administration and marketing. Marcel controls production in the utilitarian, corrugated-iron and timber winery, with the assistance of Rudi Bauer, an Austrian-born winemaker who worked at Rippon Vineyard at Wanaka from 1989 to 1992. For both winemakers German is the first language, but they speak 'mostly English, except quite late at night'.

Marcel Giesen has strong views on New Zealand Rieslings, which in the past couple of years have finally found popularity with wine lovers. 'Most of it is an up-market version of Müller-Thurgau: fruity, mild wine which tones down Riesling's natural racy acidity.' Why? 'Because many winemakers sell their Rieslings too young, and the public don't like sharp acidity. And if the wine is light in body, a high level of acidity can be unbalanced.'

Giesen delights in Riesling's mouth-wateringly crisp acidity: 'You need steely acidity for structure, elegance and longevity. With the low yields we get in Canterbury, our grapes have high extract [stuffing] and the wines can carry more acidity. Riesling like this ages longer than Chardonnay or Sauvignon Blanc.'

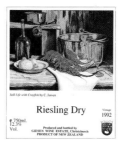

Winemaker Marcel Giesen (right) is most often in the public eye, but his brothers Theo (left) and Alexander are also heavily involved in the winery. Giesen's reputation hinges on its medium-dry and sweet botrytised Rieslings, and rich Reserve Chardonnays.

Botrytis cinerea plays a crucial role in the top Giesen Rieslings. Their first sweet wine, produced in 1985, was freeze-concentrated in the winery rather than naturally late-harvested; the 1987 was artificially inoculated with botrytis in the Giesens' sauna (and thereafter known unofficially as 'the sauna wine').

The quality breakthrough came in 1989, when rampantly botrytised fruit at soaring sugar levels was hand-harvested in the estate vineyard. After repeating the performance in the early 1990s, Marcel Giesen is confident: 'We're in the right spot for botrytis.' Giesen Botrytised Riesling is a ravishingly full-bloomed, golden, honeyish beauty, its sweetness balanced by tense acidity. This is an outstanding wine by any standards, one of New Zealand's most glorious sweet whites.

Giesen's most widely-seen Rieslings are the Riesling Dry, which displays fresh green-apple flavours and steely acidity, and the markedly sweeter, quite Germanic medium Riesling. Giesen has also marketed a Riesling Dry Reserve with strong marmalade-like flavours and a honeyish, yet surprisingly dry finish; and a Riesling Late Harvest Reserve, a delightful botrytised wine with pure, intense lemon and honey flavours and a crisp, medium-sweet finish.

Apart from Riesling, Chardonnay is the key wine in the Giesen range. The Giesens' two labels afford a clear style contrast. The Marlborough School Road wine is full-flavoured and soft, with strong, spicy German oak characters; the Reserve wine, based on a selection of the finest Canterbury fruit, which is entirely barrel-fermented and given a full malolactic fermentation, is more complex. This is an eye-catchingly powerful, chewy wine with rich, succulent stone-fruit flavours and a long, buttery-soft finish.

The Giesens' charming winery shop is lined with German wine bottles; outside, oval German wine casks (foudres) also lend a distinctly German feel. Not surprisingly, apart from Riesling the Giesens produce other traditional German varieties like the moderately spicy Gewürztraminer, the pale, plentifully sweet Müller-Thurgau, and the floral, fine, very Riesling-like Ehrenfelser.

Glenmark

Glenmark Wines
McKenzies Road, Waipara

Owner: John McCaskey

Key Wines: Riesling Dry, Riesling Medium, Gewürztraminer, Chardonnay, Waipara White, Pinot Noir, Waipara Red, Port

John McCaskey, proprietor of Glenmark, with a senior staff member in his North Canterbury estate vineyard. Glenmark early tasted success with Riesling and more recently with its rich, herbal, smooth, Cabernet-based Waipara Red.

John McCaskey, proprietor of the small Glenmark winery at Weka Plains, Waipara, is one of the Canterbury region's principal standard-bearers for Riesling.

McCaskey (55) has spent his lifetime farming his family's 392-hectare property, which originally formed part of George Henry Moore's 60,000-hectare Glenmark sheep station. The former Glenmark homestead, 'where peacocks roamed free and swans drifted on the man-made lake', is today the focal point of the winery's labels.

McCaskey's interest in winemaking was first fired twenty-five years ago, but, he says, it was not until the Glenmark irrigation scheme was under way that diversification could start. His first vines were planted in 1981 and the first Glenmark wines flowed in 1986.

Today, McCaskey's three-hectare vineyard at Weka Plains, planted in light silt loams over a base of clay and gravels, features Riesling as the principal variety with smaller plantings of Müller-Thurgau, Gewürztraminer, Chardonnay, Sauvignon Blanc, Pinot Noir, Cabernet Sauvignon and Merlot.

McCaskey in 1992 sold a half share in his winery building, a converted haybarn on the main highway at Waipara, to Torlesse Wines; Glenmark and Torlesse now share the production facility. They also share a winemaker: Kym Rayner, an Australian who formerly worked for Penfolds at Gisborne and Montana at Blenheim, before becoming a shareholder in Torlesse. Glenmark previously lacked a resident winemaker, but since the 1991 vintage Rayner has settled in as Glenmark's winemaker.

Glenmark's annual output is small — about 1750 cases. Gold medals for the robust, savoury 1991 Chardonnay and generously flavoured, herbal, Cabernet Sauvignon-based 1991 Waipara Red have recently ushered the winery into the limelight, although in the past Glenmark wines have oscillated in quality. Ex-winery sales are now at the 'Weka

Plains Weingarten', next to the vineyard in McKenzies Road.

Glenmark's Rieslings, with several labels ranging in sweetness from bone-dry to medium, are at their best light, floral and vibrantly fruity, with touches of honey and spine-tingling acidity. The Gewürztraminer is light, fresh and positively spiced. The barrel-aged Pinot Noir displays strong raspberryish flavours in an easy-drinking style.

Larcomb Wines
Larcombs Road, Rolleston

Owners: John Thom and Julie Wagner

Key Wines: Riesling, Gewürztraminer, Pinot Gris, Breidecker, Breidecker Dry, Pinot Noir

'The most important part of our business,' says John Thom of Larcomb Wines, 'is providing attractive restful surroundings where people can drink wine, have something to eat and talk to a friend. It's a very pleasant way to sell wine.'

Larcomb is a small Canterbury winery — its label originally depicted the first four Canterbury ships — lying in Larcombs Road, near Rolleston, south of Christchurch. Thom (46), a veterinarian-turned-viticulturist, and his wife, Julie, planted their first vines in 1980. Four years later they had their debut, 100 cases of 1984 Breidecker.

The tall, bearded Thom is convinced that a vet's knowledge of chemistry and hygiene is of much value in the winemaking arena. In 1988 he stopped work as a vet and plunged into full-time winemaking. 'Most vets are frustrated farmers,' he says with a hint of triumph.

In the five-hectare estate vineyard, planted in light sandy loams over river gravels, Riesling, Pinot Gris, Chardonnay and Pinot Noir are the principal varieties, with smaller plots of Gewürztraminer and Breidecker. Frost is a hazard, but the Riesling is consistently infected with 'noble rot'. A grower contracted to Larcomb has recently planted five hectares of Chardonnay and Merlot on a sheltered, north-facing site in the Port Hills; when this fruit comes on stream, Larcomb's output will double.

The corrugated-iron, polystyrene and plastic-lined winery erected in 1983 currently has an annual output of 1500 cases. 'We aim to keep the place at a size where we can sell most of the wine here,' says Thom, 'and have a small presence in the local trade.'

The majority of Larcomb's wine is sold in the delightful vineyard bar, where lunches featuring locally grown food, and wine by the glass or bottle, are served during summer. 'We want to treat the property as a farm, a garden,' says Thom, 'growing many of the ingredients — fruit, walnuts, cherries — on the property. Wine is the basis of our profitability; the rest is a logical and pleasant development.'

Larcomb wines can be frustratingly hard to buy unless you call at the winery or get on the mailing list. It's worth the effort. John Thom has slipped into the winemaker's role with apparent ease — his wines are often impressive and always hugely drinkable. The Breidecker variety, which grows easily and crops heavily in the Larcomb vineyard, is the foundation of his two enjoyable vin ordinaires. The medium Breidecker and Breidecker Dry are both light, fruity, undemanding wines, clean and fresh.

Thom sees Canterbury-grown Rieslings as 'crisper, harder than Marlborough's', yet displaying 'very good Riesling varietal character'. Larcomb Riesling has a pure, lifted bouquet and a vibrant, deep palate, brimming with grapefruit-and-lemon flavours. Dry but not austere, this is a delightful wine, all delicacy and perfume.

Gewürztraminer under the Larcomb label can also be delicious. The delicacy and clarity of flavour so captivating in the Riesling can also be found in the Gewürztraminer, a well-spiced and honeyish wine as good as any Canterbury-grown Gewürztraminer I've tried.

Pinot Gris, says Thom, is 'an orphan grape — people don't recognise the name.' Larcomb's briefly oak-matured Pinot Gris is weighty, savoury and crisp — an ideal Chardonnay alternative. The Pinot Noir is equally worthwhile. Fragrant, deep-hued, robust and supple, it is wood-aged for only three months and is highly enjoyable in its youth.

'We're very insular,' says Thom. 'I love to go fishing and diving and don't get too involved with the rest of the wine industry.' He enjoys being able to avoid the 'hustle-bustle commercialism' of the retail wine trade by selling most of his output on the estate, and has no urge to enter wine competitions: 'The locals enjoy my wine.'

John Thom adopts a low profile, selling most of his wine over the vineyard counter and in his winery café. Larcomb's Gewürztraminer, Pinot Gris, Riesling and Pinot Noir are consistently scented, delicate and satisfying.

Omihi Hills

Omihi Hills Vineyard
Reeces Road, Waipara

Owners: Danny and Mari Schuster and Brian and
Shelley McCauley

Key Wines: Pinot Noir, Pinot Blanc, Chardonnay

With his bristling black moustache and and torrent of wine conversation, Danny Schuster is one of the most distinctive figures on the Canterbury wine scene. Schuster's goal is to live at his North Canterbury vineyard, producing 'Pinot Noir to the exclusion of everything else. Pinot Noir is more challenging than Cabernet Sauvignon in the vineyard and cellar, and allows more expression of the winemaker's will. Sure, there's some ego in it — more notice is taken of great Pinot Noir.'

The eight-hectare Omihi Hills vineyard lies in rolling hill country east of the main highway at Waipara. To the north lie the snow-capped Kaikouras; sheep graze nearby fields; magpies swoop and soar overhead. On this secluded site Schuster, harnessing all the knowledge he has gained during his long love affair with Pinot Noir, has set out to produce a great Canterbury red.

German-born, Schuster (44) gathered his early wine-making experience in Europe, South Africa and Australia.

In their rolling Waipara vineyard, Danny and Mari Schuster are thirsting to produce a great North Canterbury Pinot Noir. If passion has its reward, Schuster will surely achieve his high ambition.

After arriving in New Zealand in the late 1970s, he helped set up the grape trials and microvinification cellar at Lincoln University. As winemaker at St Helena from 1980–85, he carved out a high profile based on the winery's impressively perfumed, mouthfilling and supple gold-medal winning Pinot Noirs.

The Omihi Hills vineyard — originally called Netherwood Farm — was initially a partnership between Schuster and Christchurch restaurateurs Russell and Kumiko Black, who farm the 1200-hectare Netherwood property in the Omihi Valley. After the Blacks' withdrawal in late 1989, Schuster and his wife, Mari, formed a new partnership with Brian and Shelley McCauley. Apart from his involvement in Omihi Hills, Schuster also acts as a viticultural consultant to such Californian wineries as Robert Mondavi and Stag's Leap, and is co-author, with Dr David Jackson, of the textbook *The Production of Grapes and Wines in Cool Climates*.

The elevated, sloping vineyard site, where the first vines were planted in 1986, is well protected from cool easterly and southerly winds. 'Canterbury Pinot Noir gets more concentrated the further north you go from Christchurch,' says Schuster. The vines — half Pinot Noir, the rest Chardonnay and Pinot Blanc — are planted in moderately fertile heavy clay loams, rich in ironstone and chalk. 'The structure and chemical nature of the soils have distinct parallels with Burgundy.'

The close-planted vines are remarkably spindly for their age; many have perished. Schuster doesn't believe in irrigation: 'Some say I'm suicidal, but the surviving vines have roots twenty feet [6.1 metres] deep, and once they're four years old they're safe from water stress. If you want quality, you have to accept you won't get a decent crop until the vines are five years old.'

About 2500 cases of wine flow each year from Schuster's small coolstore winery. Wines bearing the Omihi Hills label are not necessarily entirely estate-grown; grapes have been bought from various regions. Schuster's approach to winemaking is based on 'minimal intervention', with little use of sulphur dioxide, infrequent racking, and egg-white fining rather than filtering of the reds.

The wines — which are marketed by mail order and through the retail trade, but not sold directly to the public at the winery — have proved of varying quality. Schuster champions Pinot Blanc as a 'Chardonnay alternative for those who don't want to wait; it looks like a five-year-old Chardonnay when it's one-year-old and drinks well early.' Omihi Hills Pinot Blanc, which is fermented and matured in small Burgundian oak casks, at its best is impressively chewy and full-flavoured, but in lesser years lacks weight and flavour richness.

The Pinot Noir in its youth has displayed impressive body and rich, savoury, cherryish flavours, but has not always matured well over the long haul. Whether Schuster in the future can produce a great Pinot Noir will be absorbing to follow.

Sherwood Estate Wines
Weedons Ross Road, West Melton

Owners: Dayne and Jill Sherwood

Key Wines: Riesling Dry, Riesling Medium,
Chardonnay, Müller-Thurgau, Sauvignon
Blanc, Pinot Noir, Cabernet Franc/Merlot,
Port

Sherwood

The small Sherwood Estate winery in West Melton is one of the most handsome in Canterbury, with its steep-pitched roof and colour of natural wood.

Dayne Sherwood and his wife, Jill, who handles the company's administration and sales, planted their first vines in 1986. In free-draining silt loams overlying gravels, four hectares of Chardonnay, Riesling, Müller-Thurgau, Sauvignon Blanc and Pinot Noir have now been established.

Sherwood (32), is a Cantabrian who after graduating with a BA in business administration and history, initially worked for an accountancy company and a bank. After working for Torlesse and Hunter's, and gaining a postgraduate diploma in viticulture and oenology from Lincoln University, in 1990 Sherwood plunged into full-time winemaking. 'It all started from drinking the stuff,' he recalls, 'and then wanting to know more about it. Now it's a lifestyle and a business.'

The first Sherwood Estate wine was the 1990 Riesling, made at Gatehouse Wines. At about 1500 cases the output is still small, but the Sherwoods plan to eventually double their production. The whites are clean and crisp and the Pinot Noir savoury and softly mouthfilling.

St Helena Wine Estate
Dickeys Road, Coutts Island, Belfast

Owners: Norman and Robin Mundy

Key Wines: Riesling, Pinot Blanc, Pinot Gris,
Chardonnay, Müller-Thurgau, Pinot Noir

St Helena

St Helena electrified the Canterbury wine community in 1983 when its 1982 Pinot Noir scored a gold medal at the National Wine Competition. The province's oldest commercial vineyard, St Helena, had put Canterbury firmly on the New Zealand wine map.

Robin (47) and Norman Mundy (45) are down-to-earth, energetic men. After nematodes rendered their potato farm unprofitable, they early took heed of the results of Lincoln University's pioneering viticultural research. By 1978 St Helena's first vines were planted at Coutts Island, near Belfast, twenty minutes' drive north of Christchurch.

Bounded by branches of the Waimakariri River, the vineyard needs no irrigation, since the water table rises during drought conditions. The river flow also encourages air

Robin (right) and Norman Mundy (centre) pioneered commercial viticulture in Canterbury, and today head one of the province's largest wineries. The impact of winemaker Petter Evans, a Cantabrian who joined St Helena in 1991 after a successful stint at Pleasant Valley in Auckland, will be intriguing to trace.

movement, reducing the risk of frost damage to the vines.

The twenty-hectare vineyard has been established principally in Riesling, Chardonnay, Pinot Blanc and Pinot Noir, supplemented by smaller plots of Müller-Thurgau, Pinot Gris, Bacchus (for blending) and Optima (for sweet styles). 'We're putting a lot of effort into reducing vine vigour,' says winemaker Petter Evans.

With winemaker Danny Schuster at the helm from the first vintage, 1981 until 1985, St Helena achieved gold medal status with its 1982 and 1984 Pinot Noirs. His white wines, as a rule, showed excellent body but sometimes were very austere. Under Mark Rattray, winemaker from 1985 until 1990 — who now has his own Mark Rattray Vineyard label — St Helena evolved more commercial, fractionally sweeter and softer wines.

Petter Evans, a Roseworthy College oenology graduate, joined St Helena in January 1991. A Cantabrian, Evans gained his early winemaking experience at Cloudy Bay and Limeburners Bay and has also worked at Château Remy in Victoria and in Germany. Evans significantly sharpened the wine quality at Pleasant Valley in Henderson between 1988 and late 1990.

A diverse array of wines has flowed from the brick-fronted, insulated aluminium winery erected in 1981, where the annual production has reached 10,000 cases. The bargain-priced, lightly wooded Pinot Blanc (now exported to the United Kingdom) is a hearty, not quite bone-dry white with a subtle bouquet and pleasing depth of savoury/earthy flavour. The Pinot Gris is equally appealing.

The Riesling style has oscillated over the years but the latest vintages display fresh, green-apple flavours, a dryish finish and spine-tingling acidity. The Chardonnay, fermented and matured in French and American oak barriques, is a fleshy wine with ripe, peachy fruit flavours and a lingering, crisp finish.

For Petter Evans, 'Pinot Noir is the big challenge. I'm spending much of my energy on the Pinot Noir. There are so many levels of quality — to get the extra is the hard part.' The 1982, 1984, 1985 and 1988 vintages of St Helena Pinot Noir were powerful, deep-hued and supple, but of late the wine has grown markedly lighter. This wine is now good, but not outstanding, although the 1991 vintage is maturing well. Evans's impact on the wine that remains St Helena's flagship will be watched closely.

Torlesse

Torlesse Wines
State Highway One, Waipara

Owners: Dr David Jackson, Kym Rayner, Andrew Tomlin and partners

Key Wines: Riesling Dry, Riesling Medium, Chardonnay, Gewürztraminer, Chello White, Cabernet Sauvignon

The short, six-year history of the Torlesse winery has proved turbulent. Founded at West Melton by twenty shareholders, half of whom were grower-suppliers, Torlesse processed its first vintage in 1987. Three years later, however, the winery slid into receivership.

Winemaker Kym Rayner, a shareholder in Torlesse, also produces wine on a consultancy basis for other Canterbury wineries. Rayner, an Australian with lengthy winemaking experience on both sides of the Tasman, is particularly excited by the potential of North Canterbury.

Where did it go wrong? 'It was basically a production rather than market-led company,' says Kym Rayner, the current winemaker. 'The growers brought the grapes in — like Müller-Thurgau and Pinot Gris — and then said: "Let's sell the wine". Every vintage there was a different winemaker; there was no continuity of winemaking style. And until near the end, no one was trying to make a fulltime living out of Torlesse.'

Following the 1990 receivership, Torlesse was purchased by two of its original shareholders: Andrew Tomlin, an accountant, and David Jackson, a horticultural specialist at Lincoln University. Kym Rayner, who made the 1990 vintage, soon after also became a shareholder. Torlesse's new strategy, says Rayner, is to 'draw grapes from a wider range of districts within the Canterbury region, and while maintaining most of the existing wines, add a new, up-market tier.'

Rayner (41) is a lanky Australian whose parents were grape-growers in McLaren Vale. After graduating from Roseworthy College in 1975, Rayner spent five vintages at the Southern Vales Co-operative Winery in South Australia, and two vintages at the Stanley Wine Company, before he arrived in New Zealand in 1983 to run Penfolds' new winery in Gisborne. After Montana's 1986 takeover of Penfolds, Rayner spent the next three years at Montana's Blenheim winery.

After the receivership, the original coolstore winery in West Melton was sold to Gatehouse. In 1992 Torlesse purchased a half share in the Glenmark winery building; the two companies now share the production facility at Waipara, plus the services of winemaker Kym Rayner. 'This is not a company merger,' stresses Andrew Tomlin, 'and it doesn't involve any vineyards. We're just sharing ownership of a winery, and the services of a winemaker, with Glenmark.'

Torlesse draws its grapes from Andrew Tomlin's two-hectare vineyard at Yaldhurst, David Jackson's 3.5-hectare West Melton vineyard, and Kym Rayner's 1.2-hectare vineyard at Waipara (previously owned by Glenmark); fruit is also purchased from other Canterbury and Marlborough growers. Rayner is excited by his own North Canterbury grapes: 'Mark Rattray, formerly of Waipara Springs, and I are old mates. When I saw the quality of his Pinot Noir and Chardonnay, I thought: That's good enough for me.'

Highlights of the Torlesse range have included a crisp Riesling Dry with penetrating lemony flavours, and a fresh, tangy Riesling Medium; both need two or three years' cellaring to unveil their potential. With bottle-aging, the Gewürztraminer is satisfyingly full-flavoured and peppery. 'Chello' is a slightly sweet, lightly spicy blend of several grape varieties; no-one remembers how it got its name, but Rayner thinks it's a mix of 'hullo' and 'cheers'. From 1991 also flowed the winery's first Chardonnay and Cabernet Sauvignon. The 1992 Marlborough Chardonnay is impressively powerful, with a rich, creamy texture.

Waipara Springs Wine Company
State Highway One, Waipara

Owners: Andrew and Beverley Grant,
Bruce and Jill Moore

Key Wines: Chardonnay, Riesling, Sauvignon Blanc, Pinot Noir

Behind a roadside row of gum trees at Waipara nestles a collection of small grey buildings. Waipara Springs, the source of some of Canterbury's finest wines, is named after a spring which rises in the nearby hills.

The company, which processed its first vintage in 1989, is owned by Christchurch investors Andrew and Beverley Grant; and local grapegrowers Bruce and Jill Moore.

One of the founding partners, Mark Rattray, sold his shareholding in 1993. Christchurch-born, Rattray (44) studied at the Geisenheim Institute for two years in the early 1970s and worked at the fabled Schloss Johannisberg in the Rheingau. After spending five years with Montana and six years with Penfolds in Auckland, in 1985 Rattray came south to join St Helena. 'We thought the future was in the South Island,' says Rattray. 'The climate is more agreeable for the vines.'

The founding partners in Waipara Springs all own vineyards. After buying former Cabinet Minister Derek Quigley's house and vineyard, just across the highway from the winery, Rattray uprooted the established Gewürztraminer vines in 1986 and replanted with Pinot Noir and Chardonnay. The twenty-hectare Hutt Creek vineyard behind the winery, which belongs to the Moores and Grants, is planted in Chardonnay, Riesling, Sauvignon Blanc, Pinot Noir and Cabernet Sauvignon. Other fruit is drawn from Marlborough.

An old woolshed and stable have been converted into Waipara Springs' cellar and vineyard restaurant, where during summer thousands of visitors are served 'not just a ploughman's lunch', but local salmon, eel and asparagus. Waipara Springs' annual output is about 5000 cases of wine. 'We plan to reach 15,000 cases by 1995,' says Andrew Grant. 'That way we can satisfy the New Zealand market and expand our foothold in the United Kingdom.'

Waipara Springs produces a tight range of wines. 'We want to produce distinctive wines,' says Grant, 'reflecting our cool-climate conditions and the high limestone content of the soils.' Following Mark Rattray's departure to set up his own Mark Rattray Vineyard, Kym Rayner was appointed winemaker.

Waipara Springs Chardonnay displays intense fruit flavours overlaid with barrel-ferment and lees-aging complexity and an authoritatively crisp, trailing finish. This is a tautly structured wine, powerful and savoury — ideal for the cellar.

Two styles of Sauvignon Blanc have been marketed: a lush, ripe, very appealing Marlborough-grown wine, and a more complex, barrel-fermented wine based on Waipara fruit. Both have been top-flight. Waipara Springs Riesling is stop-fermented with a touch of sweetness; this is a full wine with lingering lemony flavours.

At its best the Pinot Noir is the finest in Canterbury. Matured for nine months in new Vosges oak barriques, this is a deep-scented, dark-hued, seductively soft and mouth-filling wine, delicious in its youth. If the exciting quality of the 1990 vintage is repeated — it was not in 1991 — this will be a highly sought-after label.

Waipara Springs

Founding partners in the Waipara Springs winery (left to right): Mark Rattray, Andrew Grant, Bruce Moore and Michelle Rattray. The Rattrays sold out in 1993 to found Mark Rattray Vineyard, across the road from Waipara Springs.

CENTRAL OTAGO

So eye-catching were the top Central Otago Pinot Noirs from the 1990 vintage, with their mouthfilling body, suppleness and cascading fruit flavours, the Wairarapa's Pinot Noir producers surely had a sleepless night or two.

When claims first surfaced that parts of the cooler, and frequently drier, South Island enjoyed suitable climates for viticulture, they met in the north with a frosty reception. Tom McDonald once discounted the possibility of large-scale winemaking south of Cook Strait.

Montana researched and pioneered commercial wine production in the South Island at Blenheim twenty years ago — and then abruptly silenced its critics by releasing a stream of gold-medal-winning Marlborough Rhine Rieslings and Sauvignon Blancs. Northern negativism regrouped when 'boutique' wineries later emerged in Canterbury — until St Helena electrified South Island winemakers by scoring a gold award with its 1982 Pinot Noir. Now, a small knot of enthusiasts are pioneering

winemaking at the southern frontier — Central Otago.

Otago's inland basins and valleys yielded some of the earliest New Zealand wines. Jean Desiré Feraud was cultivating 1200 vines at his Monte Cristo farm near Clyde by 1870 and his wines, bitters and liqueurs commanded high prices on the goldfields. Known as 'Old Fraud' — as Mayor of Clyde he discovered the town's water right was not legally secured, so he resigned his office, and having taken legal action to have it transferred to himself, promptly embarked on an irrigation scheme — his Constantia wine captured a First Class of Merit award at the Dunedin Industrial Exhibition, and his Burgundy a Third Class of Merit at Sydney in 1881.

Then in 1895 Romeo Bragato, on loan from the Victorian government, toured the country to advise the government on the possibilities for winemaking. At Arrowtown, he tasted his first glass of New Zealand wine — made by a Mrs Hutcheson — and 'although made after the most primitive

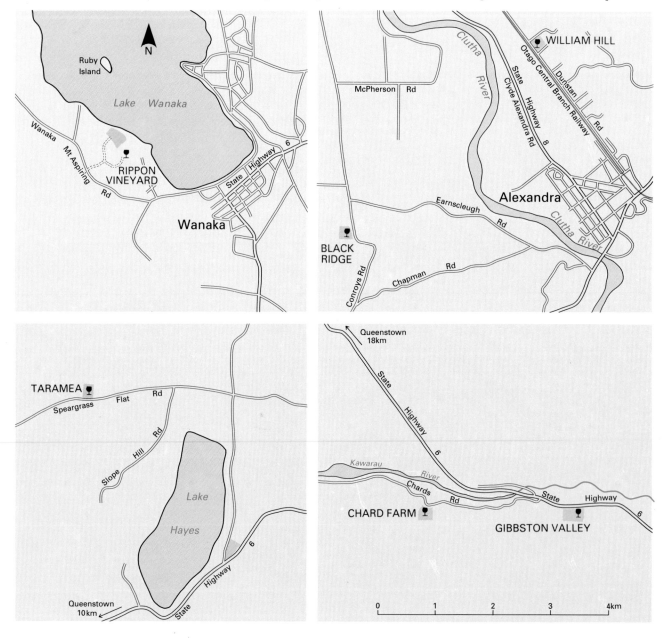

fashion, it reflected great credit on the producer . . . '

Bragato also found grapevines, tended by miners working the Clutha, flourishing outdoors at Cromwell and Clyde, and grapes fully ripened by February, 'a convincing fact to me that the summer climatic conditions here are conducive to the early ripening of the fruit.' Central Otago, he affirmed, was 'pre-eminently suitable' for winemaking.

At a public meeting in Dunedin, Bragato stirred up such enthusiasm that a Central Otago Vine and Fruitgrowers' Association was born. The word 'Vine', however, was later dropped from the title, and the eagerly anticipated new wine industry never burgeoned in the interior.

The present resurgence of activity dates from the late 1950s with Robert Duncan, of Gilligan's Gully, who planted thousands of vine cuttings near Alexandra, but failed to protect his plants from birds and frost. Trials later conducted on the Earnscleugh orchard of R.V. Kinnaird, across the Clutha from Alexandra, soon proved that fully ripe grapes could be harvested by the end of April, notably Müller-Thurgau and Chasselas.

Another trial block established in the 1970s by the DSIR under frost protection sprinklers at the research orchard at Earnscleugh, successfully ripened the prized Pinot Noir and Gewürztraminer varieties.

Central Otago is further inland than any other wine region in New Zealand. This is a region of climatic extremes; the country's highest and lowest temperatures were both recorded near Alexandra. Summer is typically hot, autumn short and winter icy-cold. The marked diurnal (day–night) temperature variation is believed to enhance the skin colours in Pinot Noir.

The crunch question is whether this region is sufficiently warm to support commercial wineries. To properly ripen, grapes must receive a certain amount of heat during the growing season. Meteorological readings confirm that the region's climate is extremely marginal for viticulture, cooler even than Germany, the most hazardous of European wine countries (and the source of its most elegant white wines). It is clear that only the most painstakingly selected sites with sheltered, sunny meso-climates — heat-traps — will enable winemakers to succeed here.

Frosts pose another real danger, threatening tender spring growth as well as the ripening bunches in autumn: it is crucial here to choose elevated vineyard sites with good cold-air drainage. Alan Brady, of Gibbston Valley, believes the major hurdle facing newcomers to the region is to secure a site which combines the prerequisites for successful viticulture — a sunny aspect for ripening, sufficient water for irrigation, suitable soils and a low frost risk.

Being so far south, however, the hours of sunlight are long. 'On a warm day it stays above 10°C for a long time, so the fruit carries on ripening,' says Rob Hay of Chard Farm. Dry autumn weather is another of the region's viticultural assets, encouraging the winemakers to leave their grapes late on the vines, into May or even June, ripening undamaged by autumn rains. Owing to the lack of humidity, the grapes are rarely infected by 'noble rot'.

The fledgling Central Otago industry is centred on Queenstown where the Gibbston Valley, Chard Farm and Taramea vineyards are based (and the tourist market beckons); Alexandra, which boasts the two southernmost vineyards, Black Ridge and William Hill; and Wanaka, where Rippon produced a rich and savoury 1990 Pinot Noir, and gold medal Barrel Selection Pinot Noir 1991, that stand comparison with any in the country.

There are currently about thirty-five hectares of vines scattered around Central Otago (with vineyards, but not wineries, having spread to Cromwell and Omarama). The most widely planted grape varieties, according to the 1992 vineyard survey, are (in descending order): Pinot Noir, Chardonnay, Riesling, Gewürztraminer and Sauvignon Blanc.

Not surprisingly for a region that boasts the most southerly commercial vineyard in the world (at Alexandra 45° 15′ south), it is felt by many that Riesling will excel here. According to Dr David Jackson of Lincoln University and Torlesse — who is often referred to as 'the father of Canterbury wine' — 'Central Otago is closest to the German situation and, I predict, will be the area where Riesling excels.'

Many of the region's white wines — particularly the Rieslings — have a bracing level of acidity and slightly green, grassy fruit aromas that bear more than a passing resemblance to the Sauvignon Blancs grown further north. Yet the Chardonnays grown on superior sites — like Chard Farm — display ample body and ripe, stone-fruit flavours.

Above all, with several outstanding Pinot Noirs — Gibbston Valley 1990 and Rippon 1990 and 1991 — safely bottled, backed up by lighter but promising Pinot Noirs from other wineries, Central Otago looks poised to emerge as one of New Zealand's top regions for Pinot Noir.

Not all of the winemaking ventures have succeeded in 'Central'. Alan Brady worries about some of the new grapegrowers 'who don't have their eyes open — they've no knowledge of viticulture'. Rob Hay tells of the 'horror stories, like the people who planted 10,000 vines without irrigation or proper weed control. Three hundred survived.'

Hay, a young winemaker trained in Germany and at Babich and Ruby Bay, planted his vineyard at Chard Farm, high above the Kawarau River, and processed his first vintage in 1989. Tony Bish, winemaker at Rippon Vineyard from 1986 to 1989, and his successor, Rudi Bauer, who departed in 1992, were both qualified winemakers. This recent influx of experienced winemakers is starting to capitalise on Central Otago's promise.

Lake Hayes Vineyards — whose shareholders include Rob and Grey Hay of Chard Farm, and the actor Sam Neill — in 1993 unveiled plans to establish up to seventy-five hectares of vines, principally Pinot Noir and Chardonnay, adjacent to Lake Hayes and at Nevis Bluff in the Gibbston Valley. The rivulet of promising Central Otago wine may soon swell into a sizeable stream.

Black Ridge

Black Ridge Winery
Conroy's Road, Alexandra

Owners: Verdun Burgess and Sue Edwards

Key Wines: Riesling, Gewürztraminer, Chardonnay, Pinot Noir

'This land is so hard, it rips the soles off your boots,' says Verdun Burgess, co-owner of the Black Ridge vineyard. Black Ridge lies on a north-facing slope in a dramatic, rocky, primeval landscape south-west of Alexandra. This is the most southern winery in the world.

Burgess (43) and his partner, Sue Edwards, planted their first vines in 1981. Born in Invercargill, Burgess came to Alexandra as a builder, but when the Clyde dam was finished, work dried up. Grapegrowing looked to be 'the most economic thing to do with the land', so in 1989 he moved into wine on a full-time basis.

Tall, rugged, moustachioed, pipe-smoking, extroverted — Burgess is a forceful personality. 'We're cowboys around

Verdun Burgess delights in the inhospitably arid, rocky terrain in which Black Ridge vines have sunk their roots. Burgess's light, crisp wines are cultivated in the most southerly commercial vineyard in the world.

here,' he grins. 'I love shooting rabbits. They're good fun and they don't shoot back.'

The six-hectare vineyard is planted in a few centimetres of top soil overlying a layer of clay, with a hard, almost impenetrable rock subsoil. For Burgess, this 'rough, rocky terrain has a particular beauty I love'. A visiting wine merchant took one look and labelled it a 'hero's vineyard'.

Riesling, Gewürztraminer, Chardonnay and Pinot Noir are the featured varieties at Black Ridge, with Breidecker also coming on stream for a quaffing wine. Cabernet Sauvignon is planted on the top slope, where, Burgess says, 'it's too hot for anything else.'

Only a trickle of wine flows from Black Ridge — about 1,000 cases a year. Until recently the wine was processed at Rippon Vineyard, but a barn-style winery was erected for the 1992 vintage, and Michael Wolter, formerly of Taramea, was appointed winemaker.

Black Ridge Riesling is pale lemon-green in hue, with a fresh, green-appley bouquet. The palate displays light citric and apple-like flavours, spine-tingling acidity and an appealing, Mosel-like delicacy.

Burgess is especially excited by Gewürztraminer, claiming Black Ridge's is 'the best in Central Otago'. That's debatable — there are several promising Gewürztraminers in the region — but Black Ridge Gewürztraminer is a full, crisp wine with a restrained spiciness.

Pinot Noir looks to have an interesting future. With its fragrant fruit aromas and light, red-berry-fruit flavours, Black Ridge lacks the eye-catching depth of the top Central Otago Pinot Noirs but is clean, gently oaked and fresh.

Black Ridge wines are not cheap — their production is currently low and the demand outstrips the supply. Their varietal flavours also lack the richness of wines from the north. They are intriguing because Burgess, one of only two viticultural pioneers in the Alexandra district, is erecting signposts to the future.

By the end of the decade, Burgess envisages having an eight-hectare vineyard and exporting one-third of his output. 'But if I'd known ten years ago just how much work and money would be involved, I think I'd have planted cherries.'

Chard Farm

Chard Farm Vineyard
Gibbston, Queenstown

Owners: Rob and Greg Hay

Key Wines: Riesling, Chardonnay, Judge and Jury Chardonnay, Gewürztraminer, Sauvignon Blanc, Pinot Gris, Pinot Noir

On a north-facing ledge seventy metres above the Kawarau River, with a riveting view across the gorge to the snow-draped Cardrona Range and Coronet Peak, Rob and Greg Hay have planted one of the country's most beautiful vineyards. 'We picked the site solely with quality wine in

mind,' says Rob Hay, 'not for its tourist potential — which is what many people think.'

Chard Farm lies just off the main highway, twenty kilometres east of Queenstown. The precipitous access route to the vineyard, skirting sheer bluffs with a steep plunge to the river for the unwary, was once the Cromwell–Queenstown road; Chard Farm starts at the cattle-stop.

The land was originally worked by Richard Chard in the 1870s as a market garden, supplying food to the miners heading for the goldfields. Later, Chard Farm became a dairy farm and an orchard. The Hay brothers uprooted the fruit trees before they planted their first vines in 1987.

Born in Motueka, Rob Hay (34) graduated from Otago University with a BSc, and then embarked on a three-year study and work course in Baden and Württemberg: 'I went to Germany to study winemaking in a genuinely cool climate,' he says. After working at Babich and Ruby Bay, in 1986 Hay came to the deep south. 'I could feel it would be good,' he recalls. 'I looked at the temperatures and tasted Alan Brady's grapes — the proof of the potential was there on the vines.'

Greg Hay (32), a marketing graduate, is Chard Farm's vineyard manager while Rob oversees the administration and winemaking. In moderately fertile silt loams with some clay bands overlying shingly sub-soil, the Hays have planted twelve hectares of Pinot Noir, Chardonnay, Riesling, Gewürztraminer and Sauvignon Blanc. A few kilometres away at Lake Hayes, another six-hectares of vines were planted on a joint venture basis in 1992.

The 1989–92 wines were produced along the road at Gibbston Valley, but Chard Farm's own winery was erected for the 1993 vintage. The annual production level is low — about 3500 cases.

The brothers plan to focus on Pinot Noir and Chardonnay. 'The wines will be a different style,' says Rob Hay. 'With the grapes' higher acidity, there'll be a need to de-acidify and use malolactic fermentation. But it's warmer here than many people think.'

The oak-matured Chard Farm Chardonnay is showing distinct promise, its ripe-tasting stone-fruit flavours underpinned by bracing acidity. The top end of the range Judge and Jury Chardonnay (named after prominent outcrops of rock across the gorge) is a predominantly barrel-fermented style.

Chard Farm Riesling is green-edged and racy in its youth, but with a couple of years' bottle age it unveils

appealing honeyed characters. The Sauvignon Blanc is intensely herbaceous ('If you like Marlborough Savvys try this little beauty!' suggests the winery newsletter); the Gewürztraminer fresh, crisp and lightly spiced.

Chard Farm Pinot Noir, matured for six months in two-year-old oak casks, is a light style with strong, ripe-tasting cherry and raspberry fruit flavours and a supple, soft finish. By pruning their vines more severely — to produce a more intensely flavoured crop — and maturing the wine in new oak, the Hays are confident they can produce a distinguished Pinot Noir.

Rob (left) and Greg Hay produce lively, mouthwateringly crisp white wines and a promisingly supple Pinot Noir on a ledge above the tumbling Kawarau River. The view to the Cardrona Range and Coronet Peak raises the question: will Chard Farm's wines match its glorious mountain setting?

Gibbston Valley Wines
Gibbston, Queenstown

Owners: Alan Brady and shareholders

Key Wines: Estate Reserve Pinot Noir, Pinot Gris, Sauvignon Blanc, Riesling, Gewürztraminer; Southern Selection Riesling, Chardonnay

'My dream is an unbroken wall of vines along the Gibbston Valley,' says Alan Brady, the bearded, Irish-born founder of one of Central Otago's largest wineries.

Brady (57), who purchased his briar-covered, craggy property in the Kawarau Gorge, twenty-five kilometres from Queenstown, in 1976, concluded 'it was an interesting piece of dirt.' He elected to do something a bit different with it and planted 350 vines in 1981 and 1982, to 'prove they would grow'. The first vintage of Gibbston Valley wine was bottled in 1984.

After coming to New Zealand as a twenty-three-year-old, Brady worked on the *Manawatu Evening Standard* and *Otago Daily Times*, became a television news editor in Dunedin, and later freelanced as a journalist, producer and

director. In early 1990, after taking in partners ('I've never had any dynastic intentions as regards the company,' he says) Brady plunged into wine on a fulltime basis. 'Wine's my obsession,' he says. 'It's so exciting and diverse. It's given me three new careers — financial entrepreneur, grapegrower and restaurateur.'

On a north-facing schist ledge at the foot of rocky bluffs, the three-hectare estate vineyard flanking the winery has been 'planted in everything — Pinot Noir, Riesling, Sauvignon Blanc, Pinot Gris, Gewürztraminer, etc.'

The majority of Gibbston Valley's fruit is drawn from other South Island regions. 'This poses a big problem,' admits Brady. 'We want to build our reputation on Central Otago grapes, so we clearly label the origin of the other wines as Canterbury or Marlborough.'

The estate-grown wines are labelled as Estate Reserve; those grown elsewhere are distinguished by a 'Southern Selection' label. Brady expects Gibbston Valley's range to be predominantly based on Central Otago grapes by 1997.

The smart red horseshoe-shaped winery and restaurant complex, erected prior to the 1990 vintage, attracts up to 600 visitors daily. Fresh salmon, venison and pasta dishes are served and the 'wine taster's platter', featuring a selec-

Gibbston Valley

Alan Brady, a pioneer of commercial viticulture in 'Central', now heads one of its key wineries. Gibbston Valley's estate-grown, slowly evolving Riesling, generously flavoured Pinot Gris and perfumed, vibrantly fruity, supple Pinot Noir are consistently absorbing.

tion of New Zealand cheeses, is popular.

What style of wine is produced in this cool mountain environment? Gibbston Valley wines display the 'crisp, fresh varietal flavours you'd expect from the highest altitude vineyards in New Zealand', says the winery's promotional brochure. '[The] aromas and flavours are more delicate and subtle than similar wines from warmer regions.' These are light white wines with slowly evolving flavours and bracing acidity.

In its youth, the Estate Reserve Riesling is crisp and shy, but at four years old the 1987 Riesling opened with a Mosel-like harmony of lemony, honeyish flavours and racy acidity. The Estate Reserve Sauvignon Blanc is weighty

and crisp, with vigorous herbal flavours and a long, dry finish. The Estate Reserve Pinot Gris, which Brady says 'loves our conditions', is full, ripe-tasting and savoury.

Gibbston Valley's exciting wine, however, is the Pinot Noir. 'Could it be that we have this affinity with Burgundy?' ponders Brady. The 1990 vintage — with Rippon Vineyard's — finally proved that Central Otago is capable of producing wines that are not just pleasant, but distinguished. Matured for nine months in French oak casks, the deliciously fruity, supple and rich-flavoured 1990 is capable of footing it in the quality stakes with any Pinot Noir in the country. The lighter, but subtle and complex 1991 is also impressive.

Rippon

Rippon Vineyard
Mt Aspiring Road, Wanaka

Owners: Rolfe and Lois Mills

Key Wines: Pinot Noir, Barrel Selection Pinot Noir, Gamay Rosé, Hotere White Wine, Osteiner, Gewürztraminer, Sauvignon Blanc, Fumé Blanc, Riesling, Chardonnay

'Burgundy is for kings, Chardonnay for duchesses and claret for gentlemen,' according to a French proverb approvingly quoted by Rolfe Mills, whose Rippon Vineyard runs down a schist slope to the shores of Lake Wanaka.

In this sublime setting Mills and his wife, Lois, have produced one of the finest Pinot Noirs in the country.

The blue-water view to the majestic snow and cloud-capped peaks of the Buchanan Range at the head of the lake is glorious, but for Rolfe Mills it is wine that is 'all consuming — that's the way it has to be currently'. The Mills' small winery sits at the crest of the slope, their house in the middle, and their little wine sales building down by the lake-shore.

Rolfe Mills planted his first vines, Seibels and Albany Surprise, at Wanaka in 1976; one of the Albany Surprise vines survives 'just for nostalgia'. A white-haired, gentlemanly, fit seventy-year-old, he is the grandson of Sir Percy Sargood, who once owned Wanaka Station. After a career

as the sales director of the family clothing and footwear company, Sargoods, Mills came to Wanaka not knowing what he wanted to do.

'My only link with wine was drinking it. I'd been to Portugal and seen land in the Douro Valley that looked like the family land at Wanaka. I knew vines would grow here, but didn't know if they would ripen fruit.' In 1981 the Millses went to Bergerac, in south-west France, to learn 'how to grow grapes. We went for the way of life in the vineyards, not to learn how to make wine. Winemaking needs a professional.' The first, tiny batch of Rippon Vineyard wine flowed in 1984, produced by Rainer Eschenbruch at the Te Kauwhata research station.

Mills has long ago buried his doubts about his site's ability to ripen grapes. 'Frost is not a problem; the cold air drains through the vineyard to the lake. I've been taking my own heat summation readings since 1977. That year the figure was 1350 degree days — comparable to Auckland — and since then we've averaged 1000 degree days. In cool years we can cut off some of the bunches to ensure that those remaining achieve full ripeness.'

The Millses have planted their seven-hectare, north-facing vineyard in a glacial moraine. 'We have terrific drainage — all that gravel beneath us, covered by good soil brought down by erosion,' says Mills. Irrigation is 'essential' to get young vines established and alleviate water stress. The principal varieties in the Rippon vineyard are Pinot Noir, Chardonnay, Riesling, Gewürztraminer and Sauvignon Blanc.

Tony Bish, now at French Farm, made Rippon Vineyard's wine from 1986 to 1989 and Rudi Bauer, now at Giesen, produced the 1990–92 vintages. Clotilde Chauvet, scion of a French Champagne-producing family, joined Rippon for the 1993 vintage.

'Central Otago wines really need eighteen months before you start assessing them,' Rudi Bauer emphasises. Nevertheless Rippon produces a trio of 'drink young, everyday wines': a pale, off-dry, light and fruity Osteiner (a crossing of Riesling and Sylvaner); a tangy, grassy, medium-dry blend labelled as Hotere White Wine; and a soft, pale pink, raspberryish Gamay Rosé — the 1991 vintage won the trophy for the champion rosé at that year's Air New Zealand Wine Awards.

Both the Sauvignon Blanc (the gold medal 1992 vintage is stunning) and barrel-fermented Fumé Blanc are lively wines with vigorous, green-edged flavours. The Riesling in its youth displays fresh, delicate lemony flavours and spine-tingling acidity. The very impressive barrel-fermented Chardonnay is a mouthfilling, powerful wine with creamy, lingering flavours and a steely acid undertow.

'I wish we had twice as much Pinot Noir,' says Lois Mills. 'It sells so quickly.' Rippon Vineyard Pinot Noir is a top-flight red, challenging the more acclaimed Pinot Noirs of the Wairarapa. This is a robust and savoury red, bursting with sweet-tasting cherryish flavours, rich, supple and sustained; the Barrel Selection Pinot Noir is particularly intense.

Taramea

Taramea Wines
Speargrass Flat Road, Queenstown

Owner: Ann Pinckney

Key Wines: Gewürztraminer, Müller-Thurgau, Riesling

Ann Pinckney lives on a stunning site above her vineyard at Speargrass Flat — straight across the poplar-lined valley tower the white flanks of Coronet Peak.

Here, the forty-year-old Lincoln University graduate planted her first Gewürztraminer and Müller-Thurgau vines in 1982. A Southlander, with winemaking experience

Ann Pinckney bottled five commercial vintages of Taramea wine from 1987 to 1991, but recently frosts have frustrated her bid to establish Taramea as a boutique winery.

in Germany, France and Italy, Pinckney was 'sick and tired of hearing bad things about the local climate' and out to prove her point that 'the climate is there' to produce premium quality wines. For the early vintages, before her equipment was upgraded, she and her friends donned gumboots and crushed the fruit by foot in buckets.

Taramea's main block of vines — a hectare of Gewürztraminer and Müller-Thurgau on the valley floor — is vulnerable to frost; most of the buds were killed when a late frost struck on 21 November 1991. The 'top block', a half-hectare of Riesling, is frost-free.

In her tiny winery, a converted farm shed, each year Pinckney produces only a trickle of wine — about 200 cases. The three small tanks at the rear, which hold Taramea's entire production, freeze in winter. Mike Wolter, who has worked at several North Island wineries and in Australia, was deeply involved in Taramea's production from 1987 to 1992.

The Müller-Thurgau is pale and dry, its biting acidity unusual for the variety. The Gewürztraminer is mildly floral and spicy; the Riesling — grown in Marlborough — slightly sweet, lemony and crisp.

The Taramea winery's future is uncertain. Cellar door sales are now 'by appointment only', and the 1992 and 1993 vintages were sold in bulk. 'We've been having such a lot of trouble with frosts,' says Pinckney. 'I'm not sure what's happening in 1994.'

William Hill

William Hill Vineyard
Dunstan Road, Alexandra

Owner: Bill Grant

Key Wines: Gewürztraminer, Chardonnay, Riesling, Pinot Noir

Bill Grant (62) produces only a trickle of Alexandra wine: 'Four thousand bottles a year, sold in a couple of months at my back door. I just want to prove it can be done; those who come after me can expand it.'

William Hill Vineyard lies on a sandy terrace above the Clutha River, on the western edge of Alexandra. Grant —

Winemaking in Central Otago's risky climate is not for the chicken-hearted. Bill Grant of Alexandra planted his first vines two decades ago, but the first serious vintage of William Hill Vineyard wines flowed in 1988.

whose full name is William Hill Grant — was raised in Dunedin and for many years was a schoolteacher. Stone quarrying and building have been his major activities in Alexandra: 'Wine is only a hobby.' While travelling in Europe, however, he became convinced there were climatic parallels between the Rhine and Alexandra.

Grant planted his first two rows of vines in 1973. 'I waited for everyone else to do something, then when I heard Ann Pinckney was going ahead at Taramea, I thought I'd better get cracking too.' By the early 1980s his vineyard had expanded to a half-hectare. After the first, experimental wines had been 'either drunk or disposed of', William Hill's first 'commercial' wines flowed in 1988.

The recently much expanded, four-hectare vineyard is planted in Chardonnay, Riesling, Gewürztraminer, Sauvignon Blanc and Pinot Noir. The 1992 and 1993 vintages were made at Black Ridge, but William Hill's own winery will be operating by 1994.

The top William Hill wines are dramatically packaged in very tall, slender bottles with vertical labels running their full height. These are uniformly light, clean wines, delicately perfumed and flavoured. The Gewürztraminer is pale and fruity, with a touch of sweetness and spiciness. The Chardonnay, a non-wooded style with plenty of weight, is tart and appley. The Pinot Noir, which is also not barrel-aged, displays appealing raspberryish aromas and a light, fresh, buoyant palate.

A SELECT BIBLIOGRAPHY

This bibliography is based on the most valuable published material used by the author during research for successive editions of this book.

All listings have been organised into one of several categories: Books, University Theses, Government Publications, Wine Magazines and Newsletters, and Miscellaneous Items. The multitude of wine columns also appearing in magazines and newspapers throughout the country constitutes another rich mine of wine material. (Any reader dedicated enough to subscribe to a press clipping service will rapidly be inundated with more words on wine than anyone can comfortably digest.)

Alternative starting-points are S.M. Bradbury's *Wines and Winemaking in New Zealand: A Bibliography*, Wellington Library School, 1970, and my own A Bibliography of New Zealand Wine to 1981, Auckland Public Library, 1983.

Books

Beaven, D., *Wines for Dining*, 1977.
Beaven, D., Donaldson, I. and Watson, G., *Wine — A New Zealand Perspective*, 2nd ed., 1988.
Beaven, D. and Schuster, D., *Wine: Care and Service*, 1985.
Bradley, R., *Australian and New Zealand Wine Vintages*, 10th ed., Melbourne, 1992.
Brooks, C., *Marlborough Wines and Vines*, 1992.
Buck, J., *Take a Little Wine*, 1969.
Cooper, M., *Michael Cooper's Buyer's Guide to New Zealand Wines*, 1992.
Cooper, M., *Michael Cooper's Pocket Guide to New Zealand Wines and Vintages*, 1990.
Evans, L. (ed.), *Australia and New Zealand Complete Book of Wine*, NSW, 1973.
Graham, J.C., *New Zealand Wine Guide*, 1971.
Graham, J.C., *Know Your New Zealand Wines*, 1980.
Graham, J.C., *Jock Graham's Wine Book*, 1983.
Halliday, J., *Pocket Guide to the Wines of Australia and New Zealand*, NSW, 1992.
Halliday, J., *Wine Atlas of Australia and New Zealand*, 1991.
Jackson, D. and Schuster, D., *Grape Growing and Wine Making: A Handbook For Cool Climates*, 1981.
Jackson, D. and Schuster, D., *The Production of Grapes and Wines in Cool Climates*, 1987.
MacQuitty, J., *Jane MacQuitty's Pocket Guide to New Zealand and Australian Wines*, London, 1990.
Reid, J.G.S., *The Cool Seller*, 1969.
Robinson, J., *Vines, Grapes and Wines*, London, 1986.
Saunders, P., *A Guide to New Zealand Wine*, published yearly since 1976.
Scott, D., *Winemakers of New Zealand*, 1964.
Scott, D., *A Stake in the Country: A.A. Corban and Family 1892-1977*, 1977.
Scott, D., *Seven Lives on Salt River*, 1987 (see Chapter 4, 'He Made the Vine Flourish', pp. 57–73).
Southern, E., *New Zealand Wine and Cheese Guide*, 1969.
Stewart, K., *The Wine Handbook*, 1988.
Thorpy, F., *New Zealand Wine Guide*, 1976.
Thorpy, F., *Wine in New Zealand*, 1971, revised ed., 1983.
Trlin, A.D., *Now Respected, Once Despised: Yugoslavs in New Zealand*, 1979, (see Chapter 4, 'The Winemakers', pp. 81–98).

University Theses

Bradding, R., 'Future Directions for MAF Viticultural Research in New Zealand', University of Waikato Bachelor of Management Studies Report, 1987.
Cooper, M.G., 'The Wine Lobby: Pressure Group Politics and the New Zealand Wine Industry', University of Auckland M.A. Thesis, 1977.
Corban, A., 'The History, Growth and Present Disposition of the New Zealand Wine Industry', Dissertation for a Diploma in Biotechnology, Massey University, 1974.
Corban, A.C., 'An Investigation of the Grape Growing and Winemaking Industries of New Zealand', University of New Zealand M.D. Thesis, 1925.
Forder, P.G., 'The Te Kauwhata Viticultural Research Station', University of Auckland M.A. Research Essay, 1977.
Gray, A., 'A Technical, Marketing and Financial Overview of the Canterbury Wine Grape System', Lincoln University B. Hort. Sc. Dissertation, 1991.
Marshall, B.W., 'Kauri-Gum Digging 1885-1920', University of Auckland M.A. Thesis, 1968.
Moore, M., 'Factors Influencing Demand for Wine in New Zealand', Victoria University M.B.A. Dissertation, 1991.
Moran, W., 'Viticulture and Winemaking in New Zealand', University of Auckland M.A. Thesis, 1958.
Norrie, B., 'The Development of Viticulture and Winemaking in Marlborough', Canterbury University M.A. Thesis, 1990.
Read, H., 'The Impact of Government on the New Zealand Wine Industry: The Grapevine Extraction Scheme', University of Auckland M.A. Thesis, 1988.
Serralaich, J., 'The Wine Industry and the Consumer', Massey University Ph.D. Thesis, 1984.
Townsend, P., Location of Viticulture in New Zealand', University of Auckland M.A. Thesis, 1976.
Trlin, A.D., 'From Dalmatia to New Zealand', Victoria University M.A. Thesis, 1967.
Vandersyp, C.J., 'A Case Study for Protection — the New Zealand Wine Industry', University of Auckland Thesis, 1979.
Zuur, D., 'Ampelographic Studies of New Zealand Grape Varieties', University of Waikato Ph.M. Thesis, 1987.

Government Publications

Berrysmith, F., 'Viticulture' (rev. ed.), 1973.
Bragato, R., 'Report on the Prospects of Viticulture in New Zealand', 1895.
Conference of New Zealand Fruitgrowers and Horticulturists, Dunedin, June 1901, NZ Department of Agriculture.
Garrett, R. and Smith S., 'Wine Industry Assistance Package', Ministry of Agriculture and Fisheries, Economics Division, 1986.
Interdepartmental Overview Committee, Mid Term Review of the Wine Industry Development Plan to 1986, 1985.
New Zealand Department of Agriculture and Fisheries, Annual Reports.
Oenological and Viticultural Bulletins, Ruakura Agricultural Research Centre.
Report and Evidence of the Royal Commission on the Kauri Gum Industry in New Zealand, AJHR, A-12, 1898.
Report No. 76C to the Tariff and Development Board, Certain Wines and Spirits, NZ Department of Industries and Commerce, 9 June 1970.
Report of the Colonial Industries Commission, AJHR, H-22, 1880.
Report of the Flax and Other Industries Committee on the Wine and Fruit Industry, AJHR, I-6B, 1890.
Report of the Industries Commission, AJHR, I-12, 1919.
Report of the Industries Development Commission: The Wine Industry Development Plan to 1986, 1980.
Report of the Licensing Committee, AJHR, I-17, 1960.
Report of the Licensing Control Commission, 31 March 1955, AJHR, A-3.
Report of the New Zealand Commission to Inspect and Classify the Kauri Gum Reserves in the Auckland Land District, AJHR, C-12, 1914.
Report of the Royal Commission on Licensing, 1946.
Report of the Royal Commission on the Sale of Liquor in New Zealand, 1974.
Report of the Winemaking Industry Committee, AJHR, I-18, 1957.
Report to the Prime Minister on Investigation of Grape Surplus by Officials Committee, 1983.
'The Sale of Liquor in New Zealand', Report of the Working Party on Liquor, 1986.
'Taxation of Alcohol and Tobacco', Treasury Report to the Minister of Finance, 22 April 1991.

Wine Magazines and Newsletters

Australia and New Zealand Wine Industry Journal (a quarterly magazine).
Campbell, B., *New Zealand Wine Annual* (published annually by *Cuisine*).
Cuisine, (a bimonthly magazine).
Saunders on Wine (a newsletter).
The New Zealand Wineglass (latterly *Pacific Wineglass*), (a monthly magazine published from 1980–86).
The Wine Report (a monthly newsletter).
Wine Review (a quarterly magazine published from 1964–78).
Wine Taste (a quarterly newsletter).

Miscellaneous

Ayto, J., 'A Review of the Policy and Economics of Selective Taxation on Wine', New Zealand Institute of Economic Research, July 1986.
Cotton, A., 'The New Zealand Wine Industry', Westpac Banking Corporation, June 1989.
Dunphy, C., 'Grape Expectations: Issues Facing the New Zealand Wine Sector', Buttle Wilson, April 1990.
'Government Revenue and the New Zealand Wine Industry 1984–1990, Coopers and Lybrand, February 1990.
A Hawke's Bay Regional Wine and Grape Industry Study, Hawke's Bay Vintners, July 1985.
Industry Study and Development Plan 1978, Wine Institute of New Zealand, 1979.
Lucas, D., 'The Influence of the Distributive Mechanism, the Sales Tax Structure and Container Prices on the Cost and Availability of New Zealand Wines', Institute of Economic Research, 1979.
McMenamin, A., and Moran, W., 'New Zealand Vineyard Survey 1989', Auckland University Geography Department mimeograph, 1989.
'New Zealand Trade Development Board Excise Tax Study', Ernst and Young, June 1990.
'The New Zealand Wine Industry', Bank of New Zealand Report, 1991.
'The New Zealand Wine Industry', Porter Project Report, 1990.
Perry, P. and Norrie, B., 'The Origins and Development of a New World Vignoble: Marlborough, New Zealand, 1970–90'. *Journal of Wine Research*, 1991, Vol.2, pp. 97–114.
'Recent Trends in the Production and Sales of Wine', Coopers and Lybrand, January 1992.
'Report to the Wine Institute of New Zealand Inc. on the Effect of Imports on New Zealand Wine Producers', Trade Consultants, October 1989.
'Wine Excise Tax Study', Southpac Corporation, 20 November 1990.
Wine Institute of New Zealand, Annual Reports.

INDEX

Entries in bold type indicate vineyards. Page numbers in bold refer to main entries while page numbers in italics refer to pictures.